QUANTUM FLUIDS AND SOLIDS

QUANTUM FLUIDS AND SOLIDS

EDITED BY

SAMUEL B. TRICKEY, E. DWIGHT ADAMS,
AND JAMES W. DUFTY
University of Florida, Gainesville

PLENUM PRESS · NEW YORK AND LONDON

Library of Congress Cataloging in Publication Data

International Symposium on Quantum Fluids and Solids, 2d., Sanibel Island, Fla., 1977.
Quantum fluids and solids.

Includes index.
1. Quantum liquids—Congresses. 2. Quantum solids—Congresses. I. Trickey, Samuel B. II. Adams, Ernest Dwight. 1933- III. Dufty, James W. IV. Title.
QC174.4 AI 157 1977 500.4'2 77-21307

ISBN-13: 978-1-4684-2420-1 e-ISBN-13: 978-1-4684-2418-8
DOI: 10.1007/978-1-4684-2418-8

Proceedings of the Second International Symposium on Quantum Fluids and Solids held on Sanibel Island, Florida, January 24–27, 1977

Softcover reprint of the hardcover 1st edition 1977
A Division of Plenum Publishing Corporation
227 West 17th Street, New York, N.Y. 10011

PREFACE

The second International Symposium on Quantum Fluids and Solids came to pass during 23-27 Jan. 1977 as the fourth and concluding part of the seventeenth consecutive running of the Sanibel Symposium Series. With approximately 120 participants from eleven countries (including, for the first time, the USSR), we found it easy to obtain a selection of papers which was fairly comprehensive. Indeed, our problem was an embarrassment of riches; in spite of our solemn vows not to crowd the schedule, we ended up with an intense program! By far, the majority of the papers presented are represented in this volume.

We are indebted to many persons and organizations for their contributions to the Symposia. First, we thank Prof. Per-Olov Löwdin, Director of the Quantum Theory Project and originator of the Sanibel Symposia. Without his patient, indulgent cooperation our task would have been vastly more difficult. We are grateful to Prof. F. Eugene Dunnam, Chairman of the Dept. of Physics and Astronomy, for providing Departmental support of our initial organizing expenses. Approximately one-half of the total cost of the Symposium was borne by a joint grant from the National Science Foundation and the U. S. Air Force Office of Scientific Research. We thank the program officers, Dr. C. Satterthwaite and Dr. D. Finnemore of NSF and Dr. T. Collins of AFOSR for this essential contribution. We are also grateful to the National Science Foundation for support of several National Academy of Sciences Travel Awards (for domestic and foreign participants) arranged by the Chemistry and Chemical Engineering Division of the National Research Council.

Much of the program resulted from suggestions by the Advisory Committee for the Symposium: V. Ambegaokar, E. Andronikashvilli, G. Baym, D. Brewer, W. Brinkman, L. Corruccini, A. Dahm, A. Fetter, W. Kirk. J. Krumhansl, A. Landesman, D. Lee, O. Lounasmaa, W. Massey, H. Meyer, L. Mezhov-Deglin, F. Pobell, and J. Wheatley. We thank them as well as the other workers in these areas who suggested papers or even entire sessions. We owe a particular debt of gratitude to those Quantum Theory Project members who had responsibility for fiscal and organizational aspects of the entire Symposium series: Prof. N. Y. Öhrn, Prof. M. Hehenberger, Prof. J. R. Sabin, and Mr. Henry Kurtz.

The typing for this volume was done by Ms. Maureen Flanagan, whose cheerful and able cooperation we acknowledge gladly.

Finally, we are grateful for the cooperation and stimulus given us by all the participants; with these gifts all the rest is made worthwhile.

Sam Trickey
Dwight Adams
Jim Dufty

Gainesville, 11 April 1977

CONTENTS

OPENING REMARKS

Robert A. Bryan, Vice-President, Academic Affairs

University of Florida

Gainesville, Florida 32611

It is my pleasant duty to bring you greetings from the University of Florida, a partner in the UF-Uppsala University Series of Winter Symposia on Quantum Science held each year on Sanibel Island. This partnership has flourished now for 17 years - and it is our hope that it will prosper for 17 more years and 17 more after that. If it can last for 51 years, surely most of the important questions facing quantum science will be answered.

Since I am a professor of English, you can imagine that I only dimly perceive the substance of the talks and discussion that go on here. I dutifully attend a few of the scientific talks, and attend the welcoming and farewell ceremonies. While I cannot recite for you the scientific concepts to which I have been exposed this year - or any of the previous three years it has been my privilege to attend as a representative of the UF - I can tell you some of the other things I have learned.

It has been borne home to me again and again that knowledge knows no nationality; that all of us who are students of nature are students first and citizens of particular countries second. I know this is a cliche; I know we have all been taught this homely truth over and over again in our various educational experiences. But nothing makes a truism - a cliche - come alive so much as to witness it in action. And each year that I come here, I witness the true spirit of knowledge-seeking. People from all over the world gather here on this strange little island to exchange ideas that are the results of years of thought and work. It is a mind-stretching experience to witness it; I can only wonder how exhilarating it would be to be an active participant.

The only thing in my experience that comes close to what you do here each year lies in the fact that English scholars attempt - as do you - to penetrate the metaphor - either literary or natures's - that surround the ultimate truth. Literature lives and thrives on metaphor; science, I think, attempts always to destroy metaphor and get at what lies beyond. But science, I observe, must constantly invent new metaphors to describe what it has learned; and literary scholarship must constantly strive to destroy old metaphors to find out what really happened.

But our paths of inquiry, so separate yet so closely related, are nourished by activities such as these that take place on Sanibel. The University of Florida is proud to be a partner with Uppsala University in this annual adventure of the mind. We pay special tribute to the guiding genius of this activity, Dr. Per-Olov Löwdin, whose vision and energy have made these conferences possible. And I am particularly pleased that this particular symposium has been established by Dr. Sam Trickey and Dr. Dwight Adams, two of our most promising young faculty members. In closing, let me wish each and every one of you a happy and productive conference.

SURFACE SINGULARITIES AND SUPERFLOW IN ^{3}He-A*

N. D. Mermin

Laboratory of Atomic and Solid State Physics
Cornell University
Ithaca, N. Y. 14853

Abstract: Except in a toroidal container, the order parameter in ^{3}He-A must have singularities at the surface. Examples of such singularities are given, and a relation between the surface topology and the kinds of singularities is derived. Quantization of circulation along surface contours is discussed, and examples of the decay of superflow through the motion of surface singularities are given.

I. INTRODUCTION

I should like to discuss two related matters: (1) the connection between the topological character of the containing vessel and the singularities in ^{3}He-A at the surface,[1] and (2) some implications of these results for the quantization of circulation in ^{3}He-A.[2] These considerations suggest the kinds of structures one may have to contend with in interpreting experiments on resonance shifts and sound attenuation; they are also important in investigating the stability of persistent currents.[3]

The problem of superflow and singularities is more intricate in ^{3}He-A than in ^{4}He-II because the superflow is not determined by a scalar phase, but by the orientation of a pair of orthonormal

---

*Work supported in part by the National Science Foundation under Grant No. DMP 74-24394 and through the Materials Science Center of Cornell University (Technical Report No. 2775).

axes $\underset{\sim}{\phi}^{(1)}$ and $\underset{\sim}{\phi}^{(2)}$, whose plane, characterized by its normal $\underset{\sim}{\ell} = \underset{\sim}{\phi}^{(1)} \times \underset{\sim}{\phi}^{(2)}$, need not be fixed. As a result, allowed values of the integral of v_s about closed contours depend on the spatial variation of $\underset{\sim}{\ell}$, and the circulation about a fixed closed contour can vary continuously in time, given the appropriate variation in $\underset{\sim}{\ell}$.

The problem is simplest at the surface because a boundary condition constrains the plane of the $\underset{\sim}{\phi}^{(i)}$ to be the tangent plane. The most general motion of $\underset{\sim}{\ell}$ one need consider at the surface is therefore the motion of the singular borders between regions in which $\underset{\sim}{\ell}$ is parallel or anti-parallel to the outward normal $\underset{\sim}{n}$.

II. THE ORBITAL ORDER PARAMETER AND $\underset{\sim}{v}_s$

We first review the definition of $\underset{\sim}{v}_s$ in ^{3}He-A.[4] The pair of orthonormal axes $\underset{\sim}{\phi}^{(1)}$ and $\underset{\sim}{\phi}^{(2)}$ characterizing the orbital part of the order parameter can be regarded as the local x- and y-axes with respect to which the order parameter is proportional to $y_{11} \propto x+iy$. The velocity field $\underset{\sim}{v}_s$ is defined by taking $-2M\delta\underset{\sim}{r}\cdot\underset{\sim}{v}_s/\hbar$ as the component along the axis $\underset{\sim}{\ell} = \underset{\sim}{\phi}^{(1)} \times \underset{\sim}{\phi}^{(2)}$ of the infinitesimal rotation vector taking the $\underset{\sim}{\phi}^{(i)}$ at $\underset{\sim}{r}$ into the $\underset{\sim}{\phi}^{(i)}$ at $\underset{\sim}{r} + \delta\underset{\sim}{r}$. It follows from this that

$$(\underset{\sim}{v}_s)_i = \frac{\hbar}{2M}\, \underset{\sim}{\phi}^{(1)} \cdot \nabla_i\, \underset{\sim}{\phi}^{(2)} \ . \tag{1}$$

Because multiplying the order parameter by $e^{-i\theta}$ is equivalent to rotating the $\underset{\sim}{\phi}^{(i)}$ through θ about the local $\underset{\sim}{\ell}$, the variable $-2M\underset{\sim}{v}_s/\hbar$ is the generalization to ^{3}He-A of the gradient of the phase in an L=0 superfluid. If two configurations of the $\underset{\sim}{\phi}^{(i)}$ differ only by a multiplicative phase factor then the associated velocity fields v_s will differ by $-\hbar/2M$ times the gradient of that phase. It follows from this that $\underset{\sim}{v}_s$ behaves like a velocity under Galilean transformations. However, $\underset{\sim}{v}_s$ itself is not in general the gradient of a phase, because of the fundamental relation

$$(\nabla\times\underset{\sim}{v}_s)_R = \frac{\hbar}{4M}\, \varepsilon_{kij}\underset{\sim}{\ell}\cdot(\nabla_i\underset{\sim}{\ell}\times\nabla_j\underset{\sim}{\ell}) \ , \tag{2}$$

which follows directly from (1) and the orthonormality of the $\underset{\sim}{\phi}^{(i)}$.

III. $\mathbf{v}_s$ AT A SURFACE

In the London limit, the boundary condition[5] requires the $\boldsymbol{\phi}^{(i)}$ to lie in the tangent plane at a surface; i.e. $\boldsymbol{\ell}$ must be parallel or anti-parallel to the outward normal $\mathbf{n}$: $\mathbf{n}\cdot\boldsymbol{\ell} = \pm 1$. The surface will thus be divided into regions of positive and negative $\mathbf{n}\cdot\boldsymbol{\ell}$, separated by line singularities in the $\boldsymbol{\phi}^{(i)}$ which we shall call "borders."

Within a region of definite $\mathbf{n}\cdot\boldsymbol{\ell}$ the only degrees of freedom of the order parameter are associated with proper rotations of the $\boldsymbol{\phi}^{(i)}$ in the plane of the surface, and the topologically stable singularities[6] will be as in ^{4}He-II; singular points about which the axes rotate through an integral multiple of 2π. These can be regarded as the terminal points of quantized vortices of strength m. We shall count m as positive if the sense of rotation of the $\boldsymbol{\phi}^{(i)}$ is clockwise as the point is encircled in a clockwise sense. With this convention the sense of $\mathbf{v}_s$ in a positive vortex is counterclockwise when viewed along the direction of $\boldsymbol{\ell}$ (Fig. 1).

In addition to these point singularities, $\mathbf{v}_s$ will in general have singularities at the borders separating the regions of positive and negative $\mathbf{n}\cdot\boldsymbol{\ell}$. In considering these border singularities it is useful to exploit some properties of $\mathbf{v}_s$ in surface regions that contain neither point nor border singularities.

Let the surface be smooth enough that in the neighborhood of any point P it is given to second order by $z = 1/2ax^2+bxy+1/2cy^2$, in the coordinate system in which the x- and y-axes lie in the tangent plane. The Gaussian curvature of the surface is the determinant of the coefficients in this expression: $K = ac-b^2$. Now the normal to the surface is a unit vector in the direction of $\boldsymbol{\nabla}(1/2ax^2+bxy+1/2cy^2-z)$, and is therefore given to first order by $\mathbf{n} = (ax+by,\ bx+cy,\ -1)$. Consequently $K = (\nabla_x n_x)(\nabla_y n_y)-(\nabla_y n_x)(\nabla_x n_y)$, which can be written in the coordinate independent form:

$$K = 1/2\ n_k \varepsilon_{kij} \mathbf{n}\cdot(\nabla_i \mathbf{n} \times \nabla_j \mathbf{n}). \tag{3}$$

Since $\boldsymbol{\ell} = \pm\,\mathbf{n}$ at the surface, and all derivatives in (3) are in the tangent plane, it follows from (2) that at the surface

$$\boldsymbol{\ell}\cdot\boldsymbol{\nabla} \times \mathbf{v}_s = \frac{\hbar}{2M} K, \tag{4}$$

i.e. the normal component of the curl of $\mathbf{v}_s$ is just $(\hbar/2M)(\mathbf{n}\cdot\boldsymbol{\ell})$ times the Gaussian curvature of the surface.[7]

$$(a) \hspace{5cm} (b)$$

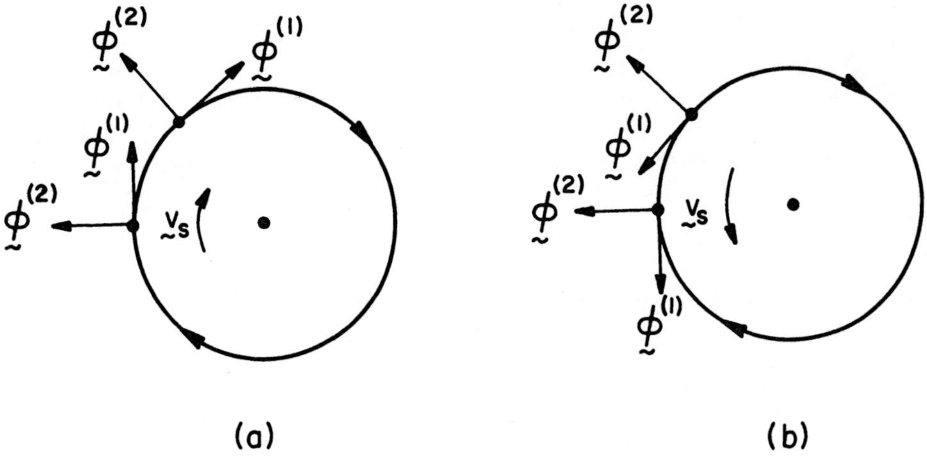

Figure 1. Vortices of strength m = +1. In (a) $\underset{\sim}{\ell}$ is out of the page; in (b) $\underset{\sim}{\ell}$ is into the page. In either part of the figure the circulation of $\underset{\sim}{v}_s$ (Eq. (2)) is counterclockwise when viewed along the direction of $\underset{\sim}{\ell}$.

In addition to this expression for $\underset{\sim}{\nabla}x\underset{\sim}{v}_s$ in terms of surface geometry, it is useful to have a closely related expression for surface line integrals of $\underset{\sim}{v}_s$ in terms of the contour geometry. Let the contour (which we take to lie in a surface region of definite $\underset{\sim}{n}\cdot\underset{\sim}{\ell}$) be parametrized by its arc length s, with unit tangent vector $\underset{\sim}{u}(s)$. We specify the $\phi^{(i)}$ along the curve in terms of the orthonormal pair of surface vectors $\underset{\sim}{u}$ and $\underset{\sim}{\ell}x\underset{\sim}{u}$:

$$\underset{\sim}{\phi}^{(1)} = (\underset{\sim}{u}x\underset{\sim}{\ell})\cos\theta + \underset{\sim}{u} \sin\theta \; ,$$

$$\underset{\sim}{\phi}^{(2)} = -(\underset{\sim}{u}x\underset{\sim}{\ell})\sin\theta + \underset{\sim}{u} \cos \theta. \hspace{2cm} (5)$$

The component of $\underset{\sim}{v}_s$ along the curve is then given by

$$\frac{2\mu}{\hbar} d\underset{\sim}{s}\cdot\underset{\sim}{v}_s = ds\underset{\sim}{\phi}^{(1)} \cdot \frac{d}{ds} \underset{\sim}{\phi}^{(2)}$$

$$= ds[\frac{d\theta}{ds} + \underset{\sim}{u}\cdot\underset{\sim}{\ell} \times \frac{d}{ds} \underset{\sim}{u}] \; . \hspace{2cm} (6)$$

If the contour is a geodesic then the rate of change of its tangent vector, $d\underset{\sim}{u}/ds$, has no component in the surface, and the second term on the right side of (6) vanishes. The quantization of circulation along a closed geodesic surface contour in a region of definite $\underset{\sim}{n}\cdot\underset{\sim}{\ell}$ is therefore the same as in ^{4}He-II (except for the factor of 2 associated with the pair mass).

More generally, it is useful to note that the quantity

$$\gamma = \underset{\sim}{\mu}\cdot\underset{\sim}{n} \times \frac{d\underset{\sim}{\mu}}{ds} \tag{7}$$

(known as the geodesic curvature) gives the angular rate (with respect to arc length) at which the curve turns (to the right in the direction of increasing s as viewed from outside the surface) in the flat local coordinate system. If a closed curve is small enough (or, more generally, if the region of surface it contains is flat enough) then the integral around the curve of its geodesic curvature will be 2π, and we again have the conventional quantization condition.

As an illustration of these results, consider a torus at the surface of which $\underset{\sim}{n}\cdot\underset{\sim}{\ell}$ has a single sign (Figure 2). Around the inner and outer geodesics, $(2M/h)\oint \underset{\sim}{v}_s\cdot d\underset{\sim}{s} = 2\pi\nu$; the same integer characterizes the circulation around both geodesics because the integral of the Gaussian curvature over an entire torus vanishes. On the other hand, the curves on the top and bottom of the torus turn through $\pm 2\pi$ in the flat coordinate system: the circulation about one of these contours has an extra quantum of vorticity; that about the other, has a quantum less. In general, the circulation about a circular contour specified by a latitude θ is given by

$$\frac{2\mu}{\hbar} \oint v_s\cdot ds = 2\pi(\nu + \sin\theta) \tag{8}$$

IV. BORDERS, VORTICES, AND SURFACE TOPOLOGY

As the border of a surface region of definite $\underset{\sim}{n}\cdot\underset{\sim}{\ell}$ is encircled, the angle the $\phi^{(i)}$ make with the border must change by an integral multiple of 2π. That integer is determined by the number of vortices reaching the surface within the region and the value of a certain topological invariant of the region (its Euler characteristic). To derive this relation [Eq. (12) below] we shall assume that in the interests of reducing the local line energy of the borders there will never be any borders with kinks, or (Figure 3) any intersections of borders. The borders will then form a set of differentiable non-intersecting closed surface curves. We parametrize each such curve with its arc length, choosing the sense of increasing s and the direction of the tangent vector $\underset{\sim}{\mu}$, so that

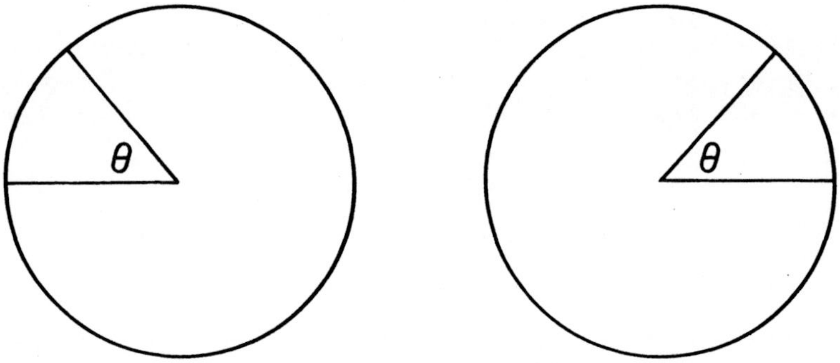

Figure 2. Cross section of a torus. The geodesic contours are given by $\theta = 0$ and $\theta = \pi$. The circulation quantization condition for a contour at general θ (assuming a single sign for $\mathbf{n}\cdot\boldsymbol{\ell}$ on the surface) is given by Eq. (8).

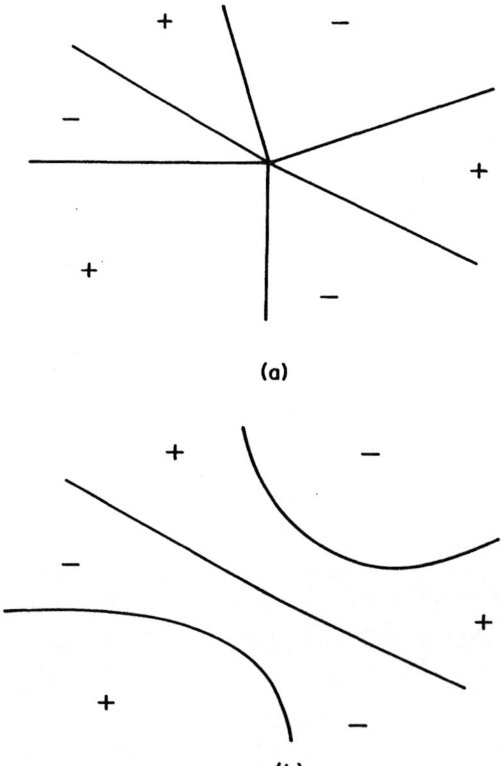

Figure 3. (a) A point where several borders separating regions of positive and negative $\mathbf{n}\cdot\boldsymbol{\ell}$ intersect. (b) Configuration of lower border line energy produced from (a) by separating the borders at their point of intersection.

$\underset{\sim}{u}\times\underset{\sim}{n}$ points to the side of positive $\underset{\sim}{n}\cdot\underset{\sim}{\ell}$ (Figure 4).

Consider now a region of uniform $\underset{\sim}{n}\cdot\underset{\sim}{\ell}$, bordered by closed curves C_i, i=1...N. Using the above sign convention and the expression (4) for $\underset{\sim}{\nabla}\times\underset{\sim}{v}_s$, we find from Stokes' Theorem that

$$\frac{2\mu}{\hbar} \sum \oint_{C_i} d\underset{\sim}{s}\cdot\underset{\sim}{v}_s = 2\pi m - \int dAK, \tag{9}$$

where m is the algebraic sum of the strengths of the vortex lines meeting the surface within the region (and the sign convention is as specified in Section III).

A second expression for the circulation of $\underset{\sim}{v}_s$ about the border is given by integrating (6):

$$\frac{2\mu}{\hbar} \sum \oint_{C_i} d\underset{\sim}{s}\cdot\underset{\sim}{v}_s = \sum \oint_{C_i} ds \frac{d\theta}{ds} + \sum_{C_i} \oint ds\, \underset{\sim}{u}\cdot\underset{\sim}{\ell} \times \frac{d\underset{\sim}{u}}{ds} \ . \tag{10}$$

To extract from these relations a condition on the change in θ about the border, we use the Gauss-Bonnet theorem.[8] This asserts that:

$$\int dAK + \sum_{C_i} \oint ds\, \underset{\sim}{u}\cdot\underset{\sim}{\ell} \times \frac{d\underset{\sim}{u}}{ds} = 2\pi E \ , \tag{11}$$

where E is the Euler characteristic of the region, defined as follows: Draw on the region a network of smooth edged polygons with simply connected interiors, the border of the region being made up of alternating edges and vertices of the network; then E = f−e+v, where f is the number of polygons ("faces") in the network, e is the number of edges, and v, the number of vertices (Fig. 5). The Fuler characteristic of a region can easily be shown to be independent of the particular network drawn, and is a topological invariant of the region.[9]

If. Eq. (9) is subtracted from (10), and (11) is used to evaluate the sum of the surface integral of the Gaussian curvature and the border integral of the geodesic curvature, then we arrive at a simple condition for the total change in the angle θ over the

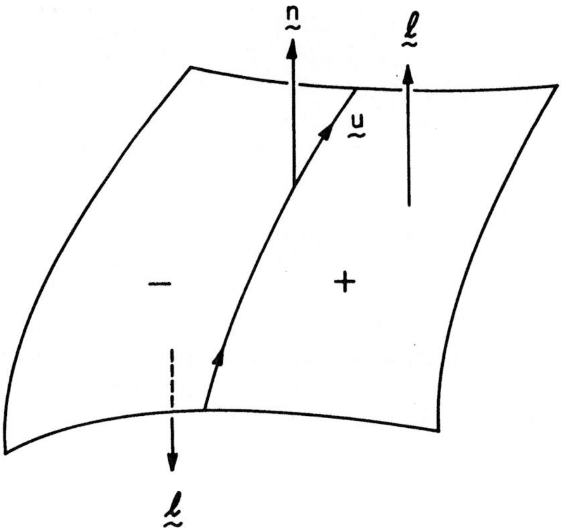

Figure 4. Direction convention for a surface border separating regions of positive and negative $\mathbf{n}\cdot\boldsymbol{\ell}$. The tangent vector $\mathbf{u}$ is oriented so that $\mathbf{u} \times \mathbf{n}$ points to the side of positive $\mathbf{n}\cdot\boldsymbol{\ell}$ (or, equivalently, so that on either side $\mathbf{u} \times \boldsymbol{\ell}$ points away from the border).

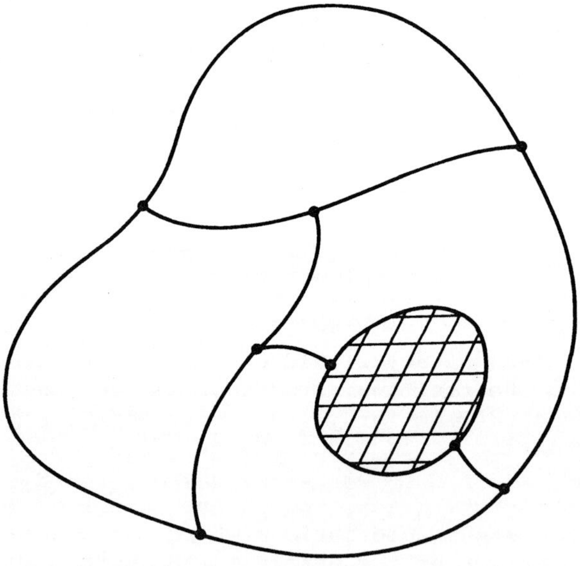

Figure 5. Construction for determining the Euler characteristic of a simple plane region with a hole (cross-hatched). The network of "polygons" has 4 faces, 12 edges, and 8 vertices, so $E = f-e+v=0$. It must be possible to shrink the circumference of each face to a point within the face; thus at least two faces must make contact with the inner hole.

closed borders of the region:[10]

$$\frac{1}{2\pi} \sum_i (\Delta\theta)_i = \frac{1}{2\pi} \sum_i \oint_{C_i} \frac{d\theta}{ds} ds = m - E \qquad (12)$$

This relation is useful in suggesting the kinds of singularities that will necessarily be present in specimens of ^{3}He-A in vessels of a given topology. We consider below several examples.

V. CONFIGURATIONS WITH $\underset{\sim}{\ell}$ REGULAR AT THE SURFACE

If $\underset{\sim}{n}\cdot\underset{\sim}{\ell}$ has a single sign over the entire surface, then the region for which (12) holds can be taken to be the whole surface; there are then no borders, and (12) reduces simply to:[1]

$$m = E .$$

Now any connected closed two-sided surface can be mapped continuously onto a sphere with n handles (Fig. 6) and its Euler characteristic is related to the number of handles by $E = 2(1-n)$.[11] Thus the unadorned sphere has E=2 and must have two quanta of vorticity at its surface if $\underset{\sim}{n}\cdot\underset{\sim}{\ell}$ is uniform over the surface. The torus (sphere with one handle) uniquely has E=0 and is the only container that can support ^{3}He-A free of any singularities whatever. The double torus (sphere with two handles) has E = -2. The constricted geometries of 4th sound experiments have enormous negative values of E.

The result for the sphere is illustrated by the solution[2]

$$\underset{\sim}{v}_s = \frac{\hbar}{2M\rho}(\nu - \ell_z)\underset{\sim}{\phi} \qquad (13)$$

to Eq. (3) appropriate to vessels with cylindrical symmetry, where ν is an integer and z, ρ, and ϕ are the cylindrical coordinates. If $\underset{\sim}{\ell}$ is everywhere outward, then ℓ_z will be +1 at the top of the vessel and -1 at the bottom. Any choice of the integer ν will therefore result in two net quanta of surface vorticity: a doubly quantized vortex at the top if $\nu = -1$, singly quantized vortices (of the same sign) at top and bottom if $\nu = 0$, a doubly quantized vortex at the bottom if $\nu = +1$, etc.

Except for vessels topologically equivalent to a torus, it may pay to have surface singularities in $\underset{\sim}{\ell}$ to avoid these singularities in $\underset{\sim}{v}_s$. The cost in energy will be proportional to the length of the borders. If all the borders surround simply connected regions, this can be reduced by shrinking down the borders,

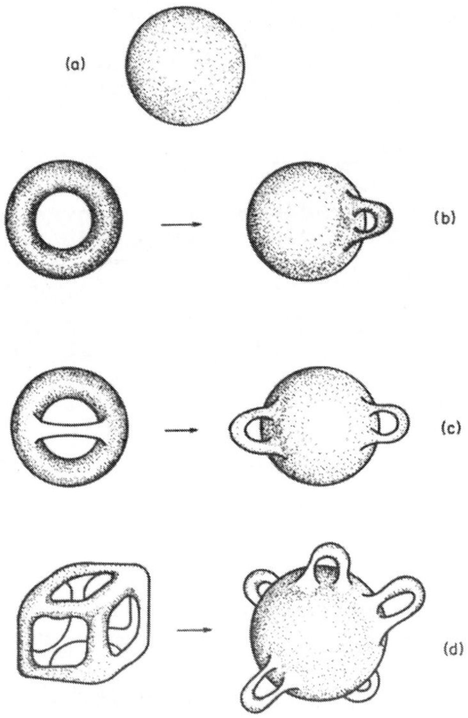

Figure 6. (a) The sphere (without handles); E=2. (b) The torus is equivalent to a sphere with one handle; E=0. (c) The bridged torus is equivalent to a sphere with two handles; E=-2. (d) The "tube cube" is equivalent to a sphere with five handles; E=-8.

without altering the topology. We therefore consider next a surface on which the sign of $\mathbf{n}\cdot\boldsymbol{\ell}$ is reversed on a single simply connected patch -- an "island" -- within a region of uniform $\mathbf{n}\cdot\boldsymbol{\ell}$.

VI. ISLANDS AND THEIR BORDERS

When $\mathbf{n}\cdot\boldsymbol{\ell}$ is not uniform over the whole surface then there will be separate constraints (12) for both the positive and negative regions. If we count the total number of vortices at the surface (regardless of the sign of $\mathbf{n}\cdot\boldsymbol{\ell}$) with the convention that a vortex is positive if $\mathbf{v}_s$ turns counterclockwise when viewed along $\mathbf{n}$ (i.e. clockwise, when viewed from outside the surface), then $m_{total} = m_+ - m_-$, and we have

$$m_{total} = E_+ - E_- + \frac{1}{2\pi}[\Delta\theta_+ - \Delta\theta_-] \qquad (14)$$

Suppose that $\underset{\sim}{n}\cdot\underset{\sim}{\ell}$ is uniform (and positive) over the whole surface except for a number of islands of negative $\underset{\sim}{n}\cdot\underset{\sim}{\ell}$. Each such island reduces E_+ by one (since the reversed region can be viewed as a single face, removed from an appropriately drawn network of polygons). In addition, each island raises E_- by one, since it augments the negative network by a single isolated polygon (easily verified to have an Euler characteristic of unity). The effect of N islands is therefore to give a value of $E_+ - E_-$ related to the Euler characteristic E of the container by

$$E_+ - E_- = E - 2N. \tag{15}$$

Now E is negative or zero for all connected containers other than those topologically equivalent to spheres. Therefore, except for the case of a single island in a spheroidal container, the effect through (15) of introducing islands to increase the magnitude of the number of quanta of vorticity in (14), and the islands will have to have borders with appropriately discontinuous $\Delta\,\theta$, if their presence is to result in a net decrease in the total number of vortices.

The discontinuity at a border in the angle θ (defined in Eq. (5)) depends on the detailed structure of the border. Consider (Fig. 7a) a little semicircle centered on the border that penetrates into the liquid in the plane perpendicular to the border. As this semicircle is traversed $\underset{\sim}{\ell}$ must reverse its direction. A particularly simple way to do this is by a rotation through 180° in the plane of the semicircle. Such borders (which we shall call "twistless") are the cores of circular or hyperbolic disgyrations (Figs. 7b-7e) depending on whether $\underset{\sim}{\ell}$ turns in the same or opposite sense as the circumference of the semicircle.

A divergence in the magnitude of $\underset{\sim}{v}_s$ at the border can only be avoided if the $\underset{\sim}{\phi}^{(i)}$ undergo no rotation about the local $\underset{\sim}{\ell}$ as the semicircle is traversed. For twistless borders this requires (Fig. 8) continuity of the components of the $\phi^{(i)}$ in the surface parallel to the border, and a sign reversal of the perpendicular components. Comparing this with Eq. (5), we find that θ is continuous across twistless borders.

As a result of the continuity of θ, islands with twistless borders will <u>increase</u> the number of surface vortices except for the case of a single such island in a container with E=2. Such an island is shown in Fig. 9. The reversal of $\underset{\sim}{n}\cdot\underset{\sim}{\ell}$ over part of the surface permits ℓ_z to have a single sign over the entire axis and Eq. (13) (with $\nu = -1$) gives no vortex singularities in $\underset{\sim}{v}_s$. Note, though, that (13) does give a discontinuity in $\underset{\sim}{v}_s$ at the border, unless the border is at the equator, where $\ell_z = 0$. This is required quite generally at non-geodesic twistless borders by Eq. (6),

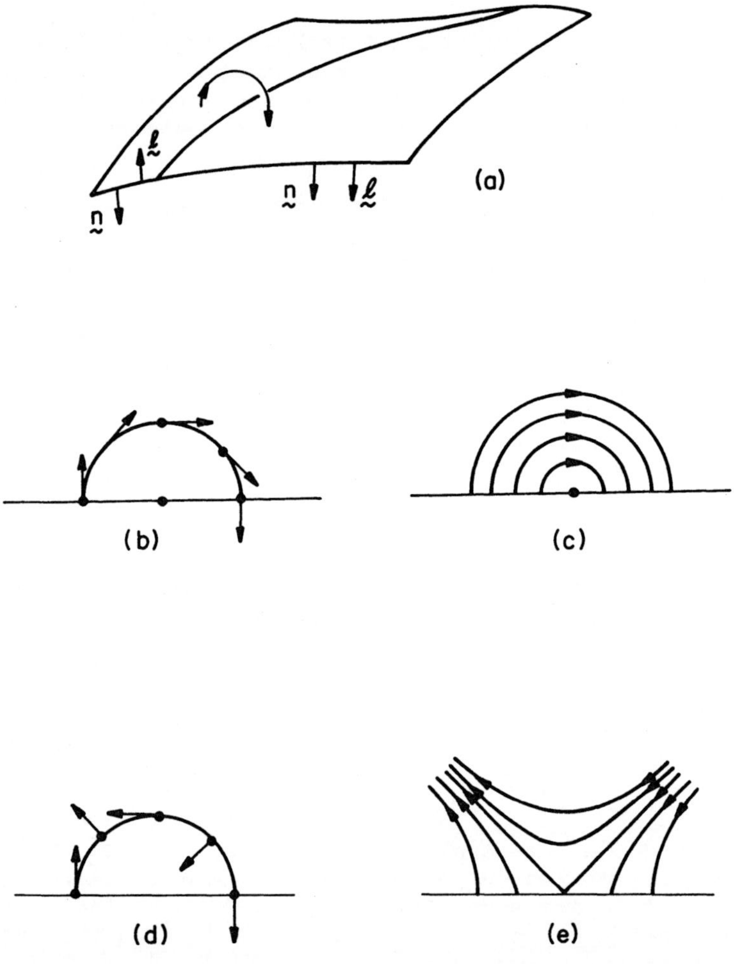

Figure 7. (a) Portion of a surface with a border separating regions of positive and negative $\underset{\sim}{n}\cdot\underset{\sim}{\ell}$. A semicircular path through the liquid joins the two sides. (b) If $\underset{\sim}{\ell}$ rotates in the same sense as the circumference of the semicircle about an axis parallel to the border the circular disgyration pattern (c) results. (d) If $\underset{\sim}{\ell}$ rotates in the opposite sense as the circumference of the semicircle about an axis parallel to the border, the hyperbolic disgyration pattern (e) results.

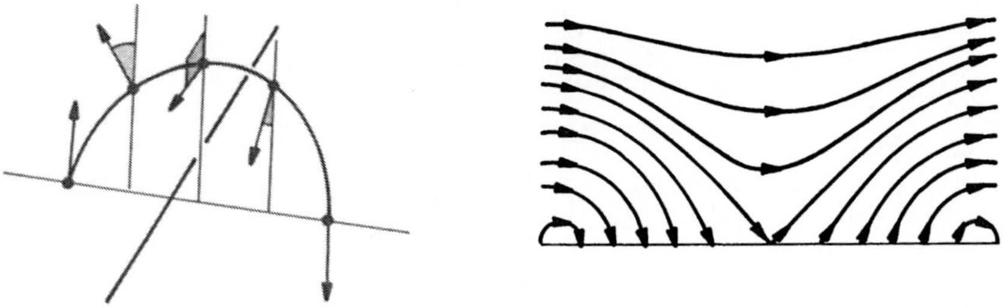

(f) (g)

Figure 7 (cont.) (f) A border with twist, in which ℓ rotates about an axis perpendicular to the border as the border is crossed. (g) Alternating circular and hyperbolic disgyrations can be used to match alternating surface regions of positive and negative $n\cdot\ell$ onto a uniform texture farther in the interior.

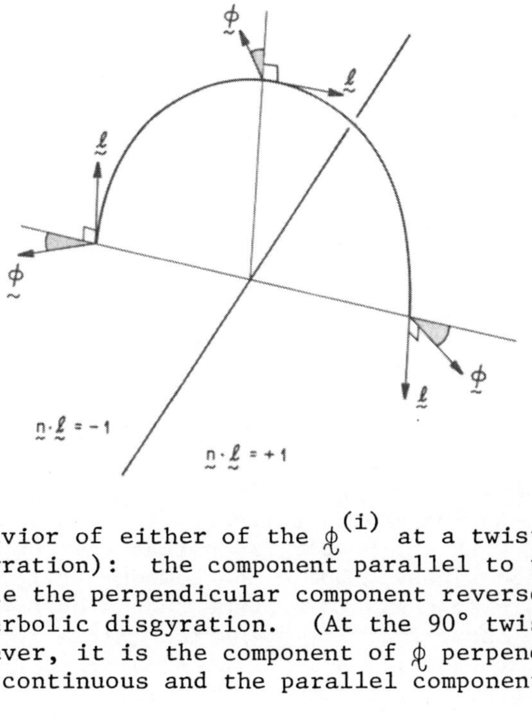

Figure 8. Behavior of either of the $\phi^{(i)}$ at a twistless border (circular disgyration): the component parallel to the border is continuous while the perpendicular component reverses. The same holds at a hyperbolic disgyration. (At the 90° twist border of Fig. 2(f), however, it is the component of ϕ perpendicular to the border that is continuous and the parallel component that changes sign.)

since continuity of θ across the border implies a discontinuity in the tangential component of v_s proportional to the geodesic curvature. The island of reversed $n\cdot\ell$ can reduce its border energy by shrinking down to a point island at the north pole (Fig. 10). In this limit the discontinuity in v_s across its border increases

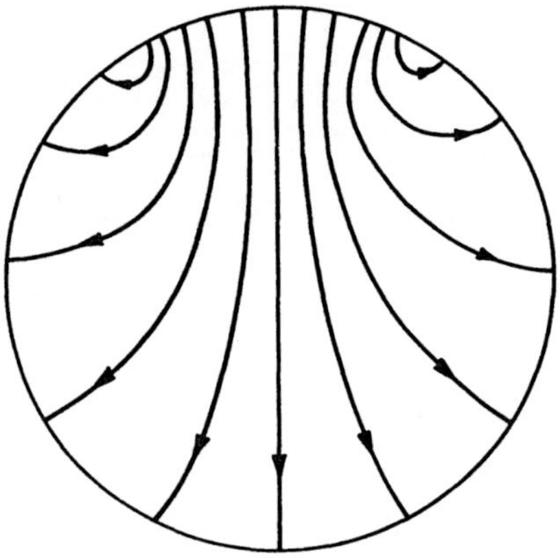

Figure 9. A simply connected surface region of reversed $\underset{\sim}{n}\cdot\underset{\sim}{\ell}$ eliminates the need for surface vortices in a sphere. The vortex free texture is given by (13) with $\nu = -1$.

and it is evident from (13) (with $\nu = -1$) that the circulation just outside of the island approaches that about a doubly quantized surface vortex point. Thus the border energetics of the island have resulted in the restoration of the two quanta of surface vorticity that the island was introduced to eliminate, with the advantage that there is no longer any associated singularity in the bending energy.[12]

If the container does not have the topology of a sphere, then if the number of vortices required by (14) is to be decreased by the inroduction of islands, their borders cannot be twistless. A border with twist can be characterized by a surface unit vector $\underset{\sim}{a}$ making an angle β with the border, such that the components of the $\phi^{(i)}$ along $\underset{\sim}{a}$ are continuous, while the components perpendicular to $\underset{\sim}{a}$ change sign. A simple way to produce such a border is for $\underset{\sim}{\ell}$ to turn through 180° about the axis $\underset{\sim}{a}$, as the circle in Fig. 7 is traversed. If, for example, $\underset{\sim}{a}$ is perpendicular to the border, the result is the twisted disgyration of Fig. 7f.

If β is constant along the border (a condition I cannot refrain from characterizing as one of fixt twist) then the discontinuity in θ across the border is also constant, and the discontinuity in $\Delta\theta$ vanishes: borders of fixt twist play the same topological role as twistless borders. To reduce the number of surface vortex points in containers other than the sphere or torus one requires islands with borders of mixt twist.

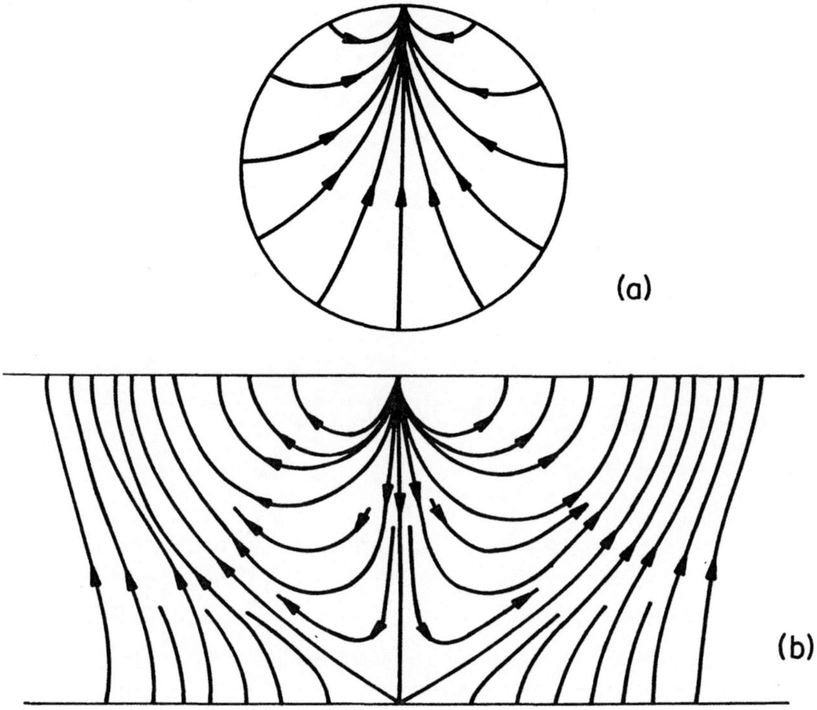

Figure 10. (a) The region of reversed $\underset{\sim}{n}\cdot\underset{\sim}{\ell}$ in Fig. 9, contracted to a point island at the north pole. This pattern is probably the most stable in a spherical container. Its many independent discoverers have proposed that it be called the flower, the fountain, the bouquet, and the boojum. (b) Two types of point islands at plane parallel surfaces normal to the page. The figure has axial symmetry about the central vertical line. The upper pattern is the flower; the lower one is made by shrinking to a point a circular border made from the hyperbolic disgyration of Fig. 7e. It no longer looks reminiscent of flowers, fountains, or bouquets, so by default it must be a hyperbolic boojum. T. L. Ho has shown (private communication) that both point singularities have convergent bending energies, by taking the lines of $\underset{\sim}{\ell}$ in the neighborhood of the singular points to be given by $z^2 = \pm A\rho^2 + B\rho^{\pm 1}$. The combined pattern in the bulk with v_s given by (13) with $\nu = -1$, constitutes a simple Anderson-Toulouse vortex texture joining the two surfaces.[12] (The term "boojum" was introduced because it was all that remained after the unstable "monopole" singularity "softly and suddenly vanished away."[13] The case for this nomenclature is enormously enhanced by the fact that the supercurrent itself may suffer this sad and celebrated catastrophe, should it encounter such a singularity. See Figs. 13 and 14.)

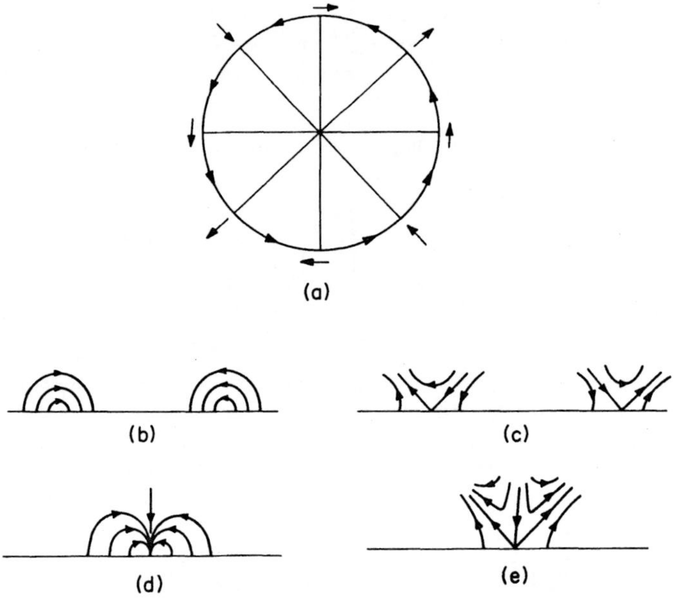

Figure 11. The simplest island of mixt twist. (a) The border is a circle. As the border is crossed $\underset{\sim}{\ell}$ rotates about an axis (shown outside the circle at 45° intervals) whose orientation turns in the opposite way to the border as the island is circumnavigated. (b) View of $\underset{\sim}{\ell}$ in a plane perpendicular to the horizontal line in (a). (c) View of $\underset{\sim}{\ell}$ in a plane perpendicular to the vertical line in (a). The border crossings along the diagonal lines in (a) have the twisted character of Fig. 7f. (d) and (e) The configuration of $\underset{\sim}{\ell}$ in planes perpendicular to the horizontal and vertical lines of (a) after the border has been shrunk to a point. This type of surface point singularity is likely to be important in any container whose topology is more complicated than a sphere or torus.

A very simple island of mixt twist is shown in Fig. 11. As the border is encircled the axis $\underset{\sim}{a}$ undergoes one complete revolution in the sense <u>opposite</u> to the tangent vector. As a result, the angle β changes by 4π as the border is encircled, and the discontinuity in θ around the border is 8π. Such an island, therefore, contributes +4 to the term in $\Delta\theta$ in (14) which more than compensates for the contribution of −2 to the term in $E_+ - E_-$. Thus a double torus (Fig. 12) with E = −2, can avoid any vortices by having one such point island, and a very simple configuration with no singular bending energy for a sphere with n handles, will have n−1 such point islands of mixt twist.

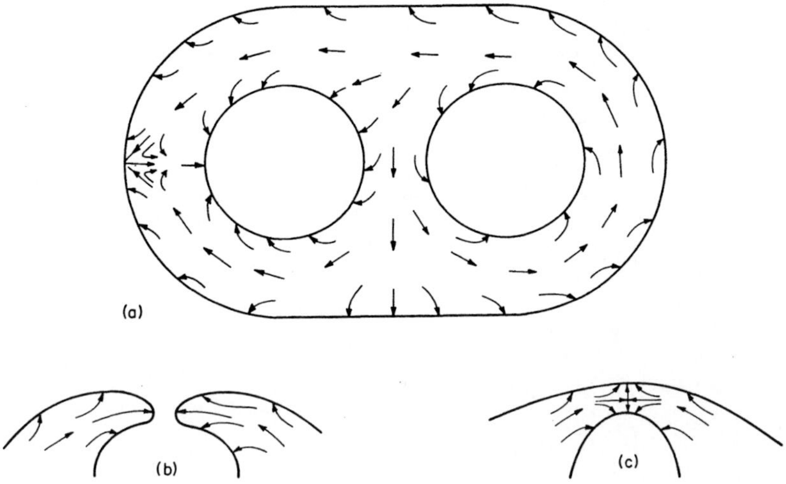

Figure 12. (a) ^{3}He-A inside of a double torus: E = -2. A single surface point singularity of mixt twist of the type shown in Figure 11 will satisfy the surface singularity condition (14). It appears at the bottom of the figure, and is the only singularity, surface or bulk. An N+1 hole torus (i.e. a sphere with N+1 handles) can be provided with all its necessary singularities by N such surface point singularities. The first two steps in the production of the singularity attendant upon the creation of the N+1th handle are shown in (b) and (c). When the semi-hyperbolic hedgehog produced in (c) moves to the surface it becomes the singularity at the bottom of (a).

VII. FLOATING ISLANDS AND THE DECAY OF SUPERFLOW

Anderson and Toulouse[12] have pointed out that the motion of non-singular vortex textures can lead to dissipation without the need for nucleating highly singular cores. Some particularly simple examples of this are furnished by considering the motion of surface singularities. If (6) is integrated around the exterior of a twistless island (with negative $\underset{\sim}{n}\cdot\underset{\sim}{\ell}$ and no singularities in its interior) then the net circulation is that of two negative vortices. If, instead, such an island has the mixt twist described in the preceding section, the circulation will be that of two positive vortices. If either type of island is present (and for all but the simplest containers the second type is quite likely to be present) then its periodic motion about the walls of a channel carrying superflow will result in the loss of two quanta

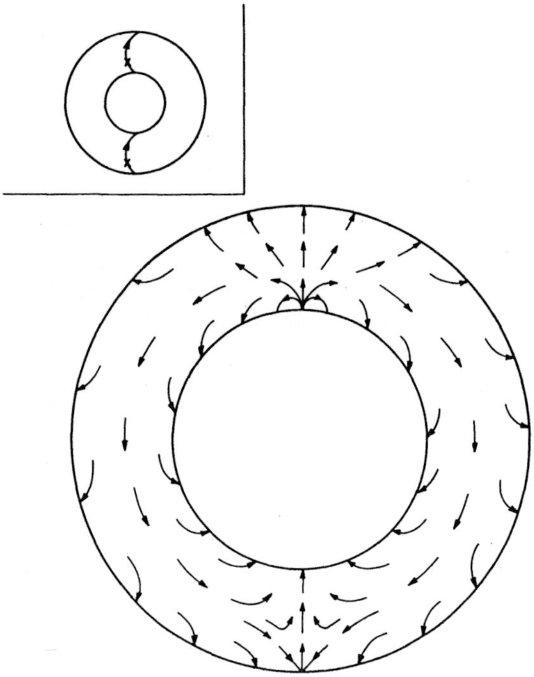

Figure 13. A torus need have no surface singularities. If it should, however, have a simple twistless circular boojum (top of torus) then this must be compensated for by at least one other surface singularity. If the second singularity is also a point island, it must be the one of mixt twist shown in Fig. 11. This twisted boojum is shown at the bottom of the torus in the cross section in which it resembles the hyperbolic twistless boojum of Fig. 10 (but note that the cross section perpendicular to the page will resemble the circular boojum). If either of the booja encircling its part of the torus (see inset) then two quanta are subtracted from the circulation about the entire torus. Such singularities can therefore catalyze the decay of a supercurrent until they drift together and annihilate. In containers of higher connectivity than the torus a certain number of such booja will be forbidden on topological grounds from so vanishing.

of circulation with each circuit of the walls (Fig. 13). Such singularities act as catalysts for the decay of superflow, without the need for any nucleation processes.

This last example should indicate the importance of a thorough study of surface singularities in the A-phase, a subject whose topological (and nomenclatural) possibilities deserve a thorough exploration.

"...beware of the day,
If your Snark be a Boojum!"

Figure 14. An early disquisition on booja (after Peter Newell). The cat apparently heeded the warning. The baker did not.

References

(1) For a discussion of the special case in which the only surface singularities are in v_s, see N. D. Mermin, "Games to Play with ^{3}He-A," Physica, to be published (Proceedings of the Sussex Symposium on Superfluid ^{3}He).

(2) The analysis here is based on the discussion of v_s given by N. D. Mermin and Tin-Lun Ho, Phys. Rev. Lett. 36, 594 (1976).

(3) The importance of these considerations for the stability of persistent currents was brought home to me by several very stimulating remarks of P. W. Anderson, and a ferocious lunch-time discussion with M. E. Fisher.

(4) For a more detailed exposition of these points, see Ref. 2. It should be emphasized that $\underset{\sim}{v}_s$ and the components of the gradient of $\underset{\sim}{\ell}$, together with the constraint (3) constitute the standard set of order parameter gradients for the hydrodynamic description of a system whose order is characterized by an orthonormal triad of vectors. The constraint on the curl of $\underset{\sim}{v}_s$ is simply the local integrability condition insuring that the $\underset{\sim}{\phi}^{(i)}$ can be reconstructed from a knowledge of

(4) (cont.) $\mathbf{v}_s$ and the gradients of $\boldsymbol{\ell}$ (in analogy to $\nabla \times \mathbf{v}_s = 0$ in ^{4}He-II, which is the constraint necessary to permit the reconstruction of the phase). The field $\mathbf{v}_s$ appears in the classical differential geometry of surfaces (see, for example, the discussion of the Gauss-Bonnet theorem in Ref. 7). It has been used to discuss singularities in nematics by M. Kleman, Phil. Mag. 27, 1057 (1973), and J. de Physique 34, 931 (1973).

(5) V. Ambegaokar, P. G. de Gennes, and D. Rainer, Phys. Rev. A 9, 2676 (1974), Phys. Rev. A 12, 345 (1975).

(6) G. Toulouse and M. Kleman, J. de Physique Lettres, 37, 149 (1976).

(7) A result apparently first proved by Gauss himself. For a particularly clear discussion see Louis Brand, "Vector and Tensor Analysis," Wiley, N.Y., (1947). Brand characterizes Gauss's relation between curl $\mathbf{v}_s$ and K as "perhaps the most important result in the theory of surfaces!" Note also that the boundary condition requires $\nabla \cdot \boldsymbol{\ell}$ to approach the mean curvature, a+c, provided that $\nabla \times \boldsymbol{\ell}$ is non-singular at the surface.

(8) See, for example, Brand, Ref. 7, p. 308ff.

(9) A very readable discussion of the Euler characteristic and the topology of two dimensional manifolds is given by William S. Massey, "Algebraic Topology: An Introduction" (Harcourt, Brace & World, N. Y., 1967).

(10) A simpler but more verbose argument leading directly to this conclusion without the differential geometry, can be constructed along the lines of the argument given in Ref. 1.

(11) Ref. 9, Theorem 5.1, Chapter 1. The number of handles, n, is called the "genus" of the surface.

(12) Such twistless islands are just the surface termination points of the doubly quantized coreless vortices discussed by P. W. Anderson and G. Toulouse, Phys. Rev. Lett. 38, 508 (1977).

(13) See Lewis Carroll, "The Hunting of the Snark," Second fit (last verse), Third fit (verses 10 and 14), and Eighth fit (verse 9).

TEXTURES OF ^{3}He-A IN A SPHERE: TOPOLOGICAL THEORY OF BOUNDARY EFFECTS AND A NEW DEFECT*

P. W. Anderson† and R. G. Palmer

Joseph Henry Laboratories of Physics
Princeton University
Princeton, New Jersey 08540

Brinkman[1] in 1974 was the first to comment on the possibility of point defects in the ^{3}He anisotropic superfluids, and he gave what is surely the correct answer: There are point defects in the B liquid when dipolar energy is considered, but not in the interior of A because of the well-known topological difficulty of "sign-posting the sphere": mapping the orientation of rigid triads continuously onto a sphere, so that if, for example, the pair angular momentum vector $\hat{\ell}$ points radially outward from a supposed defect, no nonsingular assignment of phases is possible. By phase we mean the orientation of the two gap parameters vectors $\vec{\Delta}_1$ and $i\vec{\Delta}_2$ about $\hat{\ell}$. This follows from the complete topological theory of order parameter singularities given by Toulouse and Kleman,[2] who show that the only topologically stable defect of the orbital portion of the order parameter in the interior of ^{3}He-A is one type of line, which can be deformed continuously from 2π vortex to disgyration to -2π vortex.

More recently, several papers have appeared[3,4] describing point defects ("monopoles") in the A liquid, and claiming that the normal physical boundary condition that $\hat{\ell}$ be perpendicular to the surface would enforce a point defect in a sphere filled with the A liquid. Two possible geometries commonly suggested are (1) A point defect with a 4π vortex attached (Fig. 1a), (2) A point defect connecting two opposite sign 2π vortices (Fig. 2a).

---

*Supported in part by the National Science Foundation Grant DMR 76-00886.

†Also at Bell Laboratories, Murray Hill, New Jersey 07974.

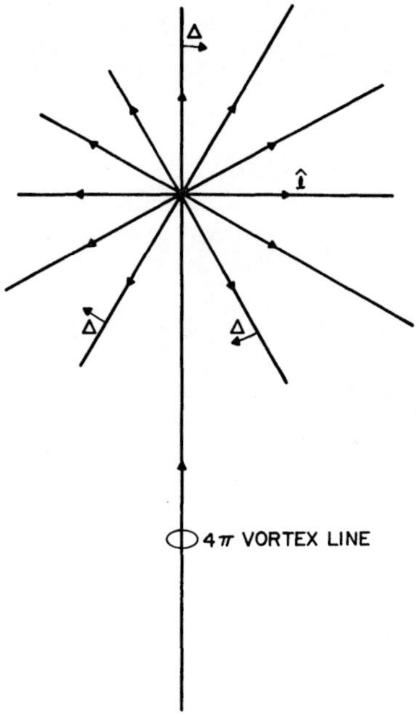

Figure 1a. Point with 4π vortex line.

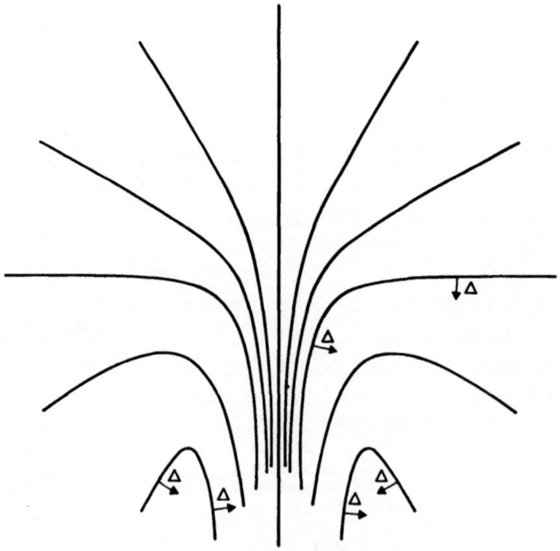

Figure 1b. 4π vortex texture terminated by hedgehog texture.

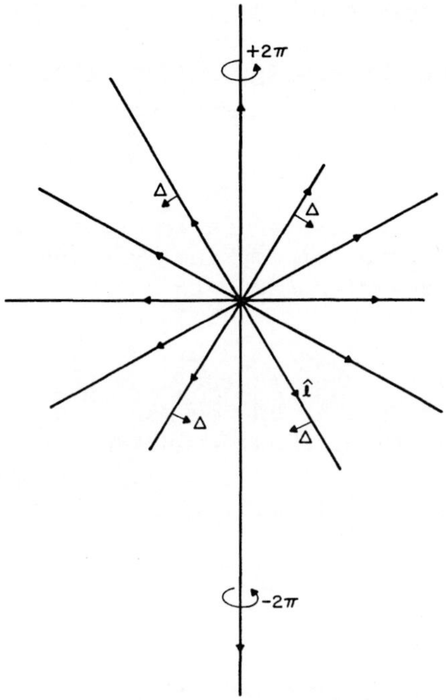

Figure 2a. Point singularity with ±2π vortex lines attached.

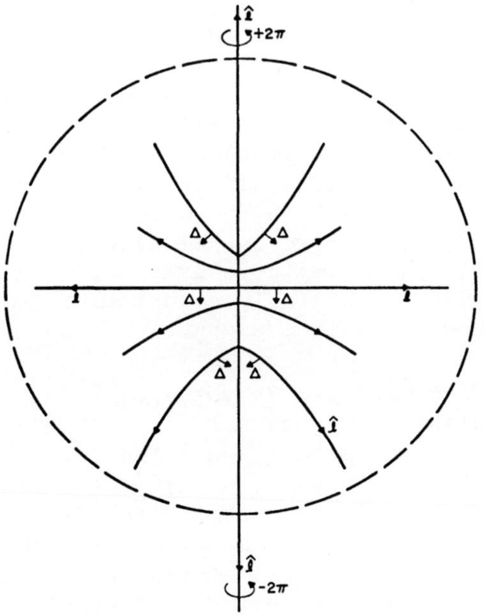

Figure 2b. Vortex line changing continuously from +2π to -2π by rotation of axis of 2π rotation.

The purpose of this paper is:

(a) To show that the configurations (1) and (2) above can indeed by deformed continuously into textures with no interior singular points, and are thus indeed dynamically as well as topologically unstable;

(b) To suggest general methods for topological classification of surface defects allowed or implied by given boundary conditions, and to show how these surface defects are related to bulk defects or textures.

(c) To apply these ideas to the classification of boundary defects of ^{3}He-A in a sphere, and by matching these to bulk singularities or textures, to find all solutions the "^{3}He-A in a sphere" problem. Many of these solutions have been proposed speculatively by Mermin[5] at the Sussex Symposium.

INSTABILITY OF "MONOPOLE" SOLUTIONS

The texture with no singularities which mimics a point defect with a single 4π vortex tail (Fig. 1a) has already been given in a previous letter.[6] In that letter it was shown that a 4π "vortex texture" can be created which can terminate in a "hedgehog" texture in which $\hat{\ell}$ simply splays out in radial fashion. The texture is created by starting with $\hat{\ell}$ in say the +z direction and rotating around the radial direction as one goes out along any radius: by π as one goes out from the "stem" or 4π vortex, by $\pi/2$ at the termination, and of course, not at all at the top of the diagram (Fig. 1b). Figure 1b clearly contains no $R\ell n\ (R/\xi)$ energy term for the 4π vortex because the gradients and currents are everywhere finite, of order $1/R$ (where R is the size of the vessel) and one see easily that it is therefore dynamically stabler than the point defect from which it is a simple relaxation. Thus a point with 4π vortex in free space can simply deform into (1b) continuously along a path of decreasing free energy.

The point defect with two 2π vortices (Fig. 2a) can be continuously deformed into a segment of disgyration line (Fig. 2b). If one has $\hat{\ell}$ in the +z direction above the point, and in the -z direction below, one sees that $\vec{\Delta}_1$ and $i\vec{\Delta}_2$, the real and imaginary parts of the gap, rotate in the same sense about the vortex line (let us say to the right) by 2π as one makes a complete circuit. At the point $\hat{\ell}$ is perpendicular to the line and it, in turn, rotates to the right by 2π around the vortex. Clearly, instead of turning from +z to -z <u>discontinuously</u> at the point, one may rotate it continuously about (say) $i\Delta$, and in that case the $+2\pi$ vortex will deform <u>continuously</u> into a 2π disgyration near the original point and then into the -2π vortex line. Again, we are merely replacing infinite gradients of the order parameter by finite ones, and not otherwise changing the configuration: this must lead to a lowering of the current and other $(\Delta d_{ij})^2$ energies.

CLASSIFICATION OF SURFACE DEFECTS

There are three distinguishable problems in the classification of surface defects. Firstly, one must decide what types of defect are possible (topologically stable). Secondly, one must ask what defects, or set of defects, are necessary, in the sense that they are required by the global character (topology) of the boundary. Thirdly, one must discuss how various surface defects are to be matched to bulk singularities or textures.

The boundary conditions imposed on the order parameter at the surface restricts the space of allowable order parameters to a subspace of that in the bulk. In general, the particular subspace singled out will depend on the local character - usually orientation - of the surface. Thus, in ^{3}He-A, the $\hat{\ell}$ - perpendicular condition depends on the local direction of a surface - normal. This coupling of order parameter space to real space complicates the application of simple topological principles to the problem. However, the Toulouse-Kieman theory may be applied locally on a surface, since any physical surface may be regarded as locally flat. We therefore surround a point defect by a closed loop (1-dimensional sphere), and line defect by two points (0-dimensional sphere), and then ask for the homotopy group for mappings from these surrounding spaces onto the manifold of the order parameter, restricted by the appropriate boundary condition. This completes the classification of possible surface defects.

To answer the second question - what defects are necessary - one must consider the global nature of the surface; defects needed in a spherical boundary will be quite different from those in a torus.

Geometrical intuition and the topological theorems connected with the Euler characteristic appear sufficient to treat most cases of physical interest. The Euler characteristic, χ, of a two dimensional manifold may be defined by

$$V - E + F$$

where V, E, and F are the number of vortices, edges, and faces in any triangulation of the manifold. χ is 2 for a sphere, 0 for a torus and 2n-2 for a pretzel with n holes. It can be proved that χ is equal to the sum of the indices of all point singularities of a vector field defined to be tangent to the manifold (the index of a point singularity is the number of turns of the vector field along a path encircling the point; ±1 for a 2π vortex point, etc.). This theorem, and straightforward generalizations for unsigned vectors, etc., solves the problem only for point defects, but line defects may be easily treated intuitively.

The third consideration concerns the matching of surface defects to bulk defects. If there is a singularity on the boundary, corresponding to irreducible paths in a given surrounding manifold (circle for a point, two points for a line), there may or may not be a corresponding bulk singularity (line or boundary) since the paths in the larger order parameter space of the bulk will allow more general deformations than in the boundary and may become reducible. Thus there can either be a singularity in the bulk terminating in the surface (line terminating in a point, for instance) or only a texture, leaving one with an isolated boundary singularity. We will encounter examples of both cases below.

^{3}He-A IN A SPHERE

At a surface the accepted boundary condition for ^{3}He-A is that $\hat{\ell}$ is perpendicular to the surface, but its sign and phase are arbitrary: $\vec{\Delta}_1$ and $i\vec{\Delta}_2$ can rotate at will and form either a left- or right-handed pair of vectors. Thus the manifold of the order parameter is $Z_2 \times SO(2)$: $\pm\hat{\ell}$ times a phase. The "orientation" of the SO(2) group, considered as a subset of the $SO(3)P_3$ manifold of unrestricted order parameters in the bulk, depends on the orientation of the surface at the point considered. The Toulouse-Kieman theory, or simple observation, shows that possible surface defects are lines, where $\hat{\ell} \to -\hat{\ell}$, and vortex points of any charge, $\pm 2\pi n$. The Euler characteristic theorem tells us that, in the absence of line defects, we need a total of 4π of vortex charge. Using these principles, we now discuss the possible sphere solutions.

An obvious solution on the surface is to have two 2π vortex points, with $\hat{\ell}$ everywhere outward (or everywhere inward). The vortex points are clearly terminations of vortex lines in the bulk. A single vortex line strung between the two surface points will not do (it would require one $+2\pi$ point, one -2π), but the configuration of Fig. 2b clearly satisfies all conditions. Fig. 2b is thus a solution to the sphere problem.

Another surface solution is a single 4π vortex point, with $\hat{\ell}$ everywhere outward (or inward) as before. Such a vortex point can attach to no stable line defect in the bulk, so it implies no bulk singularity, only a texture.

What texture is obvious: the 4π vortex point must terminate the Anderson-Toulouse 4π vortex texture! (See Fig. 3). It is this configuration which we have named a fountain[7] and observe that it is an example of a new class of order parameter singularities: boundary singularities of bulk textures. The rule for construction of the"fountain" must roughly be that starting at the origin and going out radially along a radius r at an angle Θ from the z axis, the order parameter is rotated about the radial direction through an angle $\varepsilon = \pi$ and $\hat{\ell}$ is always down. Clearly for large enough r we can change this rule in such a way as to fit onto a 4π vortex texture if we wish.

Other sphere solutions having only point defects on the surface (such as the vortex points of charges 4π, 2π, -2π) are clearly possible, but will certainly be higher in energy than those already considered. We therefore turn to solutions with surface line singularities. The only interesting solution of this class is one in which the sense of $\hat{\ell}$ changes across the line singularity (line singularities in the phase alone are topologically unstable according to the Toulouse-Kieman theory). Consider, for example, and equatorial line singularity with $\hat{\ell}$ outward in the upper hemisphere and inward in the lower hemisphere. A line boundary on the surface must be attached to a surface boundary in the bulk, but there are none of those in ^{3}He-A, so it implies no bulk singularity,

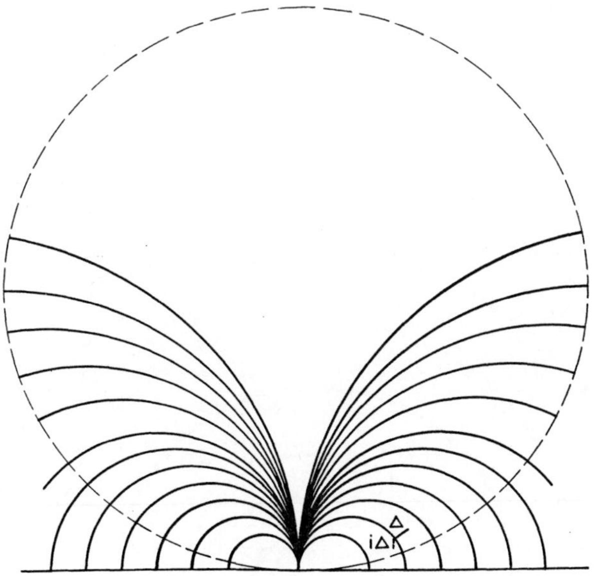

Figure 3. "Fountain": stream lines of $\hat{\ell}$.

only a texture. In fact, a $\pm\hat{\ell}$ boundary may be locally regarded as a half-disgyration on the surface (Fig. 4a). Globally a half-disgyration, with $\vec{\Delta}$, say, along the axis, cannot be used to solve the sphere problem without further point singularities, but this problem may easily be overcome (Fig. 4b) by allowing the phase to rotate along with $\vec{\Delta}$ and $i\vec{\Delta}$ alternately along the axis (making one revolution in all around the sphere). This defect cannot be deformed away except by shrinking it up to and off, say, the north pole, where it leaves behind a 4π vortex point. Such a process may well be prevented energetically in containers with a neck or with a hourglass shape.

In all, we have discussed three strong contenders for the solution to the sphere problem, shown in Figures 2b, 3, and 4b. In a true sphere it seems likely that the fountain (Fig. 3) is the best solution, being the only one entirely devoid of logarithmically divergent energies. By deforming the vessel to a shape still topologically equivalent to a sphere, it is probably possible to make any of these textures into the most energetically stable one. In a constricted hourglass shape the equatorial half-disgyration could well end up the lowest, in spite of its $\ln 1/\xi$ energy proportional to the length of the disgyration. It seems easy to design containers in which these defects could be probed by NMR or ultrasonic attenuation techniques. An observation of the fountain would be almost unique determination of the triad nature of the A-phase order parameter, if such is needed.

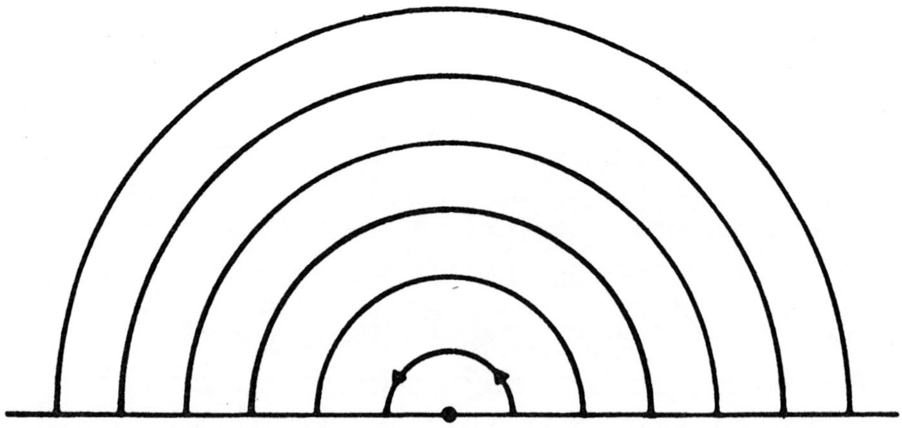

Figure 4a. Half-disgyration at a surface.

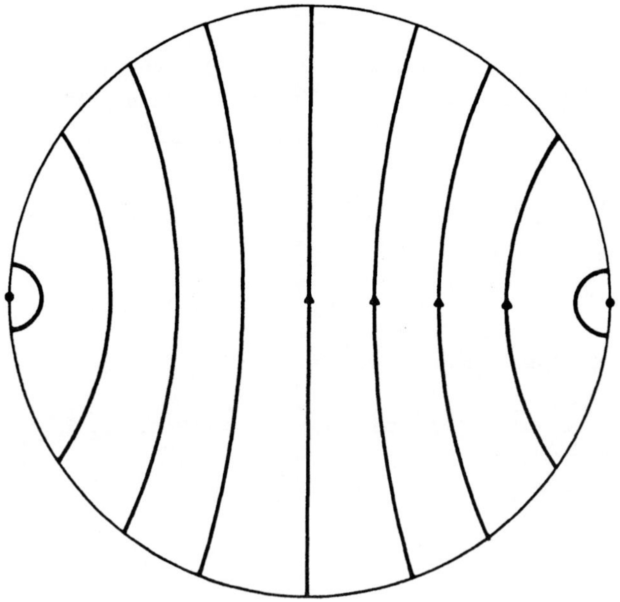

Figure 4b. Sphere with equatorial half-disgyration. A section perpendicular to this (also through the poles) would show the same stream lines of $\hat{\ell}$, but iΔ would replace Δ in the figure.

In other geometries, not topologically equivalent to a sphere, the defect configurations may be quite different. The methods of this paper may be easily applied to show that no surface or bulk defects are needed in a torus, while a pretzel with two holes (χ=-2) requires defects similar to those in a sphere. One important aspect of these surface singularities is helping to understand the superfluid dynamics of ^{3}He-A. In bulk there is no rigorous vorticity quantization, and phase slippage can occur by texture motion without the help of any true singularities.[6] The surface of ^{3}He-A, on the other hand, behaves almost like a normal superfluid, with quantized vortices and a simple phase slippage theorem. It is literally true that the chemical potential difference between two surface points is equal to the rate of passage of vortex points across any path joining them. It seems likely that under some circumstances the vortex points will be double ones (4π) carrying a fountain along with them through the bulk liquid.

References

(1) W. F. Brinkman, in lectures at the Scottish Universities Summer School, St. Andrews, 1974; published in P. W. Anderson and W. F. Brinkman, in The Helium Liquids (Farquhar, ed.), 1975, p. 315.

(2) G. Toulouse and M. Kleman, J. de Physique Letters 37, 2149 (1976).

(3) S. Blaha, Phys. Rev. Letters, 36, 874 (1976).

(4) K. Maki, various preprints (1976).

(5) D. Mermin, Proceedings of "Simon" Symposium, Sussex, Aug. 1976.

(6) P. W. Anderson and G. Toulouse, submitted to Phys. Rev. Letters.

(7) D. Mermin suggests "flower."

STATIC AND DYNAMIC TEXTURES IN SUPERFLUID ^{3}He-A[†]

Louis J. Buchholtz[*] and Alexander L. Fetter

Institute of Theoretical Physics, Department of Physics
Stanford University
Stanford, California 94305

Abstract: Static textures in confined superfluid ^{3}He-A near T_c are studied with the Ginzburg-Landau free energy. Plane boundaries induce a uniform texture that becomes deformed at a critical perpendicular magnetic field; above that value, the mean superfluid density measured in torsional oscillations decreases monotonically from $\rho_s^{\perp}$ to $\rho_s^{||}$. Cylindrical boundaries induce a flared texture, with a circulating current and an angular momentum per particle $\approx 0.782\ \hbar\rho_s^{||}/\rho$; this configuration deforms continuously with increasing axial magnetic field up to a critical value, when a discrete transition takes place to a planar texture with neither current nor angular momentum. Leggett's dynamical equations are generalized to incorporate static inhomogeneous textures, which then represent the zero-order solutions in the presence of a weak r-f field. The first-order response characterizes the various textures, which may permit an unambiguous experimental identification.

I. INTRODUCTION

Superfluid ^{3}He may be described by a tensor order parameter[1-3] $A_{\mu i}$ that is analogous to the scalar order parameter familiar in superconductivity.[4] In both cases, the overall magnitude characterizes the strength of the pairing amplitude, but the tensor

---

† Research supported in part by NSF grant DMR 75-08516.
* Danforth Foundation Fellow.

indices μi for ^{3}He reflect the internal unit spin and orbital angular momenta associated with the triplet p-wave structure. Although these additional degrees of freedom complicate the analysis, they correspondingly enlarge the range of phenomena. To illustrate these possibilities, we here study ^{3}He-A in confined geometries where the boundaries themselves can act to orient the internal vectors. Moreover, the imposition of hydrodynamic flow or a static magnetic field alters the equilibrium configuration, producing either continuous deformations, or, more dramatically, discrete transitions to topologically distinct textures. Finally, the addition of weak oscillatory magnetic fields induces new dynamic states, in which the inhomogeneous zero-order structures affect the (already rich) NMR spectra in characteristic and striking ways.

Throughout these considerations, the free energy $\int d^3r\, F(\mathbf{r})$ plays a fundamental role. First, it must attain a local minimum for any static equilibrium configuration, with the true equilibrium representing the absolute minimum. Second, the free energy also acquires dynamical significance, for Leggett[5] identifies it as part of an effective Hamiltonian that determines the equations of motion. These ideas have their simplest applications to bulk homogeneous fluid, where the free-energy density is a spatial constant. For such uniform static configurations, the applied magnetic field orients the spin angular momentum, and the dipole energy density F_D then aligns the orbital angular momentum, partially lifting the residual internal degeneracy. As noted above, the same F_D also appears in Leggett's phenomenological equations, producing the well-known transverse and longitudinal shifts[3,5,6] in the NMR frequency of small-amplitude oscillations about the equilibrium.

The preceding formalism is readily generalized to incorporate nonuniform states merely by adding the appropriate bending energy density F_K. As a concrete example, this paper examines the static (Sec. II) and dynamic (Sec. III) behavior of ^{3}He-A confined by planar and cylindrical boundaries. In addition, our discussion is restricted to the Ginzburg-Landau regime near T_c, where the various contributions to the free-energy density have particularly simple forms.

II. STATIC CONFIGURATIONS

In the bulk A phase, the order parameter becomes a direct product

$$A_{\mu j} = \Delta e^{iS}\, \hat{d}_\mu (\hat{n}_1 + i\hat{n}_2)_j \ , \tag{1}$$

where Δ is a real constant, S is a spatially varying phase, $\hat{d}_\mu$ is a unit vector perpendicular to the local spin density, and $\hat{n}_1$ and $\hat{n}_2$ are real orthogonal vectors. Moreover, the vector

$$\hat{\ell} = \hat{n}_1 \times \hat{n}_2 \tag{2}$$

characterizes the orbital angular momentum of the Cooper pair. Evidently, the identification of the phase S is somewhat arbitrary, for it can be eliminated with a suitable redefinition of $\hat{n}_1$ and $\hat{n}_2$. Nevertheless, this added flexibility is often convenient; it does not affect the vector $\hat{\ell}$. For a given geometry, the parameters in Eq. (1) must be determined by minimizing the integral of the free-energy density, which consists of a bulk term F_0 and a magnetic term F_Z in addition to the dipole term F_D and bending term F_K mentioned previously. Qualitatively, the bulk term F_0 fixes the magnitude Δ in terms of the Ginzburg-Landau phenomenological constants, the magnetic term

$$F_Z = 2g_Z \, \Delta^2 (\hat{d} \cdot \underset{\sim}{H}^0)^2 \tag{3}$$

aligns $\hat{d}$ perpendicular to the applied field $\underset{\sim}{H}^0$, the dipole term

$$F_D = 2g_D \Delta^2 [(\hat{d} \cdot \hat{n}_1)^2 + (\hat{d} \cdot \hat{n}_2)^2 - \tfrac{2}{3}] = 2g_D \Delta^2 [\tfrac{1}{3} - (\hat{d} \cdot \hat{\ell})^2] \tag{4}$$

aligns $\hat{d}$ and $\hat{\ell}$, and the bending term F_K aligns $\hat{\ell}$ along an external hydrodynamic flow $\underset{\sim}{v}$,[7] maintaining the texture as uniform as possible. Furthermore, the configuration is subject to boundary conditions, and we here make the simplest choice of specular conditions,[2] which have the physical consequence that the particle current has no normal component at the walls.[8]

A. Planar Geometry

Consider a slab of width W, placed in a perpendicular magnetic field $\underset{\sim}{H}^0 = H^0\hat{z}$. The magnetic energy F_Z is minimized if $\hat{d}$ is tangential ($\perp \hat{z}$), and the boundary conditions require $\hat{\ell} \parallel \hat{z}$ at the walls. In addition, the dipole energy F_D tends to align $\hat{d}$ and $\hat{\ell}$, and the competition between these various textures produces two characteristic parameters[3,9,10] – a length $L^* \approx 6$ μm and a field $H^* \approx 25$ Oe. For the most practical case that $W \gg L^*$ and $H^0 \gg H^*$, the magnetic field enforces the condition $\hat{d} \perp \hat{z}$, and the dipole coupling requires $\hat{\ell} \parallel \hat{d}$ almost everywhere. In detail, however, the actual configuration of $\hat{\ell}$ is rather intricate, for it bends smoothly from normal at the walls to tangential in the bulk,[11] with the transition occurring in a narrow boundary layer of thickness $\approx 2^{-1/2} L^*$.

To this level of approximation, all directions in the xy plane are equivalent, but the orientational degeneracy can be lifted by other external perturbations. One interesting example is a weak hydrodynamic flow, which arises in the oscillating-disk geometry[12] used to study the anisotropy in the superfluid-density tensor[3,11]

$$(\rho_s)_{ij} = \rho_s^{\perp} (\delta_{ij} - \hat{\ell}_i \hat{\ell}_j) + \rho_s^{\parallel} \hat{\ell}_i \hat{\ell}_j \, . \tag{5}$$

The induced superflow is purely azimuthal ($\propto \hat{\phi}$), and measurement of the corresponding spatial average

$$\langle \rho_s \rangle = \rho_s^{\perp} + (\rho_s^{\parallel} - \rho_s^{\perp}) \quad (\hat{\ell}\cdot\hat{\phi})^2 \tag{6}$$

provides direct information about the azimuthal component of $\hat{\ell}$. If the applied normal field H^0 is small, then the equilibrium state[9,10] has $\hat{\ell} \parallel \hat{z}$ and $\langle \rho_s \rangle = \rho_s^{\perp}$. As H^0 increases past a critical value, however, an orientational transition to a deformed texture occurs, in which H^0 acts to bend $\hat{d}$ toward the horizontal and the dipole coupling induces parasitic bending of $\hat{\ell}$. Furthermore, the azimuthal flow aligns both vectors $\hat{d}$ and $\hat{\ell}$ along $\hat{\phi}$. As a result, $\langle \rho_s \rangle$ for the deformed state falls below its weak-field value because $\rho_s^{\parallel} - \rho_s^{\perp} \approx - 1/2\rho_s^{\perp}$. If $W > L^*$ and $H^0 > H^*$, most of the sample has $\hat{\ell} \parallel \hat{\phi}$, and $\langle \rho_s \rangle$ approaches $\rho_s^{\parallel} \approx 1/2\rho_s^{\perp}$. In particular, a model calculation[11] for $W \approx 50$ μm predicts the high-field limit $\langle \rho_s \rangle = 0.59 \, \rho_s^{\perp}$, in good agreement with the observed value[12] $0.62\rho_s^{\perp}$ at $H^0 \approx 290$ Oe. These experiments not only demonstrate the essential anisotropy of $(\rho_s)_{ij}$ but also support the particular texture described here. Systematic study of the field-dependent $\langle \rho_s \rangle$ in channels of different widths would be most desirable.

B. Cylindrical Geometry

When the plane walls are replaced by a long cylinder, the situation becomes considerably more complicated, for the specular boundary conditions now must hold along the curved surface. If the cylinder's radius R greatly exceeds the temperature-dependent coherence length $\xi(T) \approx 120$ Å $(1 - T/T_c)^{-1/2}$, however, it is still permissible to require that $\hat{\ell}$ be normal at the outer surface (i.e. $\hat{\ell} = \hat{r}$), with Δ remaining constant there. For large $R (\gg L^*)$, as is expected for most configurations, the dipole coupling again forces $\hat{d}$ to be parallel to $\hat{\ell}$ throughout the sample, just as for a planar texture.

Given these qualitative restrictions on $\hat{d}$ and $\hat{\ell}$, we must seek the arrangement that minimizes the total free energy. In the simplest case of zero applied magnetic field, the only relevant contribution is the bending energy, which tends to keep $\hat{d}$ and $\hat{\ell}$ uniform. One possible configuration is a flared state[8,13,14] with $\hat{d}$ and $\hat{\ell}$ bending from radial at the walls to axial at the center. The resulting uniform texture near the symmetry axis eliminates the local curvature energy, because Δ is unchanged and $A_{\mu j}$ reduces to $\Delta\hat{z}_\mu(\hat{y} - i\hat{x})_j$ at $r = 0$. Re-expressed in cylindrical polar coordinates (r, ϕ, z), the order parameter becomes $\Delta e^{i\phi} z_\mu(\hat{\phi} - i\hat{r})_j$ near the origin, and consideration of the boundary condition at the outer surface suggests the general trial function

$$A_{\mu j} = \Delta e^{i\phi}(z\cos\theta + \hat{r}\sin\theta)_\mu \; [\hat{\phi} + i(\hat{z}\sin\theta - \hat{r}\cos\theta)]_j \qquad (7)$$

where $\hat{d}\cdot\hat{z} = \hat{\ell}\cdot\hat{z} = \cos\theta$ depends only on the radial coordinate. Detailed calculations show that $\theta(r)$ varies approximately linearly ($\approx \pi r/2R$), with R setting the characteristic scale of length. As a result, the total free energy per unit length for the state (7) exceeds the bulk value by a constant independent of R. The axial bending of $\hat{\ell}$ has another important consequence,[8,13,14] for it induces an aximuthal current and an associated angular momentum $L_z \approx 0.782\,\hbar\rho_{s||}/\rho$ per particle.

Other plausible textures for large cylinder ($R >> L^*$) in zero magnetic field involve purely planar configurations of $\hat{d} \approx \hat{\ell}$. One possibility is de Gennes' dysgyration,[2,15] in which $\hat{d}$ and $\hat{\ell}$ remain radial with a singular region along the cylinder's axis. In this case, Δ itself vanishes near the symmetry axis within a region of radius $\approx \xi(T)$, cutting off the logarithmic divergence in F_K and producing an excess free energy proportional to $\ell n(R/\xi)$. Alternatively, $\hat{d}$ and $\hat{\ell}$ may assume a more intricate texture with two singular lines symmetrically placed on the cylindrical surface. This latter "double dysgyration" also requires Δ to vanish in the singular regions, and the excess total free energy is again proportional to $\ell n(R/\xi)$. Planar states like these carry no particle current and therefore have no orbital angular momentum. For $R >> L^* >> \xi(T)$, neither the single nor the double dysgyration is competitive with the flared state (7), which therefore is thought to represent the equilibrium texture in zero magnetic field.

The situation is modified by the addition of an axial magnetic field H^0z, which forces $\hat{d}$ toward the xy plane. For small H^0 ($<<H^*$), the state (7) becomes progressively more deformed ($\delta\theta \gtrsim 0$), still retaining the uniform central region. The corresponding free

energy and angular momentum per unit length both increase as $(H^0)^2$. Conceivably, a modulated applied field might induce synchronous torsional oscillations in a suspended cylinder of ^{3}He-A; an experimental search for such an effect would be interesting. As H^0 increases beyond H^*, the magnetic energy becomes comparable with the dipole energy, and $\hat{d}$ and $\hat{\ell}$ no longer remain parallel. At approximately the same value of H^0, however, a discrete transition occurs to a topologically distinct (planar) state, in which $\hat{d}$ lies in the xy plane; this new texture eliminates the magnetic energy (3) entirely, and the dipole energy again enforces $\hat{\ell} \approx \hat{d}$ nearly everywhere. Both the single and double dysgyrations are plausible candidates for the high-field configuration ($H^0 > H^*$), but only the former has been studied in detail. Since neither texture has an orbital angular momentum, this orientational transition should be quite striking in a suspended cylinder. As discussed subsequently, each of these two possible dysgyrations should yield characteristic NMR spectra, and experimental studies may ultimately resolve this interesting question.

III. NMR FOR INHOMOGENEOUS TEXTURES

Leggett's phenomenological equations[5] successfully describe the NMR behavior of superfluid ^{3}He in bulk states. This analysis has been extended to certain specific inhomogeneous states by Takagi[16] and by Smith et al.[17] We have generalized the formalism to include arbitrary inhomogeneous A-phase configurations. The necessity for such an extension is not hard to see. In general, the equilibrium texture is determined as a minimization of competing contributions to the free energy, including bulk, dipole, magnetic, and kinetic terms. As a rule, the spin density and order parameter describing the superfluid will then have spatial dependence. NMR studies analyze the dynamic behavior of the spin density, which is influenced by the dipole energy. Hence we expect that inhomogeneities will lead to nonuniform forces that must also be included in considering the NMR.

Leggett begins his analysis of a uniform medium by devising a phenomenological Hamiltonian, with commutation relations among the dynamical variables. The approach is semiclassical in that it generates equations for the variables as operators and then treats them as classical quantities. He suggests using $\int d^3r[(\gamma^2/2\chi)S^2 - \gamma \underset{\sim}{S}\cdot\underset{\sim}{H} + F_D]$ as the phenomenological Hamiltonian, where F_D is the dipole energy density. The dynamical variables may be taken as the spin density $\underset{\sim}{S}$ and the vector $\underset{\sim}{A}$ related to the order parameter $A_{\mu i}$ as $\underset{\sim}{A}_\mu = A_{\mu i}\hat{k}_i$; they obey the equations of motion:

$$\partial \underset{\sim}{S}/\partial t = \gamma \underset{\sim}{S} \times \underset{\sim}{H} + (i\hbar)^{-1} [\underset{\sim}{S}, F_D] \tag{8}$$

$$\partial \underset{\sim}{A}/\partial t = \gamma \underset{\sim}{A} \times (\underset{\sim}{H} - \gamma\chi^{-1}\underset{\sim}{S}). \tag{9}$$

For these equations to be acceptable, we must require that $\dot{\underset{\sim}{S}} = \dot{\underset{\sim}{A}} = 0$ for the equilibrium texture. In bulk states this condition is satisfied automatically, and the investigation proceeds directly. For confined states, however, these simple equations become inadequate, as seen in any of several examples. The obvious extension, and one that turns out to be correct, is to augment the phenomenological Hamiltonian with the kinetic-energy contributions.[17] The first problem is to show that the equilibrium configuration dervied by minimizing the free energy is indeed stationary.

Our proof proceeds in the following manner. As dynamical variables, use $A_{\mu i}(\underset{\sim}{r})$, $\underset{\sim}{S}(\underset{\sim}{r})$, and as a phenomenological Hamiltonian use

$$\mathcal{H} \equiv \int d^3r[(\gamma^2/2\chi)S^2 - \gamma\underset{\sim}{S}\cdot\underset{\sim}{H} + F_D(\underset{\sim}{r}) + F_K(\underset{\sim}{r})], \tag{10}$$

where

$$F_D = g_D(A^*_{ii}A_{jj} + A^*_{\mu i}A_{i\mu} - \frac{2}{3} A^*_{\mu i}A_{\mu i}) \tag{11}$$

and

$$F_K = K_1A^*_{\mu i;i}A_{\mu j;j} + K_2A^*_{\mu j;i}A_{\mu j;i} + K_3A^*_{\mu j;i}A_{\mu i;j} \ . \tag{12}$$

All tensor indices in this paper refer to physical components,[8] so that neither caret nor raised indices are required. We generalize Leggett's commutation relations[5] to include spatial variations

$$[S_i(\underset{\sim}{r}),\ S_j(\underset{\sim}{r}')] = i\hbar\ \varepsilon_{ijk}\ S_k(\underset{\sim}{r})\ \delta(\underset{\sim}{r}-\underset{\sim}{r}')$$

$$[A_{\mu i}(\underset{\sim}{r}),\ S_j(\underset{\sim}{r}')] = i\hbar\ \varepsilon_{\mu jk}\ A_{ki}\ \delta(\underset{\sim}{r}-\underset{\sim}{r}')$$

and then use $\mathcal{H}$ to generate the dynamical equations. We find

$$\partial S_k/\partial t = \gamma(\underset{\sim}{S}\mathrm{x}\underset{\sim}{H})_k - 2\mathrm{Re}\ \varepsilon_{k\sigma\mu}A^*_{\sigma\lambda}[g_D(\delta_{\mu\lambda}A_{jj} + A_{\lambda\mu})$$

$$- (K_1 + K_3)A_{\mu j;j;\lambda} - K_2A_{\mu\lambda;j;j}] \tag{13}$$

$$\partial A_{\mu i}/\partial t \approx \gamma\varepsilon_{\mu\nu\lambda}A_{\nu i}(\underset{\sim}{H} - \gamma\chi^{-1}S)_\lambda \tag{14}$$

To show that these equations ensure stationary behavior at equilibrium, observe first that $\gamma\underset{\sim}{S}^0 \equiv \underset{\approx}{\chi}\underset{\sim}{H}^0$. Then notice that

$$(F_s - F_n)_{\text{magnetic}} \quad -\frac{1}{2}(\chi_s - \chi_n)_{ij}\, H_i^0 H_j^0$$

$$= g_Z\, H_\mu^0 H_\nu^0\, A_{\mu\lambda}^* A_{\nu\lambda} ,$$

which identifies the shift in the susceptibility

$$2\mathrm{Re}\, g_Z\, A_{\mu i}^* A_{\nu i} = -(\chi_s - \chi_n)_{\mu\nu} .$$

One then infers the equilibrium spin density

$$\gamma S_\nu^0 = \chi_n H_\nu^0 - 2\mathrm{Re}\, g_Z\, A_{\nu i} A_{\mu i}^* H_\mu^0 ,$$

and the cross product with $\underset{\sim}{H}^0$ yields

$$\gamma(\underset{\sim}{S}^0 \mathrm{x} \underset{\sim}{H}^0)_k = 2\mathrm{Re}\, g_Z\, \varepsilon_{k\mu\nu}\, A_{\nu i} A_{\sigma i}^* H_\mu^0 H_\sigma^0$$

With these results, Eq. (14) becomes

$$\partial A_{\mu i}/\partial t = 2\gamma g_Z \chi^{-1} \varepsilon_{\mu\nu\lambda} A_{\nu i} \mathrm{Re} A_{\lambda j} A_{\lambda j}^* H_\sigma^0 .$$

Since $A_{\mu i} \propto \hat{d}_\mu(\hat{n}_1 + i\hat{n}_2)_i$ for the A phase, where $\hat{d}_\mu$ is real, it follows $\partial A_{\mu i}/\partial t = 0$ in equilibrium. To understand the significance of Eq. (13), we must refer back to the variational equation $\delta\mathcal{H}/\delta A_{\mu\lambda}^* = 0$ that determines $A_{\mu i}$ in equilibrium. Without going into details, we assert that the right-hand side of Eq. (13) becomes $-2\,\mathrm{Re}\,\varepsilon_{k\sigma\mu}A_{\sigma\lambda}^*\delta\mathcal{H}/\delta A_{\mu\lambda}^* \equiv 0$, which ensures that $\underset{\sim}{S}^0$ is indeed stationary.

The application of these equations is clearly quite difficult. It is, however, possible to set up the C. W. problem. We study the first-order perturbations about equilibrium, assuming $\underset{\sim}{S} = \underset{\sim}{S}^0 + \underset{\sim}{S}'(t)$, $A_{\mu i} = A_{\mu i}^0 + A_{\mu i}'(t)$, and $\underset{\sim}{H} = \underset{\sim}{H}^0 + \underset{\sim}{H}'(t)$. Further, a natural choice for $A_{\mu i}'$ in the A phase is $\Delta_0 \hat{d}_\mu'(\hat{n}_1 + i\hat{n}_2)_i$ where $\underset{\sim}{d}'\cdot\hat{d}_0 = 0$. No perturbations are introduced for $\hat{n}_1$ or $\hat{n}_2$ because they refer to spatial orientations, and, following Leggett's argument,[5] ought to have much slower motions than those of interest. Then under the assumption that $\underset{\sim}{H}'(t) = \underset{\sim}{H}' e^{-i\omega t}$ we may solve Eq. (14) and find $\underset{\sim}{d}'$ in terms of $\underset{\sim}{H}'$ and $\underset{\sim}{S}'$

$$\underset{\sim}{d}' = \frac{i\omega\underset{\sim}{D}+\Omega_z \hat{d}_0 x \underset{\sim}{D}}{\omega^2-\Omega_z^2} \qquad (15)$$

where $\underset{\sim}{D} \equiv \gamma(\Delta/\Delta_0)\hat{d}^0 x[\underset{\sim}{H}'(t) - \gamma\chi^{-1}\underset{\sim}{S}'(t)]$ and $\Omega_Z \equiv 4\gamma g_Z \Delta^2 \chi_{\wedge}^{-1} d^0 \cdot \underset{\sim}{H}^0$ is a frequency that depends on the local orientation of d and size of Δ^2. With this result Eq. (13) may be written solely in terms of the desired unknown $\underset{\sim}{S}'$. This expression is a matrix differential equation and suggests a variety of solutions. An immediate novel feature is the presence of new fundamental frequencies. Besides the usual Larmor and dipole frequencies $\Omega_L = \gamma H^0$, $\Omega_D{}^2 = 4g_D\Delta^2\gamma^2\chi^{-1}$, there appear the spatially dependent magnetic frequency Ω_Z and also an important kinetic frequency $\Omega_K{}^2 = 4K_2\Delta^2\gamma^2/R^2\chi$ where R is the characteristic dimension of the container; Ω_K will be comparable to Ω_D for R $\approx$ 20-100 μm.

We are currently investigating the resonant frequencies and absorption strengths for specific textures in ^{3}He-A and can here only give qualitative results. The basic nature of the problem is the following. Instead of a matrix inversion as for a uniform medium, we now have a matrix Green's function boundary value problem. For example, one realistic case is a cylindrical geometry of radius $R >> L^*$ with $\underset{\sim}{H}^0 \,||\, \hat{z}$ and $H^0 >> 30$ Oe. The $\hat{d}^0$ vector is $\perp$ to $\underset{\sim}{H}^0$ and everywhere radial. This means $\Omega_Z = 0$ and implies

$$\underset{\sim}{d}' = i\omega^{-1}\,\gamma(\Delta/\Delta_o)\hat{d}^0 \times [\underset{\sim}{H}'(t) - \gamma\chi^{-1}\underset{\sim}{S}'(t)]. \qquad (16)$$

The most reasonable set of boundary conditions on $\underset{\sim}{S}'$ seems to be: $\underset{\sim}{S}'(r = 0) = 0$, which follows from the differential equation, and $\partial_r \underset{\sim}{S}'(r = R) = 0$, which follows from Eq. (16). In particular, a longitudinal excitation $H'\hat{z}$ produces only an S_z' component of spin density. The defining equation for S_z' takes the form $\mathfrak{D} S_z' = f(r)H_z'$ where $\mathfrak{D}$ is a second-order linear differential operator in r and f is known. In contrast, a transverse excitation $H'\hat{x}$ induces both S'_r and S'_ϕ.

For more general textures, $\mathfrak{D}$ will be a second-order partial differential operator, and Ω_Z will appear with magnitude comparable in size to the other frequencies. The resulting mixing of the various vector components of $\underset{\sim}{S}'$ becomes more complicated and won't be mentioned further here.

NMR seems to offer an ideal tool for probing these inhomogeneous configurations through line broadening and the appearance of new or shifted resonances. Besides determining the basic textures, one might also hope to obtain values for the characteristic phenomenological constants associated with superfluid ^{3}He. It may even be feasible to investigate new structures like singularities and vortices. The wealth of possible phenomena promises to keep this subject exciting for many years.

References

(1) R. Balian and N. R. Werthamer, Phys. Rev. 131, 1553 (1963).

(2) V. Ambegaokar, P. G. de Gennes, and D. Rainer, Phys. Rev. A 9, 2676 (1974); A 12, 245 (1975).

(3) A. J. Leggett, Rev. Mod. Phys. 47, 331 (1975).

(4) V. L. Ginzburg and L. D. Landau, Zh. Eksp. Teor. Fiz. 20, 1064 (1950).

(5) A. J. Leffett, Ann. Phys. 85, 11 (1974).

(6) J. C. Wheatley, Rev. Mod. Phys. 47, 415 (1975).

(7) P. G. de Gennes and D. Rainer, Phys. Lett. A 46, 429 (1974).

(8) L. J. Buchholtz and A. L. Fetter, Phys. Lett. A 58, 93 (1976) and (to be published). As demonstrated in these papers, a general nonuniform texture in ^{3}He-A requires two gap functions, Δ_1 and Δ_2; this subtlety has proved unnecessary for the present considerations.

(9) V. Ambegaokar and D. Rainer (unpublished).

(10) A. L. Fetter, Phys. Rev. B 14, 2801 (1976).

(11) A. L. Fetter, Phys. Rev. B 15 (to be published).

(12) J. E. Berthold, R. W. Giannetta, E. N. Smith, and J. D. Reppy, Phys. Rev. Lett. 37, 1138 (1976).

(13) P. W. Anderson and W. F. Brinkman, in "The Helium Liquids", edited by J. G. M. Armitage and I. E. Farquhar (Academic, London, 1975), Sec. VIII, pp. 406-413.

(14) N. D. Mermin and T.-L. Ho, Phys. Rev. Lett. 36, 594 (1976).

(15) P. G. de Gennes, Phys. Lett. A 44, 271 (1973); in "Collective Properties of Physical Systems", edited by B. Lundquist and S. Lundquist (Academic, New York, 1974), pp. 112-115.

(16) S. Takagi, J. Phys. C 8, 1507 (1975).

(17) H. Smith, W. F. Brinkman, and S. Engelsberg, Phys. Rev. B (to be published).

STABILITY OF PLANAR TEXTURES IN SUPERFLUID ^{3}He-A

Pradeep Kumar*

Department of Physics
University of Southern California
Los Angeles, Ca. 90007

Planar textures in superfluid ^{3}He-A are due to the symmetry breaking terms in the free energy e.g. current, magnetic field and the energy due to the nuclear dipolar interaction.[1] The spin triplet - p wave order parameter is characterized in terms of two unit vectors $\hat{l}$ and $\hat{d}$ which respectively refer to the orbital and spin axes. The dipole energy gives two possible orientations for the $\hat{d}$ vector, one which is parallel to the $\hat{l}$ vector and the other which is antiparallel to the $\hat{l}$ vector. A pure $\hat{d}$ texture is the domain wall between the two confiurations in the case when $\hat{l}$ is uniform over the whole system.[2] In an open system $\hat{l}$ is not constrained to be uniform, the presence of a static magnetic field confines both $\hat{l}$ and $\hat{d}$ to a plane perpendicular to the field. Simultaneous rotation of both $\hat{l}$ and $\hat{d}$ (in opposite direction) gives rise to the composite structure (see Maki's review at this conference for details). A third structure has been postulated recently by Hall and Hook[3] where the current provides 2 possible directions to the $\hat{l}$ vector and in an analogous fashion the texture is the domain wall between the two configurations. Fetter[4] has also studied the possible textures in the presence of currents as well as static fields. All of these textures have a characteristic signature in the form of a coherent motion which can be resonantly excited.

These structures have been studied under one constraint or the other (by neglecting some forces). The object here is to study the stability of planar structures as the assumed constraints are relaxed. An example that we discuss in detail is the pure $\hat{d}$ texture. It is possible to make the $\hat{l}$ vector uniform by a judicious use of narrow spacing between the parallel plates or inducing flow in an open system. However in an open system where $\hat{l}$ is not constrained very strongly, the $\hat{d}$ soliton has energy[2] $4\Omega_A^2$ (in units of $\frac{1}{2}\chi_N/\gamma^2$, where χ_N is the normal state susceptibility and γ, the nuclear gyromagnetic ratio) stored in a length ξ_1. Here Ω_A is the A phase Leggett frequency and $\xi_\perp$ is the dipolar coherence length ($\sim 10^{-3}$ cm). The $\hat{l}$ vector then may rotate to reduce the dipolar energy and the pure $\hat{d}$ soliton is unstable towards the formation of composite texture. Should a pure $\hat{d}$ texture be created first in a field turn off or tipping experiment,[5] it comes to a stop relatively quickly ($\sim 10^{-4}$ seconds) due to spin diffusion effects.[2] The time taken for conversion into a composite texture (to be precise the upper limit since we neglect the effect of the original constraint here) can be estimated rather easily by calculating the growth rate of the small fluctuations about a static pure d texture.

The calculations make use of the Ginzburg Landau free energy, the same that is used to calculate the textures originally (Maki, eq. 4). We will concentrate on a time scale longer than $(\Omega_A)^{-1}$. The equations of motion then are the Cross-Anderson equation[6] for the $\hat{l}$ vector and for the $\hat{d}$ vector, we ignore the inertial term. The sole time dependence comes from the spin-diffusion term.

If

$$\hat{l} = \sin\phi\,[\sin\chi\,\hat{x} + \cos\chi\,\hat{y}] + \cos\phi\,\hat{z}$$

and

$$\hat{d} = \sin\sigma\,[\sin\Psi\,\hat{x} + \cos\Psi\,\hat{y}] + \cos\sigma\,\hat{z} \tag{1}$$

we have;

$$-\frac{4\mu\gamma^2}{\chi_N C^2}\sin^2\phi\,\frac{\partial\chi}{\partial t} = -\frac{\partial}{\partial z}\left[(1+\cos^2\phi)\frac{\partial\chi}{\partial z}\right] - \frac{4}{\xi_\perp^2}\sin(\Psi-\chi)\,F(\Psi-\chi,\sigma,\phi)$$

$$\frac{D}{C^2}\frac{\partial^2}{\partial z^2}\left[\frac{\partial\Psi}{\partial t}\right] = -\frac{1}{2}\frac{\partial}{\partial z}\left[\sin^2\sigma(1+\cos^2\phi)\frac{\partial\Psi}{\partial z}\right] + \frac{1}{\xi_\perp^2}\sin(\Psi-\chi)\,F(\Psi-\chi,\sigma,\phi)$$

where

$$F(\Psi-\chi,\sigma,\phi) = \sin\sigma \sin\phi \cos(\Psi-\chi) + \cos\sigma\cos\phi, \quad (2)$$

μ represents the Cross-Anderson viscosity, γ, the nuclear gyromagnetic ratio for ^{3}He, χ_N, the normal state susceptibility, D, the spin diffusion constant, and $C = \xi_\perp \Omega_A$ represents the spin wave velocity in the z-direction. The equilibrium configuration for the pure $\hat{d}$ texture is taken to be χ=o, $\sigma=\phi=\pi/2$ and $\Psi = 2\tan^{-1}(\exp z/\xi)$. Therefore the fluctuations of interest are also going to be the ones with same spatial dependence. Similar equations can be derived for the angles σ and ϕ as well. Indeed the σ,ϕ fluctuations turn out to be unstable as well; however, the largest instability growth rate belongs to the motion of $v\equiv\chi-\psi$. It is apparent that the instability in v points in the direction of formation of composite solitons as the final state.

In terms of v, we have

$$\frac{1}{25\xi_\perp^2}\frac{\partial}{\partial t}\left[16\,\tau v - \frac{D\xi^2}{C^2}\frac{\partial^2 v}{\partial z^2}\right] = \frac{1}{5}\frac{\partial^2 v}{\partial z^2} - \frac{1}{\xi_\perp^2}(1-2\,\mathrm{sech}^2 z/\xi_\perp)v$$

$$-\frac{1}{\xi_\perp^2}\,\mathrm{sech}\, z/\xi_\perp \tanh z/\xi_\perp$$

$$= -Lv\frac{1}{\xi_\perp^2}\,\mathrm{sech}\, z/\xi_\perp \tanh z/\xi_\perp \quad (3)$$

where $\tau=\mu\gamma^2/\chi_N\Omega A^2$, the $\hat{l}$ relaxation time measured by Paulson et al.[7] The operator L has a complete orthonormal set of eigenvectors. If v(z,t) is expanded in terms of that complete set, the exponentially growing term corresponds to the bound state for which the eigenvalue λ_b and the eigenfunction f_b are given as

$$\lambda = -.4597\,\xi_\perp^{-2} \text{ and } f_b = N(\mathrm{sech}\, z/\xi_\perp)^{(\sqrt{41}-1)/2} \quad . \quad (4)$$

The associated growth rate for the instability is,

$$\Gamma = \frac{11.4922}{16\,\tau + 1.1398\, D/C^2}$$

Very close to T_c, Γ is determined by $C_{\perp}/D$. That temperature region turns out to be extremely small. Using $D = \cdot 03\ cm^2/sec$ and $C_{\perp} = 2\cdot 54\ (1-T/T_c)^{1/2}$ meters/sec. (both at P = 27 bar) and $t^{\perp} \simeq 10\ (1-T/T_c)^{1/2}$ in msec, the temperature range $(1-T/T_c)$ over which the growth rate is dominated by spin diffusion is found to be $\sim 10^{-5}$. Therefore

$$\Gamma = \cdot 7183/t\ (m\ sec)^{-1} \qquad (6)$$

The growth rate is completely determined by the $\hat{l}$ vector relaxation time. Once the composite soliton is formed, it will show up in the N.M.R. with its characteristic satellite. Since the composite soliton energy is smaller than the pure $\hat{d}$ texture, the energy is distributed into spin waves and orbital waves. The propagation of the latter, as seen from the linearized form of Eq. (2), is analogous to the heat pulse propagation.

To conclude, the composite soliton stability has also been studied by looking at the time dependence of small fluctuations in an analogous fashion. The lowest eigenvalues of the corresponding operators L turn out to be zero, indicating linear stability. The eigenvalues correspond to simple translation or rotation of the whole structure. Details will be published elsewhere.[8]

This work was done in collaboration with K. Maki. The author also acknowledges a stimulating discussion with J. M. Delrieu which initiated this work.

References

(1) A. J. Legett, Rev. Mod. Phys. 47, 331 (1975).

(2) K. Maki and H. Ebisawa, J. Low Temp. Phys. 23, 351 (1976).

(3) H. Hall and J. Hook (unpublished).

(4) A. L. Fetter, Phys. Rev. B 14, 2801 (1976).

(5) K. Maki and P. Kumar, Phys. Rev. B 14, 3920 (1976); A. C. Scott *et al*. J. App. Phys. 47, 3272 (1976).

(6) M. C. Cross and P. W. Anderson, Proc. of L. T. - 14 Otaniemi, Finland 1975, edited by M. Krusius and M. Vuorio (North Holland, Amsterdam 1975), Vol. 1, p. 29.

(7) D. N. Paulson, M. Krusius and J. Wheatley, Phys. Rev. Lett. 22, 1322 (1976).

(8) K. Maki and P. Kumar (unpublished).

*Work supported by NSF under contract no. DMR76-21032.

SOLITARY STRUCTURES IN NONLINEAR FIELDS*

J. A. Krumhansl

Laboratory of Atomic and Solid State Physics
Cornell University
Ithaca, New York 14853

Abstract: An overview of recent work in nonlinear fields is presented, with intent to provide a pedagogical introduction to the recent literature. The ideas reach into many branches of physics, chemistry, and continuum mechanics. The significant feature of strong nonlinearity is that spatially extended waves interact and organize into well-characterized, localized objects. The result is that objects like domain walls, dislocations, and such may be found. These structures rely on the nonlinearity inherent in the problem to such an extent that they can not be generated by a perturbation theory starting with a plane-wave basis. Most important, they have static and dynamic stability in a wide number of cases, and play a significant role in both the statistical and quantum mechanical descriptions of many physical systems.

I. INTRODUCTION

This introduction to solitary structures in nonlinear fields is intended to be pedagogical rather than to deal with any specific application in a detailed technical manner. The subject of solitons, solitary waves, and objects of similar generic character is

---

*Supported by Energy Research and Development Administration under contract number EY-76-S-02-3161.

a rapidly advancing one; for physicists three important reviews are those by Scott, Chu, and McLaughlin,[1] G. L. Lamb and McLaughlin,[2] and Rajaraman.[3]

There is no more delightful way of introducing the subject than quoting from Scott-Russell,[3] who first introduced the idea of solitary waves into the field of hydrodynamics:

> "I was observing the motion of a boat which was rapidly drawn along a narrow channel by a pair of horses, when the boat suddenly stopped-not so the mass of water in the channel which it had put in motion; it accumulated round the prow of the vessel in a state of violent agitation, then suddenly leaving it behind, rolled forward with great velocity, assuming the form of a large solitary elevations, a rounded, smooth and well-defined heap of water which continued its course along the channel apparently without change of form or diminution of speed. I followed it on horseback, and overtook it still rolling on at a rate of some eight or nine miles an hour, preserving its original figure some thirty feet long and a foot to a foot and a half in height. Its height gradually diminished, and after a chase of one or two miles I lost it in the windings of the channel. Such, in the month of August 1834, was my first chance interview with that singular and beautiful phenomenon..."

That report contains the essential features of what has now been found in a wide variety of situations:

(1) A medium which is capable of carrying low amplitude harmonic wave-like excitations responds quite differently when large amplitude excitations emphasize the nonlinearity.

(2) There results a spatially compact (i.e. solitary) well-defined excitation, with highly specific characteristics (internal energy, amplitude, speed).

(3) The structure exhibits remarkable stability properties.

For the past several decades most physicists have been trained to think almost entirely in terms of linear systems, or formalisms. Superposition of normal modes, phonons, plane-wave states, etc.

have been the conceptual and working approach - whether the application be many-body theory, statistical physics, or quantum field theory. When interaction between these basic excitations occurred, perturbation and diagrammatic methods were developed to "dress" what usually remained recognizable as a qualitatively similar "quasi-particle: with renormalized properties. However, in strong perturbation situations this procedure breaks down completely,[3] as will become clear. What is fortunate and unexpected is that relatively simple new excitations are found.

These generalities about the style of recent physics are, of course, not universally true. The Landau-Ginzburg equations, whose solutions show domain and fluxoid structure, and the Bloch wall equations in ferromagnets are familiar examples where the nonlinearity plays an essential role. In addition to these applications, the reviews referenced discuss solitary phenomena in a wide variety of situations: hydrodynamic waves, ion-acoustic waves, nonlinear optical pulses, self-induced transparency, dislocations, Josephson transmission lines, anisotropic ferroelectrics, nonlinear lattice vibrations, structural phase transitions, spinodal decomposition, waves in spatially extended chemical and biological reactions, 1-d conducting materials, quantum field theory, and probably many more.

Finally, it should be noted that the main advances in this subject have been made by applied mathematicians until very recently,[5] and computer studies have played an essential role in augmenting formal development; for example, the pioneering work of Fermi, Pasta, and Ulam,[6] and the extensive studies of Zabusky.[7]

II. THE TEXTBOOK EXAMPLE: "SINE-GORDON"

The 1-d Klein-Gordon equation for a particle of rest mass m is ($\hbar = 1$, $c = 1$, $E^2 = m^2 + p^2$)

$$\frac{\partial^2\psi}{\partial t^2} - \frac{\partial^2\psi}{\partial x^2} + m^2\psi = 0 \tag{1}$$

which is a linear equation. When the equation is modified to the nonlinear "sine-Gordon" equation (dimensionless)

$$\psi_{tt} - \psi_{xx} + m^2 \sin\psi = o \tag{2}$$

the solutions are found to have remarkable properties. In 1962, Perring and Skyrme[8] showed in computer experiments that two compact solitary wave solutions of this equation would pass right through each other and emerge intact. While some nonlinear systems exhibit only single or periodic solitary solutions, those solutions for which the solitary structures pass through each other are given the special name of "solitons". In their context, quantum field theorists associate new massive elementary particles with these solutions.

This ubiquitous equation has appeared in studies of dislocations,[9] Bloch walls,[10] quantum field theory,[11] magnetic flux propagation on a Josephson transmission line,[12] and in charge density waves in models[13] of 1-d conducting organic materials. We discuss the nature of its solutions.

Before doing so, it will be useful to point out that this equation can be modeled by the continuum limit of a linear array of pendula of length ℓ fixed to an elastic bar which couples them torsionally; if θ_j is the rotation angle of the j'th pendulum and s is the torsional stiffness, then

$$mg\ell \sin\theta_j + m\ell^2\ddot{\theta}_j + s\,[2\theta_j - \theta_{j-1} - \theta_{j+1}] = 0 \tag{3}$$

Letting $\dfrac{[\quad]}{(\Delta x)^2} \to -\dfrac{d^2\theta}{dx^2}$, $\dfrac{s(\Delta x)^2}{m\ell^2} \to c^2$, $g/\ell \to \omega_o^2$

we have

$$\ddot{\theta} - c^2\,\theta_{xx} + \omega_o^2 \sin\theta = 0 \tag{4}$$

The correspondence with Eq. 2 follows if $m^2 = (\omega_o^2/c^2)$.

We are familiar with the types of solutions which are found when the pendula are all hanging at $\theta = 0$ (downward) and we give the system a small kink. For small displacements $\sin\theta \simeq \theta$ and we find the usual "linear" waves ($a \ll 1$)

$$\theta = a\cos(\omega t - kx) \tag{5}$$

with

$$\omega^2 = \omega_o^2 + c^2k^2 \tag{6}$$

One might quantize these as "phonons". This standard simplification breaks down at larger amplitude.

Most conventional approaches then resort to perturbation or "mean field" methods; for example

$$\sin\theta \simeq \theta - \frac{\theta^3}{3} \simeq \theta\,(1 - \frac{3\langle\theta^2\rangle}{3})$$

whence the dispersion relation becomes amplitude dependent

$$\omega^2 \simeq \omega_o^{\,2}\,(1 - \alpha\,\langle a^2\rangle) + c^2k^2 \quad .$$

Indeed, if the amplitude squared is thermalized, proportional to kT, one finds a soft mode behavior

$$(\omega_o^{\,2})_T \simeq \omega_o^{\,2}\,(1 - \alpha T).$$

But this is really fortuitous and the crude treatment of the nonlinearity should be obvious.

Fortunately, one can go far beyond these common approximations. First, because of the Lorentz invariance, the use of the variable $x - ct = \xi$ [in units of a characteristic length (c/ω_o)] reduces Eq. (2) to

$$\psi_{\xi\xi} + \sin\psi = 0 \tag{7}$$

The general solutions are elliptic functions; it turns out, though, there are some limiting forms of the solutions that are particularly interesting - i.e. solitary waves.

We expect that there are solutions in which the pendula crank all the way through 2π; indeed, the limit of one of the elliptic functions is

$$\theta^{(1)}_{s,a} = (\pm)\;4\,\tan^{-1}\,[\exp\,(\frac{\omega_o}{c}\;\frac{(x - v_d t)}{\sqrt{1-(v_d^2/c^2)}})] \tag{8}$$

The "s" goes with (+) and denotes soliton, the "a" goes with (-) and denotes antisoliton. The solutions are shown in Fig. 1. Several features of this solution are:

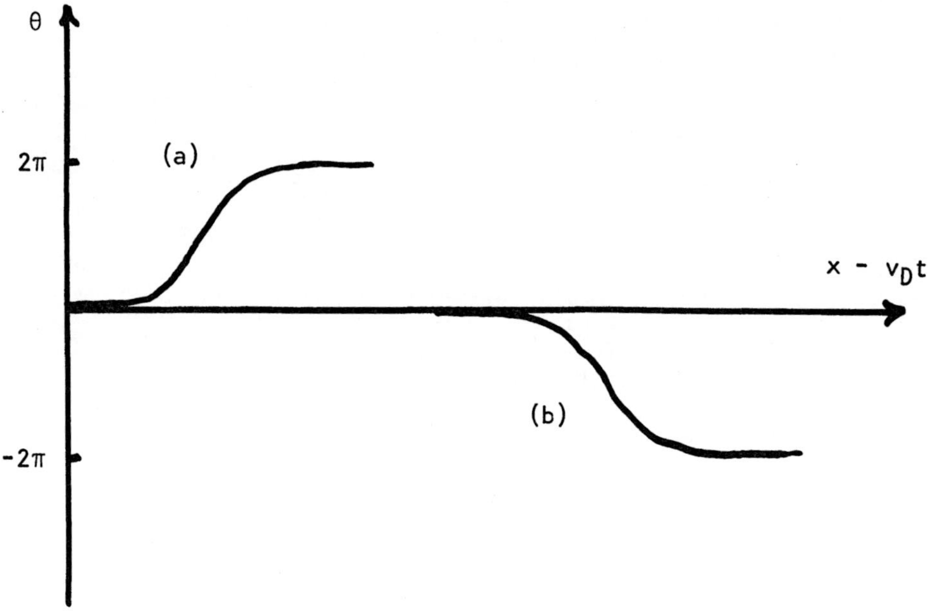

Figure 1. (a) soliton, (b) antisoliton for "sine-Gordon" equation. (Equation 8).

(1) The superscript (1) on $\theta_{s,a}^{(1)}$ denotes a single solitary wave. Note that the physical situation here find the system almost everywhere in one of two low energy "ground states", except in a transition region of thickness $\simeq c/\omega_o$. They are the 1-d equivalent of domain walls.

(2) The period of this disturbance is infinite in time and space. Thus this solution can not be reached by renormalization or self-energy corrections via the conventional perturbation theoretic machinery.

(3) The drift velocity v_d can be anything from 0 to c.

(4) Because of translational invariance this solution can be displaced by an finite amount and remain a solution.

(5) It will be shown below that otherwise this solution is stable against small distortions. As such, it represents a local minimum of the field energy - which significantly influences both the statistical and quantum mechanics of the system.

(6) It is characteristic of the solution of nonlinear equations that: (a) the amplitude of the solution (i.e. $o \to \pm 2$) is very specific, (b) solutions cannot, in general, be simply superposed.

(7) Finally, note that in some sense this "solitary wave", "kink", or "soliton" is conjugate to the low amplitude plane wave solutions; the former are localized in x-space, the latter are localized in k-space.

We next see that these are not solitary waves, but also solitons. The exact N-soliton solutions of the sine-Gordon equation are well known by now;[1] we consider the "two-soliton" solutions:

$$\tan\left[\frac{\theta}{4}\right] = u\,\frac{\sinh\,(x/\sqrt{1-u^2})}{\cosh(ut/\sqrt{1-u^2})} \qquad \text{(soliton-soliton)}$$

$$\tan\left[\frac{\theta}{4}\right] = \frac{\sinh(ut/\sqrt{1-u^2})}{u\,\cosh(x/\sqrt{1-u^2})} \qquad \text{(soliton-antisoliton)} \tag{9}$$

in dimensionless units $u = (v/c)\; x(\omega_o/c) \to x$.

Taking the soliton-soliton case, when $t \to -\infty$, we have

$$\theta = 4\,\tan^{-1}[(\pm)\;u\,\exp(\pm\,x + ut)/\sqrt{1-u^2}];\; x \;\substack{>> \\ <<}\; 0 \tag{10}$$

which describe a left-going solitary wave for large positive x simultaneous with a right-going kink at large negative x. At $t \simeq 0$ there is a collision. Then remarkably, at $t \to \infty$

$$\theta = 4\,\tan^{-1}[(\pm)u\,\exp(\pm\,x - ut)/\;1 - u^2)];\; \substack{x\,>>\,0 \\ x\,<<\,0} \tag{11}$$

and for $x >> 0$ the right-going soliton has appeared, and at $x << 0$ we find the left-going one - both intact!

For the soliton - antisoliton pair a similar behavior occurs. They might have annihilated but didn't! It now appears that there are many nonlinear systems for which similar behavior is found. In some cases it is possible to see that this spatial self-organization is a consequence of a happy cooperation between dispersion and nonlinearity, each of which would generally tend to destroy a wave packet, but for the right amplitude and sets of parameters there are "valleys in phase space" in which the dynamic trajectories are stabilized by nonlinearity.

III. SOLITON STABILITY ANALYSIS

To continue the discussion, as well as to provide very important details needed in physical applications, we now consider a stability analysis. Suppose we place one soliton in the system, and then ask about the behavior of additional disturbances; the main features can be found in a linear stability analysis, that is

$$\theta = \theta_s^{(1)} + \chi \tag{12}$$

and

$$\chi \ll \theta_s .$$

Substitute into Eq. (2) and keep terms linear in χ; the result is

$$\chi_{tt} - \chi_{xx} + m^2 (1 - \cos \theta_s^{(1)}) \chi = o \tag{13}$$

Examining solutions of this (linear) equation in the form

$$\chi(x,t) = f(x)\, e^{i\omega t}$$

we solve the eigenvalue problem

$$f_{xx} + [\omega^2 - m^2(1 - \cos \theta_s^{(1)})]f = 0$$

The eigenspectrum and eigenfunctions are:

$$\text{(a) } \omega^2 = 0, \; \chi = \frac{d}{dx} [\theta_s^{(1)}] = \frac{d}{dx} [4 \tan^{-1}(\exp x)] \tag{14}$$

This has a simple interpretation. The operator $(1 + \delta\, d/dx)$, where δ is small, simply represents a translation in x of the solution $\theta_s^{(1)}(x)$ by δ. Since the field equation is unchanged by such a transformation, obviously $\chi = d/dx\, (\theta_s^{(1)})$ is a steady state solution, i.e. $\omega = 0$.

(b) For all other solutions,

$$\omega^2 = m^2 + k^2 > 0 \tag{15}$$

$$f(x) = e^{-ikx}[k + i\, m \tanh (mx)]$$

Two properties of these solutions are: first, since all $\omega^2 > 0$, small oscillations about the soliton solutions are stable; second, small amplitude traveling waves exist, and pass right through the soliton without reflection (although they pick up a phase shift).

This example illustrates two very important features of solitions: their stability, and their "independence" of the usual phonon excitations.

Several remarks need to be made in relation to application of these solutions to real physical systems.

First of all, the stability theory just discussed is only a linear anlaysis. However, the sine-Gordon equation has simultaneously existing independent N-soliton, phonons, and "breathers" solutions. Indeed Faddeev[14] has shown that a set of action-angle variables can be constructed in which the Hamiltonian separates completely into these three types of excitations.

Although these interesting properties lead us to speculate about many possible consequences; one should exercise caution in trying to extend these consequences to three-dimensional systems and potentials other than that for the sine-Gordon case. Nonetheless, it is likely that these features will survive in an approximate form, or on a metastable basis. At least during the present era of physics these new viewpoints are worth considering.

IV. APPLICATIONS AND PHYSICAL CONSIDERATIONS

Suppose we continue to take a pedagogical viewpoint and discuss the statistical mechanics of a model Hamiltonian for a 1-d displacive phase transition. Variations of this model have been used for approximating self-consistent phonon models of soft modes, for computer simulations of "ferroelectric" phase transitions, and for field theoretical studies in two space-time dimensions.

Imagine an ion displacement $u(x_j)$ in some preferred direction in a lattice; x_j is the position of the j'th site. Schrieffer and the author[15] have discussed this case in some detail; the reader is referred to that paper for discussion of the underlying motivation and the results of both an exact statistical mechanical calculation, and a calculation in which phonons and kinks are treated phenomenologically as independent particles.

We wish to demonstrate here why, quite generally, one may expect to find kink structures influencing the statistical mechanics when strong nonlinearity is present.

The Hamiltonian, which may be an effective classical limit of a quantum mechanical theory, now is taken to be that of a continuous field u(x), and its conjugate momentum p(x).

$$H = \int dx\, \mathcal{H}\,[p(x),u(x)] = \int dx\left[\frac{p(x)^2}{2p} + V[u(x)] + \rho\,\frac{c_o^2}{2}\left(\frac{du}{dx}\right)^2\right] \quad (16)$$

It is common to take the "u^4-potential":

$$V[u(x)] = \frac{A}{2}u^2 + \frac{B}{4}u^4 \quad (17)$$

Frequently $A = a(T-T_{m.f.})$; sometimes A - const < 0. In any case, the potential is the familiar double well. By comparison the sine-Gordon potential is $\sim(\cos u-1) = -(u^2/2!) + (u^4/4!) - \ldots$ and in low order of nonlinearity corresponds to a special choice of A and B in "u^4-theory".

Studies show that: (a) the "u^4-potential" has kink solutions which are solitary waves - not solitons, (b) linear stability analysis shows the kink to be stable, and (c) that the kink interacts only weakly with phonons (providing only a phase shift but no reflection). [However, in higher order nonlinear stability analysis the exact separations found for sine-Gordon do not follow[11] for "u^4"].

The fundamental statistical mechanics problem is always to compute the partition function

$$Z(\beta) = \mathrm{Tr}\,[e^{-\beta H}] \quad (18)$$

with $\beta = (kT)^{-1}$. But here the (assumed) classical form leads to the functional integral

$$Z(\beta) = \iint \mathcal{D}[p(x)]\mathcal{D}[u(x)]\; e^{-\beta \int dx\, \mathcal{H}[p(x),\, u(x)]} \quad (19)$$

The symbols $\mathcal{D}[p(x)]$, $\mathcal{D}[u(x)]$ imply that we are to integrate over all field paths. In the classical limit kinetic and potential terms separate

$$Z(\beta) = Z_{kin}\, Z_{pot}. \quad (19a)$$

$$Z_{kin} = \int \mathcal{D}[p(x)]\; e^{-\beta\int dx\, \frac{p(x)^2}{2\rho}} \quad (19b)$$

$$Z_{pot} = \int \mathcal{D}[u(x)]\ e^{-\beta\ dx\left\{V[u(x)] + \rho \frac{c_o^2}{2}\left(\frac{du}{dx}\right)^2\right\}} \tag{19c}$$

The only complication arises in evaluating 19(c). In 1-d cases this can be done exactly, using transfer integral methods; in (15) the standard methods were applied.

However, the Feynman path integral visualization is very informative, and Fig. 2 illustrates the application to this simple classical case. The main point is that (as in any computation of the partition function) Z_{pot} will be dominated by those paths which minimize

$$\int dx \left\{V[u(x)] + \rho \frac{c_o^2}{2} \frac{du}{dx}^2\right\}. \tag{20}$$

But application of the Euler-Lagrange variational method yields a defining equation for u(x):

$$\rho\ c_o^2\ u_{xx} + A\ u + B\ u^3 = 0 \tag{21}$$

If $A = -|A|$, V[u] has a minimum $u = \pm\ u_o = \pm\ (A/B)^{\frac{1}{2}}$, and one set of minimal potential fields is to have u displaced to $\pm\ u_o$, with small oscillations about these "displaced ordered states".

But, another very low energy solution is with $u = -u_o$ for $x \ll x_o$, $u = \pm\ u_o$ for $x \gg x_o$ and a kink localized at $x \simeq x_o$; this state will not cost much energy, and thus must be considered. This kink is then a "defect" in our otherwise uniform order parameter.

Now, what does the discussion of solitary waves contribute to this situation? The answer is that by an analysis similar to that for the sine-Gordon equation one finds

$$u_{kink} = \pm\ u_o\ \tanh\ (x/d) \tag{22}$$

where $d^2 = (2\rho\ c_o^2/|A|)$, and the oscillations about this solution

$$\chi_o = [d(u_{kink})/dx];\ \omega^2 = 0 \tag{23}$$

$$\chi_1 = \sinh\ (x/d)\ \text{sech}^2(x/d)\ e^{i\omega t};\ \omega^2 = 3/4\ \omega_o^2 \tag{24}$$

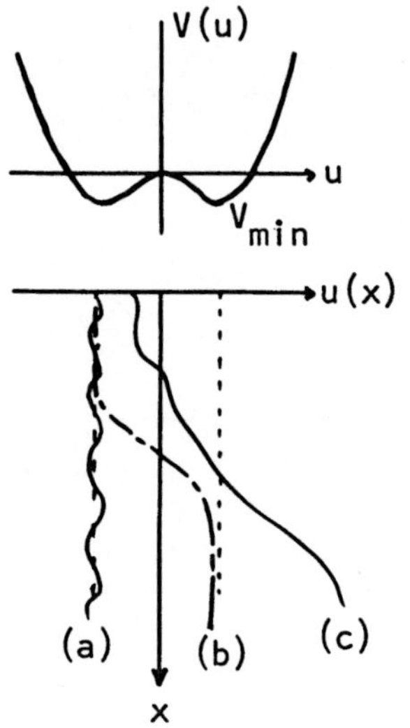

Figure 2. Possible paths contributing to the path integral expression for the partition function Eq. 19(c); (a) small amplitude oscillations about a local minimum of V (displaced from u = o); (b) a kink solution; (c) a possible general path, which contributes negligible to the thermodynamics.

$$\chi_k = e^{i(kx-\omega t)}(3\tanh^2(x/d) - 1 - k^2 - 3ik\tanh(x/d)];$$

$$\omega^2 = \omega_o{}^2 + c_o{}^2\, k^2 \tag{25}$$

where ω_o is the Einstein frequency in the bottom of the potential well.

Not only does this demonstrate the stability, but also it provides us with the information needed to compute the additional contributions to Z of paths neighboring the kink path. Thus, we can finally represent the partition function very closely by

$$Z = \exp\,[-\beta(F_{phonon} + F_{kink} + F_{kink\text{-}fluctuation})]$$

whence the free energy becomes

$$F = -k\,T\,\ln Z = F_{phonons} + F_{kinks} + (\text{fluctuations-corrections.})$$

This was shown to be exactly true, by transfer integral methods, in Ref. 15.

The advantage of the more general path integral description is that it allows us to see, at least in principle, that similar behavior could occur in three dimensions. In fact, we have looked at kink solutions of the "u^4" problem in three dimensions; we find local stability of a domain wall with respect to fluctuations, but not global stability of the macroscopic shape. The statistical mechanics then becomes that of describing arrays of grain boundaries in three dimensions. As yet, a quantitative calculation has not been done.

There is far too much going on to present here; however, it is hoped that a useful introduction to the literature of the subject of solitary waves has been given.

Just in closing, we show in Fig. 4 the results of computer simulations of the finite temperature molecular dynamics, by

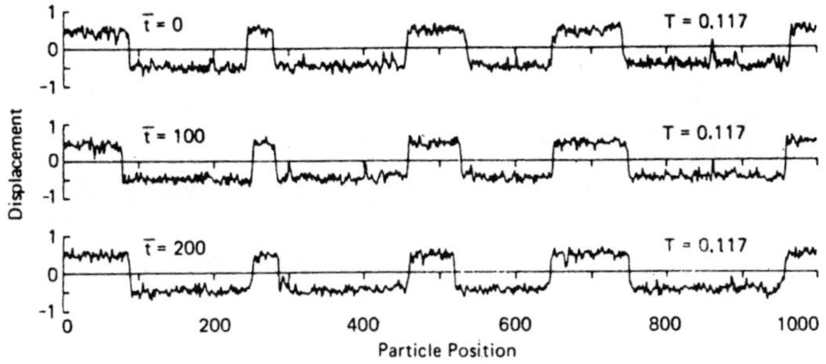

Figure 3. Displacements at 1,000 lattice sites for $\bar{T}$ = 0.117; t = 0, 100, 200.

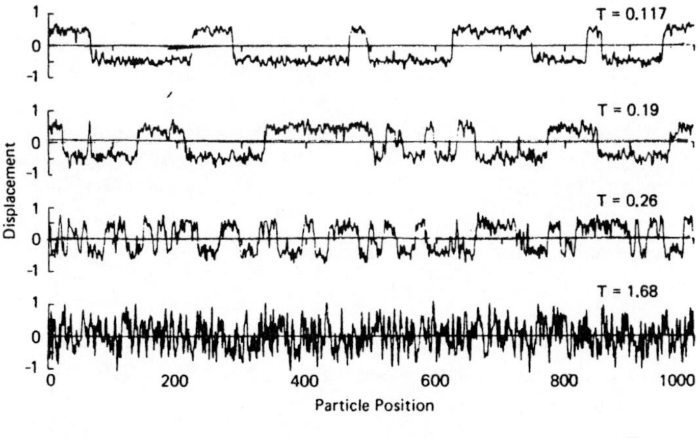

Figure 4. Displacements at 1,000 lattice sites for $\bar{T}$ = 0.117, 0.19, 0.26, 1.68.

T. R. Koehler, et al.[16] There is no phase transition in this one-dimensional system; however, the kink structure is apparent, and in quantitative agreement with the theory sketched above.

References

(1) A. C. Scott, F. Y. F. Chu, and D. W. McLaughlin, Proc. IEEE 61, 1443 (1973).

(2) G. L. Lamb and D. W. McLaughlin (unpublished).

(3) R. Rajaraman, Physics Reports 21C, 229 (1975).

(4) J. Scott-Russell, Proc. Roy. Soc. Edinburgh, 319, (1844).

(5) NSF Conference on Solitons, Tucson, Arizona (1976).

(6) E. Fermi, J. R. Pasta, S. M. Ulam, Los Alamos Sci. Lab., Rep. LA-1940 (1955).

(7) N. J. Zabusky, Comput. Phys. Commun. 5, 1 (1973).

(8) J. K. Perring and T. H. Skyrme, Nucl. Phys., 31, 550 (1962).

(9) A. Seeger, H. Donth, and A. Kochendörfer, Z. Phys., 134, 173 (1953).

(10) W. Döring, Z. Naturforschung, 31, 373 (1948).

(11) R. Dashen, B. Hasslacher and A. Neveu, Phys. Rev. D11, 3424 (1975), S. Coleman, Phys. Rev. D11, 2088 (1975).

(12) A. C. Scott, Nuovo Cimento, 69B, 214 (1970).

(13) M. J. Rice, A. R. Bishop, J. A. Krumhansl, and S. E. Trullinger, Phys. Rev. Lett. 36, 432 (1976).

(14) L. D. Faddeev, L. A. Takhatajan, Uspekhi Math. Sci. 29, 249 (1974); Theor. Math. Phys. 21, 160 (1974) (See also preprint JINR E27998 (1974)).

(15) J. A. Krumhansl and J. R. Schrieffer, Phys. Rev. B11, 3535 (1975).

(16) T. R. Koehler, A. R. Bishop, J. A. Krumhansl, and J. R. Schrieffer, Solid State Commun., 17, 1515 (1975).

SOLITONS AND RELATED PHENOMENA IN SUPERFLUID ^{3}He*

Kazumi Maki

Physics Department
University of Southern California
Los Angeles, California 90007

Abstract: Solitons (or planar textures) in superfluid ^{3}He are reviewed. In the A phase, solitons are classified into $\hat{d}$-soliton, $\hat{\ell}$-soliton and composite soliton. In an open system the last one is the only stable soliton. Creation of $\hat{d}$-soliton by magnetic means, and magnetic resonances associated with a variety of solitons are described.

The condensate of superfluid ^{3}He is described in terms of nine complex order parameters. This large degree of freedom reflects the triplet P-wave pairing central to the condensate of superfluid ^{3}He. The most remarkable manifestation of this large degree of freedom is the effect of textures, the concept originally introduced by de Gennes.[1] In this talk I would like to review work initiated with Tsuneto[2] and Ebisawa[3] and further developed in collaboration with Kumar[4] on solitons in superfluid ^{3}He. Generally speaking, I am interested in the simplest textures (planar textures) in superfluid ^{3}He, which depends on a single space variable. The term "soliton" is appropriate for the magnetic solitons, which are nothing but domain walls associated with the spin configurations of the condensate. In particular, in the A phase the $\hat{d}$ vector takes the opposite orientation in the two sides of the domain wall.[3,4] The $\hat{d}$ vector changes continuously from one direction to another within the wall. The three typical walls (solitons) are shown in Fig. 1. They can be called the splay, the bending and the pure twist solitons, respectively. Furthermore, the basic equation describing the $\hat{d}$ texture in the A phase is a sine-Gordon

* Work supported by the National Science Foundation DMR 76-21032.

equation and consequently the d solitons have properties common to solitons in other fields in physics. In the case of the B phase, the basic equations for the spin coordinates are slightly more complicated.[5] However, we may still call the planar structures solitons. For more general planar structures, the dynamic equations are quite different at least in the vicinity of the transition temperatures. Therefore, the domain walls may be more appropriate for those structures, although at low temperatures we may use the term solitons as well. In the following, we will start with the Ginzburg-Landau free energy as first written down by Ambegaokar et al.[6] In the most general form, it is given by[7]

$$F_{kin} = \frac{1}{2}\int d^3r[K_1\partial_i A_{\mu i}\partial_j A^*_{\mu j} + K_2\partial_i A_{\mu j}\partial_i A^*_{\mu j} + K_3\partial_i A_{\mu j}\partial_j A^*_{\mu i}], \quad (1)$$

where $A_{\mu i}$ (μ is the spin index and i the orbital) are nine order parameters. In the Ginzburg-Landau regime and in the weak-coupling limit we have[6,7]

$$K=K_1=K_2=K_3=\frac{6}{5}\,(7\zeta(3)/(2\pi T_c)^2)\,(8m^*)^{-1}N \quad , \quad (2)$$

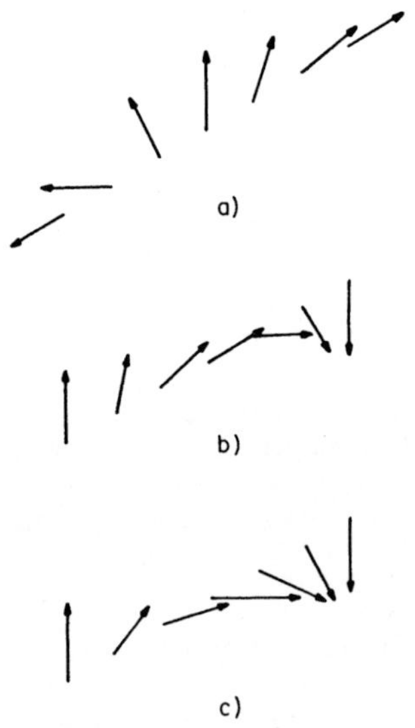

Figure 1. Three typical planer structures are shown, where arrows indicate the direction of $\hat{d}$-vector or $\hat{\ell}$-vector. (a) splay (b) bending, and (c) pure twist.

with m* the effective mass and N the ^{3}He density.

1. SOLITONS IN A-PHASE

In the A phase, where the condensate is the axial state, the vector order parameter $A_{\mu i}$ is given by

$$A_{\mu i} = \frac{\Delta_o}{\sqrt{2}} \hat{d}_\mu (\vec{\delta}_i^1 + i\vec{\delta}_i^2), \tag{3}$$

where $\hat{d}$ is unit vector describing the spin coordinates, while $\vec{\delta}^1, \vec{\delta}^2$, and $\hat{\ell} (\equiv \vec{\delta}^1 \times \vec{\delta}^2)$ are mutually orthogonal unit vectors describing the orbital coordinates. We can then rewrite Eq. (1) in terms of $\hat{d}$ and $\vec{\Delta} = \frac{\Delta_o}{\sqrt{2}} (\vec{\delta}^1 + \vec{\delta}^2)$ as

$$F_{kin} = \frac{1}{2} K \int d^3r \, \{(2+k)|\vec{\nabla}\cdot\vec{\Delta}|^2 + k(\vec{\nabla}\times\vec{\Delta})^2 + 2|\vec{\Delta}\cdot\vec{\nabla}\hat{d}|^2 + k|\vec{\Delta}|^2$$

$$\times (|\vec{\nabla}\cdot\hat{d}|^2 + |\vec{\nabla}\times\hat{d}|^2)\} \quad , \tag{4}$$

where $K = \frac{1}{2}(K_1+K_3)$, $k=K_2/K$ and we have dropped pure divergence terms, since they do not contribute in the following consideration. In the following analysis we take mostly k=1 (i.e. the weak coupling limit), except occasionally we come back to the more general situation with $k \neq 1$.

To be definite, we assume that a static magnetic field $\vec{H}$ is applied in the z direction, which limits the equilibrium configuration of $\hat{d}$ in the x-y plane. Then we take

$$\hat{d} = \sin\psi\hat{x} + \cos\psi\hat{y},$$

and

$$\hat{\ell} = \sin\chi\hat{x} + \cos\chi\hat{y}, \tag{5}$$

where ψ and χ are functions of position to be determined later. The vector order parameter consistent with Eq. (5) is given by

$$\vec{\Delta} = \frac{\Delta_o}{2} e^{i\Phi}(-\cos\chi\hat{x} + \sin\chi\hat{y} + i\hat{z}), \tag{6}$$

where Φ is another function to be determined later.

Substituting Eqs. (5) and (6) into Eq. (4) (with k=1), we have

$$F = \frac{1}{2}A \int d^3r\{|\vec{\nabla}\chi|^2 + 2[\sin\chi(\partial\chi/\partial x)+\cos\chi(\partial\chi/\partial y)]^2$$

$$+4|\vec{\nabla}\Phi|^2 - 2[\sin\chi(\partial\Phi/\partial x)+\cos\chi(\partial\Phi/\partial y)]^2$$

$$+4|\vec{\nabla}\psi|^2 - 2[\sin\chi(\partial\psi/\partial x)+\cos\chi(\partial\psi/\partial y)]^2$$

$$+2(\partial\chi/\partial z)[\sin\chi(\partial\Phi/\partial x)+\cos\chi(\partial\Phi/\partial y)]$$

$$-6[\sin\chi(\partial\chi/\partial x)+\cos\chi(\partial\chi/\partial y)](\partial\Phi/\partial z)$$

$$+4\xi_\perp^{-2}\sin^2(\chi-\psi)\}\ , \tag{7}$$

where $A = \frac{1}{2}K\Delta_0^2 = \frac{1}{4}\chi_N C_\perp^2$, and $\xi_\perp = C_\perp/\Omega_A$ is the dipolar coherence length with $C_\perp$ the spin wave velocity perpendicular to $\hat{\ell}$, and Ω_A is the resonance frequency. In Eq. (7), I have included the dipolar energy (the last term), which is given by[7] $F_d = -1/2\chi_N\Omega_A^2(\hat{\ell}\cdot\hat{d})^2$. This dipolar energy provides the crucial symmetry breaking term necessary for existence of domain walls.

a) $\hat{d}$-soliton (magnetic soliton)

When $\hat{\ell}$ is uniform (e.g. $\chi = 0$, $\Phi = 0$) we have a pure $\hat{d}$-texture given by[3,4]

$$\tan\frac{\psi}{2} = \exp[\pm s/\xi_\perp(1-\frac{1}{2}k_2{}^2)^{1/2}]\ , \tag{8}$$

with

$$s = k_1x+k_2y+k_3z = \hat{k}\cdot\vec{x} \text{ and } |\hat{k}| = 1. \tag{9}$$

The surface energy of the $\hat{d}$-soliton is given by

$$f^{\hat{d}}(\hat{k}) = E/\sigma(\hat{k}) = 2\chi_N\Omega_A{}^2\xi_\perp(1-\frac{1}{2}k_2^2)^{1/2} = 8A\xi_\perp^{-1}(1-\frac{1}{2}k_2^2)^{1/2}\ , \tag{10}$$

where $\sigma(\hat{k})$ is the surface area normal to $\hat{k}$.

The $\hat{d}$ soliton has the lowest energy when $\hat{k}||\hat{\ell}$ (i.e. $k_2=1$). Furthermore, the soliton carries spin current with the spin polarization in the z direction

$$\vec{J}^{spin} = -\frac{\overleftrightarrow{\rho}_s^{\,o}}{2m}\vec{k}\left(\frac{\partial\psi}{\partial s}\right) = -\overleftrightarrow{\rho}_s^{\,o}\vec{k}\,[2m\xi_\perp(1-\tfrac{1}{2}k_2^2)^{1/2}]^{-1}$$

$$\times\ \mathrm{sech}[s/\xi_\perp(1-\tfrac{1}{2}k_2^2)^{1/2}]\ , \tag{11}$$

where $\overleftrightarrow{\rho}_s^{\,o}$ is the reduced superfluid density tensor. The integrated spin current is nonvanishing:

$$\langle\vec{J}_{spin}\rangle = \int_{-\infty}^{\infty} ds\,\vec{J}_{spin}(s) = -\frac{\pi}{2m}(\overleftrightarrow{\rho}_s^{\,o}\vec{k})\ . \tag{12}$$

When in motion the $\hat{d}$ solitons carry magnetization pulses and they can be created magnetically.

b) $\hat{\ell}$-soliton (orbital soliton)

We may think of a uniform $\hat{d}$ texture ($\psi=0$). In this case, we have a pure $\hat{\ell}$ texture.[8] Assuming that $\chi=\chi(s)$ and $\Phi=\Phi(s)$, with s defined in Eq. (9) we have

$$\Phi_s = k_3 a\chi_s/(2-a^2)\ ,$$

and

$$\chi_s = \pm 2\xi_\perp^{-1}\sin\chi\left[\frac{2-a^2}{2+(3-2k_3^2)a^2-2a^4}\right]^{1/2}\ , \tag{13}$$

with

$$a = (k_1\sin\chi + k_2\cos\chi).$$

Here, the suffix s means a derivative in s.

The surface energy for the $\hat{\ell}$-soliton is given by

$$f^{\hat{\ell}}(\hat{k}) = F/\sigma(\hat{k}) = 2A\xi_\perp^{-1}\int_0^{\pi} d\chi\sin\chi\left[\frac{2+(3-2k_3^2)a^2-2a^4}{2-a^2}\right]^{1/2}$$

The minimum is obtained for $\hat{k}||\hat{z}$, $f^{\hat{\ell}}(\hat{z}) = 4A\xi_{\perp}^{-1}$, which describes a pure twist domain wall. In this limit, the $\hat{\ell}$-soliton carries a supercurrent associated with curl of $\hat{\ell}$,

$$\vec{v}_{irr}(z) = \frac{1}{8m}(\vec{\nabla}\times\hat{\ell}) = \frac{1}{8}\hat{\ell}\chi_s \tag{15}$$

The integrated $\vec{v}_{irr}(z)$ is given by

$$\langle\vec{v}_{irr}\rangle = \int_{-\infty}^{\infty} dz\, \vec{v}_{irr}(z) = \frac{1}{4m}\hat{x} \tag{16}$$

c) Composite soliton

Finally, we consider a planar texture, where both $\hat{\ell}$ and $\hat{d}$ fields are involved.[9] In this case, we can solve exactly only for a pure twist solution. Introducing new variables u and v by

$$u = \chi+4\psi \text{ and } v = \chi-\psi , \tag{17}$$

a solution is given by

$$u = \text{const, and } \tan\frac{v}{2} = \exp\left(\pm\sqrt{5}z/\xi_{\perp}\right) . \tag{18}$$

Here u is the center of mass coordinates and v is the relative coordinates of χ and ψ. The corresponding surface energy is given by

$$f^{\hat{\ell},\hat{d}}_{(0)} = \frac{2}{\sqrt{5}}\chi_N\Omega_A{}^2\xi_{\perp} , \tag{19}$$

which is the smallest among the planar structures so far considered. In fact, in the absence of the external constraints, the composite soliton will be the end product of all planar structures so far considered in the A-phase.

The composite soliton carries both spin current and mass current

$$\vec{j}^{spin} = -\frac{\overleftrightarrow{\rho}_s{}^o}{2m}\hat{z}\left(\frac{\partial\psi}{\partial z}\right) = \frac{\overleftrightarrow{\rho}_s{}^o}{2\ 5m\xi_{\perp}}\hat{z}\ \text{sech}(\ 5\ z/\xi_{\perp}) , \tag{20}$$

and

$$\vec{v}^{irr} = \frac{1}{8m}\hat{\ell}(\frac{\partial\chi}{\partial z}). \qquad (21)$$

The integrated values are given as

$$<\vec{J}^{spin}> = \frac{\pi}{10m}\overleftrightarrow{\rho}^{o}_{s}\hat{z}, \quad <\vec{v}^{irr}> = \frac{1}{8m}[(1+\cos(\frac{\pi}{5}))\hat{x}+\sin(\frac{\pi}{5})\hat{y}] \ . \qquad (22)$$

We are still unable to solve a more general composite soliton where both χ and ψ depend on s (Eq. (9)). However, for small $k_\perp$, the free energy is calculated to be

$$f^{\hat{\ell}\cdot\hat{d}}(k) \cong f^{\hat{\ell}\cdot\hat{d}}_{(0)} \ [1 + \frac{7}{40}(1- \frac{25}{39}\cos(\frac{\pi}{5}))k_\perp^2]$$

$$= f^{\hat{\ell}\cdot\hat{d}}_{(0)}[1 + 0.084 \ k_\perp^2] \ , \qquad (23)$$

indicating that the twist composite soliton is the one with the lowest energy.

2. SOLITONS IN B-PHASE

In the B-phase the condensate is isotropic and given by[10]

$$A_{\mu i} = \frac{\Delta_o}{\sqrt{\beta}}\{\cos\theta\delta_{\mu i}+ (1-\cos\theta)n_\mu n_i+\sin\theta\varepsilon_{\mu ik}n_k\} \ , \qquad (24)$$

where $\vec{n}$ is unit vector, designating the axis of rotation and θ is the rotation angle of the spin coordinates. Then we have three kinds of solitons, θ-solitons, $\vec{n}$-solitons, and θ-$\vec{n}$ composite solitons. Furthermore, in order to have pure $\vec{n}$-solitons, we need additional symmetry breaking energy besides the dipole interaction energy.[11] Here, we limit ourselves to θ-solitons for simplicity. Then in the presence of a magnetic field along the z axis, we can take $\vec{n}||\hat{z}$. In this limit, substituting Eq. (24) into Eq. (1), we obtain[5]

$$F = \frac{4A}{3}\int d^3r[(\partial\theta/\partial x)^2+(\partial\theta/\partial y)^2+ \frac{1}{2}(\frac{\partial\theta}{\partial z})^2+ \frac{16}{15}\xi_\perp^{-2}(\cos\theta+ \frac{1}{4})^2], (25)$$

where $A = \frac{1}{2} K\Delta_o^2$ and $\xi_\perp^{-2} = \frac{3}{8} \chi_B\Omega_B^2/A$. As before, we included the dipolar coherence length in the B-phase. Then assuming that $\theta=\theta(s)$ with

$$s = \hat{k}\cdot\vec{x} \quad \text{and} \quad |\vec{k}| = 1 , \tag{26}$$

we have

$$\tan\frac{\theta}{2} = \pm\sqrt{\frac{5}{3}} \coth(s/2\xi_\perp (1 - \frac{k_3^2}{2})^{1/2}), \tag{27}$$

and

$$\tan\frac{\theta}{2} = \pm\sqrt{\frac{5}{3}} \tanh(s/2\xi_\perp (1 - \frac{k_3^2}{2})^{1/2}) , \tag{28}$$

for the type I and the type II soliton, respectively. The surface energy associated with these solitons are given by

$$f_I^\theta = 2[1+(1/\sqrt{15}) (\theta_o-\pi)]\chi_B\Omega_B^2\xi_\perp(1- \frac{1}{2} k_3^2)^{1/2} \tag{29}$$

and

$$f_{II}^\theta = 2(1+(1/ \sqrt{15})\theta_o)\chi_B\Omega_B^2\xi (1 - \frac{1}{2} k_3^2)^{1/2} \tag{30}$$

for the type I soliton and type II soliton, respectively, where $\theta_o = \cos^{-1}(- 1/4)$ the Leggett angle. We note that Eqs. (29) and (30) imply that

$$f_I^\theta/f_{II}^\theta = (1+\frac{1}{\sqrt{15}} (\theta_o-\pi))/(1+\frac{1}{\sqrt{15}} \theta_o) = 0.449 \tag{31}$$

3. CREATION OF MAGNETIC SOLITONS

The $\hat{d}$-solitons are created magnetically either by turning off an inhomogeneous magnetic field[4] or by localized tipping of the magnetization. In particular, in the A phase, the turn-off experiment is formulated exactly in terms of the inverse scattering method developed by Ablowitz et al.[12] for the sine Gordon equation. However, in usual situations solitons are created by the hundreds. Therefore, an approximate method which gives the total number of created solitons is extremely useful. Here, I limit myself to

$\hat{d}$-soliton creation for simplicity, although the method can be easily generalized to θ-solitons in the B phase.

a) Turn-Off experiment

In this case the basic equation is given by[2]

$$\alpha_{tt}-c_{\perp}^{2}\alpha_{zz} + {}_{A}^{2}\sin\alpha\cos\alpha = 0, \tag{32}$$

where α is the angle between $\hat{d}$ and $\hat{\ell}$ in the x-y plane and suffixes t and z imply the corresponding derivatives. We note that Eq. (32) has soliton solutions

$$\tan(\tfrac{\alpha}{2}) = \exp[\pm(z-c_{\perp}ut)/\xi_{\perp}(1-u^2)^{1/2}] , \tag{33}$$

which are the time dependent version of Eq. (8). The soliton moves with velocity $c_{\perp}u$ and carries a magnetic pulse,

$$M_z(z) = \chi_N\Omega_A \frac{u}{(1-u^2)^{1/2}} \operatorname{sech}[(z-c_{\perp}ut)/\xi_{\perp}(1-u^2)^{1/2}] . \tag{34}$$

When a magnetic field $\Delta H(z)$ parallel to the z axis is turned off at t=0, the time development of α is described by Eq. (32), with the initial conditions

$$\alpha = 0, \text{ and } \alpha_t = \Delta\omega_o(z) \text{ at } t = 0 , \tag{35}$$

where $\Delta\omega_o(z) = \gamma_o\Delta H(z)$ the Larmor frequency due to the turned-off field and γ_o is the gyromagnetic ratio of ^{3}He nucleus.

The above equation can be solved exactly in terms of the inverse scattering method.[4] However, here I would like to adapt a heuristic method, which can be easily extended to the case of the tipping experiment. When $(\Delta\omega_o(z))_{max}$ is larger than Ω_A, we may assume that $\alpha(z,t)$ is approximately determined by the Ω_A - independent term,

$$\alpha_{tt} - c_{\perp}^{2}\alpha_{zz} \cong 0 \tag{36}$$

Then Eq. (36) together with the initial conditions (35) yields

$$\alpha(z,t) = \frac{1}{c_\perp}\int_{z-c_\perp t}^{z+c_\perp t} \Delta\omega_o(z')dz' \quad . \tag{37}$$

If we assume further for simplicity that the turned-off field is a hat shape function; $\Delta\omega_o(z) = \Delta\omega_o\theta(a^2-z^2)$, with $\theta(x)$ the step function, Eq. (37) gives

$$\begin{aligned} \alpha(0,t) &= 2(\Delta\omega_o)t \qquad &&\text{for } t<t_o \equiv a/c_\perp \\ &= 2(\Delta\omega_o)t_o \qquad &&\text{for } t>t_o \end{aligned} \tag{38}$$

This is easily understood from Fig. 2. As long as the point P(z,t) is within the causal diamond, $\alpha(P)$ behaves as if a homogeneous field $\Delta H = \Delta\omega_o/\gamma_o$ is turned off, since the effects of edges in $\Delta H(z)$ are not felt. Only outside of the causal diamond, do the effects of edges become of importance. Therefore, within the present approximation, we have the maximum value of $\alpha(P)$ as

$$\alpha_{Max} = 2(\Delta\omega_o)t_o = 2(\Delta\omega_o)a/c_\perp \tag{39}$$

As α increases through π, a pair of soliton and anti-soliton may be created, we may estimate the total number of pairs N_o as[13]

$$N_o = \frac{2(\Delta\omega_o)a}{\pi c_\perp} = \frac{2}{\pi}\left(\frac{\Delta\omega_o}{\Omega_A}\right)\frac{a}{\xi_\perp} \quad . \tag{40}$$

However, the exact solution[4] of Eqs. (32) and (35) tells that Eq. (40) is not the whole story. It is true that N_o corresponds to the total number of pairs. But some of the pairs are still in the bound state (i.e. the breather modes). In the bound states, these pairs never separate each other and ultimately damp out due to the spin diffusion term. The exact treatment predicts that for large N_o, the number of outgoing pairs (unbounded pairs) N has an additional dependence of $\Delta\omega_o/\Omega_A$,

$$N/N_o \cong \frac{1}{2}\left[1-\left(\frac{\Omega_A}{\Delta\omega_o}\right)^2\right]^{1/2} + \frac{1}{\pi N_o}\cos^{-1}\left(\frac{\Omega_A}{\Delta\omega_o}\right) \quad . \tag{41}$$

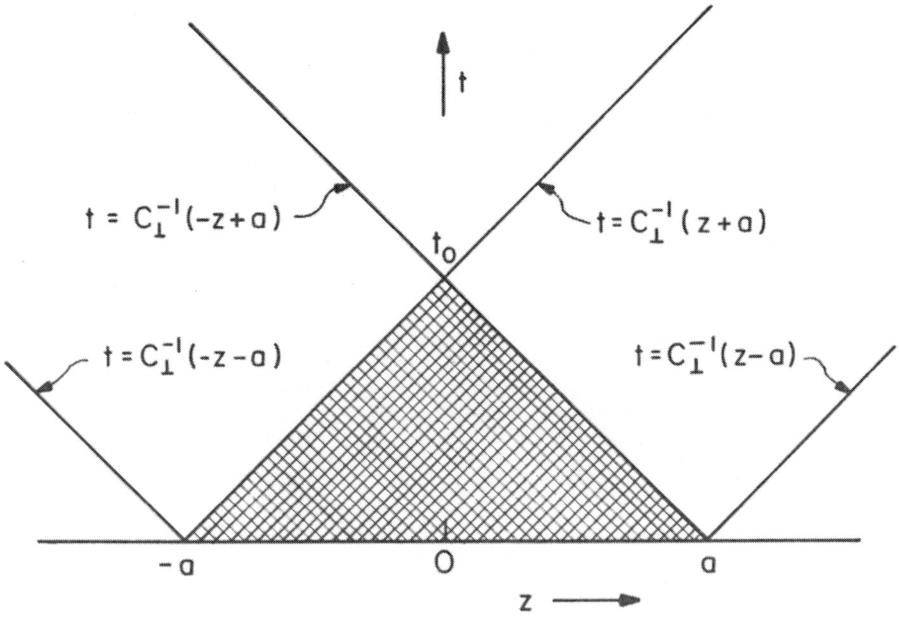

Figure 2. The t-z plane, where the winding of the $\hat{d}$ vector takes place. The shaded area is the causal diamond.

In the presence of the dipolar energy (i.e. the Ω_A-dependent term), $\Delta\omega_o$ has to be greater than Ω_A for α to increase unhindered by the dipolar potential barrier.

The above equation is plotted in Fig. 3 for $N_o = 10$, and $N_o > 10^2$. For $\Delta\omega_o = 2\Omega_A$, we have $N/N_o \simeq 0.433$.

In actual experimental situations, since $\xi_\perp \simeq 10^{-3}$ cm, the solitons are created by the hundreds in a turn-off experiment if $a \simeq 1$cm.

b) Tipping experiment

In this case a burst with the resonance frequency ($\omega \cong \omega_o$) is applied in the transverse direction, which causes tipping of the total magnetization by an angle $\theta(z)$. The basic equation is now given by

$$\frac{\partial}{\partial t}\left(\frac{\partial \mathcal{L}}{\partial \alpha_t}\right) + \frac{\partial}{\partial_z}\left(\frac{\partial \mathcal{L}}{\partial \alpha_z}\right) - \frac{\partial \mathcal{L}}{\partial \alpha} = 0 \qquad (42)$$

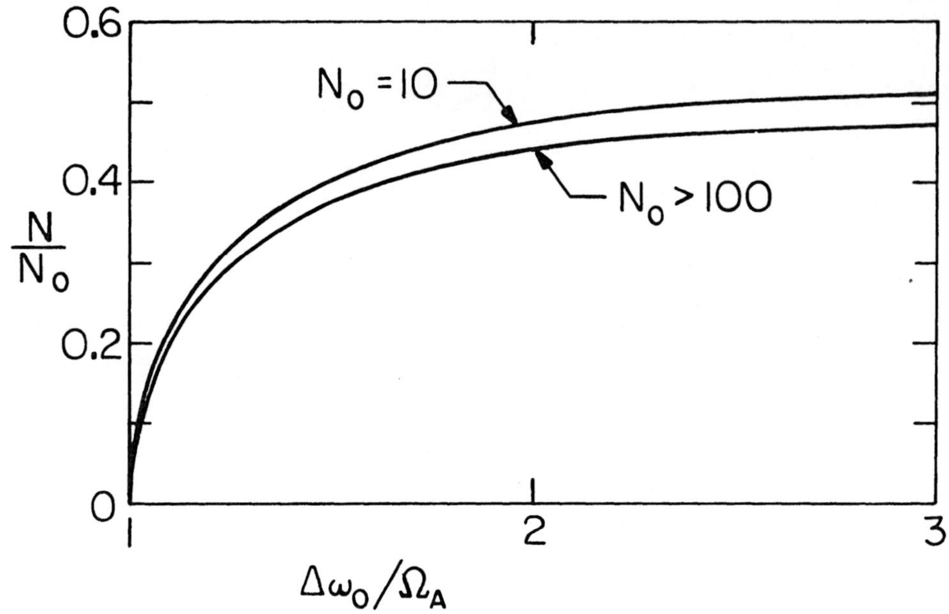

Figure 3. The total number of the outgoing soliton pairs N normalized by N_0, the total winding angle α of $\hat{d}$ vector divided by π, is shown as functions of $\Delta\omega_0/\Omega_A$ for N=10 and $N>10^2$.

etc., where $\mathcal{L}$ is the Lagrangian describing the spin dynamics in the A phase,

$$\mathcal{L} = \frac{1}{2}\chi_N\{\alpha_t^{\,2} + \beta_t^{\,2} + \gamma_t^{\,2} + 2\alpha_t\gamma_t\cos\beta - 2\omega_0(\alpha_t + \gamma_t\cos\beta)$$

$$- c_\perp^{\,2}[\alpha_z^{\,2} + \beta_z^{\,2} + \gamma_z^{\,2} + 2\alpha_z\gamma_z\cos\beta - (\gamma_z\sin\beta\sin\gamma + \beta_z\cos\gamma)^2]$$

$$- \Omega_A^{\,2}[\cos\alpha\cos\gamma - \sin\alpha\sin\gamma\cos\beta]^2 \qquad (43)$$

Here α, β, and γ are Eulerian angles describing the rotation of the spin coordinates and ω_0 is the Larmor frequency. We assumed that $\hat{\ell}$ is along the y axis and the static field in the z direction.

Even in the absence of the dipolar interaction, the above set of equations are rather unwieldy. However, in the homogeneous case and in the absence of the dipolar energy term, Eq. (42) can be solved as

$$\alpha = -\gamma = \omega_o t \quad \text{and} \quad \beta = \theta \ . \tag{44}$$

As in the case of the turn-off experiment, Eq. (44) should be valid in the causal diamond. In particular, if $\theta(z)$ is a hat shape function, at the origin (i.e. z=0) we have Eq. (44) for $t \leq t_o \equiv a/c_\perp$. On the other hand, the total rotation angle Φ of the $\hat{d}$ vector is given by

$$\cos\Phi = \cos\alpha\cos\gamma - \sin\alpha\sin\gamma\cos\beta$$

$$= \frac{1}{2}(1 + \cos\theta + (1-\cos\theta)\cos(2\omega_o t)). \tag{45}$$

Therefore for $\theta = 2\pi$ (i.e. 180° - tipping), we have $\Phi = 2\omega_o t$. The maximum Φ is then estimated to be $\Phi_{max} = 2\omega_o t_o = 2\omega_o\, a/c_\perp$. This phase winding corresponds to the number of soliton pairs N_o

$$N_o = \frac{2}{\pi}\,\omega_o a/c_\perp = \frac{2}{\pi}\left(\frac{\omega_o}{\Omega_A}\right)\frac{a}{\xi_\perp} \tag{46}$$

We note that in the case of 180° - tipping, $\Delta\omega_o$ for the turn-off experiment is replaced by ω_o, which appears quite reasonable. As in the case of the turn off experiment, the actual number of the outgoing soliton pairs may be reduced somewhat from Eq. (46). However, Eq. (46) will provide a good estimate for the soliton numbers.

These solitons spread into space from their positions of creation. Then they stop[3] somehwere due to the spin diffusion within 10^{-4} sec. Then the orbital instability sets in.[9] In an open system they are converted into the composite solitons[9] within the time scale approximately $10^{-2}(1-T/T_c)^{1/2}$ sec. due to the orbital viscosity.[14,15] In this process both spin waves and orbital waves are radiated all over the space. The former propagates in space like waves with strong dispersion, while the latter propagates as diffusive wave, say like a heat pulse in normal liquids. Therefore, the secondary radiation associated with the conversion may be accessible by both magnetic and ultrasonic measurements.

4. MAGNETIC RESONANCES

The planar structures I have described may be most readily accessible to the magnetic resonance, although existence of pure $\hat{\ell}$ solitons or composite solitons affects certainly the sound propagation in liquid as well. In general, the spin oscillation in the planar structures has a discrete spectrum associated with the localized oscillations bound at the dipolar potential well produced by the planar structures. These localized modes or bound states appear as satellites to the main resonances in superfluid ^{3}He.

a) Pure $\hat{d}$-soliton

In this case, we consider a pure twist $\hat{d}$ soliton without loss of generality. Assuming that $\hat{d}$ is given by

$$\hat{d} = [\sin(\psi+f)\hat{x} + \cos(\psi+f)\hat{y}] \cos g + \hat{z} \sin g \ , \tag{47}$$

where ψ is given by Eq. (8) with $\hat{k}=\hat{z}$, and f and g are infinitesimal fluctuations, we can expand the free energy in powers of f and g,

$$\delta f = \frac{\delta F}{\sigma_3} = \frac{1}{\sigma_3} (F-F(\psi)) = \frac{1}{2} A \int dz \ [4(f_z{}^2 + g_z{}^2)$$

$$+ 4\xi_\perp^{-2}(1 - 2 \operatorname{sech}^2(z/\xi_\perp))(f^2+g^2)] \ . \tag{48}$$

The corresponding eigen-equations are given by

$$\lambda f = -\xi_\perp^2 f_{zz} + (1-2\operatorname{sech}^2(z/\xi_\perp))f \ , \tag{49}$$

and the same equation for g. Eq. (49) allows one bound state with

$$\lambda = 0 \text{ and } f \propto \operatorname{sech}(z/\xi_\perp) \tag{50}$$

The corresponding nuclear resonance frequencies are determined by noting that δF provides the potential V in the Lagrangian (43) in this general situation, where T is still given by the kinetic energy in Eq. (43),

$$\mathcal{L} = T - V$$

and

$$T = \frac{1}{2}\chi_N(\alpha_t{}^2+\beta_t{}^2+\gamma_t{}^2+ 2\alpha_t\gamma_t\cos\beta - 2\omega_o(\alpha_t+\gamma_t\cos\beta)) \qquad (51)$$

In order to determine the resonance frequencies, we have to express f and g in terms of α, β, and γ. This is most easily done by noting that the rotation $R(\alpha,\beta,\gamma)$ brings $\hat{d} = \hat{y}$ into

$$\hat{d} = -(\sin\alpha\cos\gamma + \cos\alpha\cos\beta\sin\gamma)\hat{x}$$

$$+ (\cos\alpha\cos\gamma-\sin\alpha\cos\beta\sin\gamma)\hat{y} + \sin\beta\sin\gamma\hat{z}. \qquad (52)$$

Comparing this with Eq. (5), we have

$$\psi + f = -\alpha,\ \gamma = g \text{ and } \beta \cong \frac{\pi}{2} \ . \qquad (53)$$

With this identification, we can solve the equation of motion for small f and g. The satellite frequencies for pure $\hat{d}$ solitons are given by

$$\omega_\ell = 0$$

$$\omega_t = \omega_o \ , \qquad (54)$$

for the longitudinal and the transverse resonance respectively; the pure $\hat{d}$ soliton gives rise to unshifted resonance frequencies.[16] There are Goldstone modes associated with $\hat{d}$-soliton; the f-mode corresponds to the sliding of $\hat{d}$-soliton, while the g-mode corresponds to the rotation of $\hat{d}$-soliton along an axis perpendicular to the z axis. Therefore, in a narrow channel with a gap of order of 10^{-3} cm, we expect the existence of unshifted resonances due to pure $\hat{d}$-solitons.

b) $\hat{\ell}$-soliton

At the time of writing, I do not know how to create pure $\hat{\ell}$ solitons. Maybe such solitons exist pinned at a planar wall of the container in the presence of a static magnetic field perpendicular to the wall. Therefore, without giving details, we present here satellite frequencies associated with pure $\hat{\ell}$-solitons,

$$\omega_\ell \cong 0.829\ \Omega_A \tag{55}$$

$$(\ _t^2 - \ _o^2)^{1/2} = 0.341\ \ _A$$

for the longitudinal and the transverse resonance, respectively.

c) <u>Composite soliton</u>

As already pointed out, in an open system the composite solitons are most likely the end product of $\hat{d}$-solitons. The composite soliton is the only stable planar configuration in the A phase for $F = F_{kin} + F_d$. Therefore, in an open system, composite solitons should be most relevant. We will take now $\hat{d}$ as given by Eq. (48), with ψ and χ

$$\psi = -2/5\ \tan^{-1}[\exp(\sqrt{5}\ z/\xi_\perp)], \chi = 8/5\ \tan^{-1}[\exp(\sqrt{5}\ z/\xi_\perp)] \tag{56}$$

The fluctuation free energy is then given by

$$\delta f = \frac{\delta F}{\sigma 3} = \frac{1}{2} A \int dz\{4(f_z{}^2 + g_z{}^2) + 4\xi_\perp^2[(1-2\mathrm{sech}^2(\sqrt{5}\ z/\xi_\perp))f^2$$

$$+ (1 - \frac{6}{5}\ \mathrm{sech}^2(\sqrt{5}\ z/\xi_\perp))g^2]\}\ , \tag{57}$$

with eigen-equations

$$\lambda f = -\xi_\perp^2 f_{zz} + (1-2\mathrm{sech}^2(\sqrt{5}\ z/\xi_\perp))f \tag{58}$$

$$\lambda g = -\xi_\perp^2 g_{zz} + (1 - \frac{6}{5}\ \mathrm{sech}^2(\sqrt{5}\ z/\xi_\perp))g\ . \tag{59}$$

Both Equations (58) and (59) have bound states with

$$\lambda_f = 1/2(\sqrt{65} - 7) \quad .5311,\ f \propto (\mathrm{sech}(\sqrt{5}\ z/\xi_\perp))^{1/2(\sqrt{13/5}\ -\ 1)}$$

and

$$\lambda_g = 4/5 \quad , \quad g \propto (\mathrm{sech}(\sqrt{5}\ z/\xi_\perp))^{1/5} \tag{60}$$

respectively.

We can identify f and g with α, β, and γ as before and find the satellite frequencies associated with the composite soliton,

$$\omega_\ell = (\lambda_f)^{1/2}\Omega_A = .722\ \Omega_A$$

and

$$[\omega_t{}^2-\omega_o{}^2]^{1/2} = (\lambda_g)^{1/2}\ \Omega_A = .896\ \Omega_A \tag{61}$$

for the longitudinal and transverse resonance, respectively. About a year ago the Orsay-Saclay group[17] reported observation of a satellite peak in their longitudinal resonance experiment in superfluid ^{3}He-A. The satellite frequency was reported to be $\omega_\ell/\Omega_A \simeq 1/\sqrt{2}$. More recently, an exhaustive study of these satellites, in both longitudinal and transverse cases, has been reported by Gould and Lee.[18] Their results are summarized as

$$\omega_\ell/\Omega_A(\equiv R_\ell) = .74-.35(1-T/T_c)^{1/2},\ \text{and}\ (\omega_t{}^2-\omega_o{}^2)^{1/2}/\Omega_A(\equiv R_t)$$

$$= .835 \tag{62}$$

The longitudinal satellite frequency agrees remarkably well with the predicted one, if we ignore a small temperature dependence. This temperature dependence may be accounted qualitatively, in terms of the temperature dependent Fermi liquid corrections as discussed by Cross.[19] On the other hand, the agreement for the transverse satellite is not completely satisfactory. Furthermore, the apparent constancy of R_t is rather puzzling.

One way out of this dilemma may be to consider the case $k \neq 1$. (Note $k = 1$, for a weak coupling theory.) Then the corresponding frequencies are now given by

$$R_\ell = [\ \frac{1}{2k}\ \{(2+3k)^{1/2}(2+11k)^{1/2}-(5k+2)\}]^{1/2} \tag{63}$$

and

$$R_t = [\frac{2(1+k)}{2+3k}]^{1/2}\ , \tag{64}$$

respectively. Equations (63) and (64) are shown in Fig. 4. We note from Fig. 4, it is rather difficult to fit R_t to the observed value without destroying the good agreement for R .

Another puzzling point is that the observed R_t is so close to the predicted R_ℓ associated with the wall pinned pure $\hat{\ell}$ soliton (see Eq. (55)). Due to the rather localized nature of the bound states, the width of these satellites are principally due to the spin diffusion. Making use of the wave functions given in Eq. (60), we obtain widths

$$\Delta\omega_\ell \cong 1/2\ D\langle f_z^2\rangle = 9/4(\sqrt{5/13} - 5/9)D\xi^{-2}$$

and

$$\Delta\omega_t \cong 1/2\ D\langle g_z^2\rangle = 1/14\ D\xi_\perp^{-2} \qquad (65)$$

respectively.

The results on the satellite frequencies associated with a variety of planar textures are summarized in Table 1.

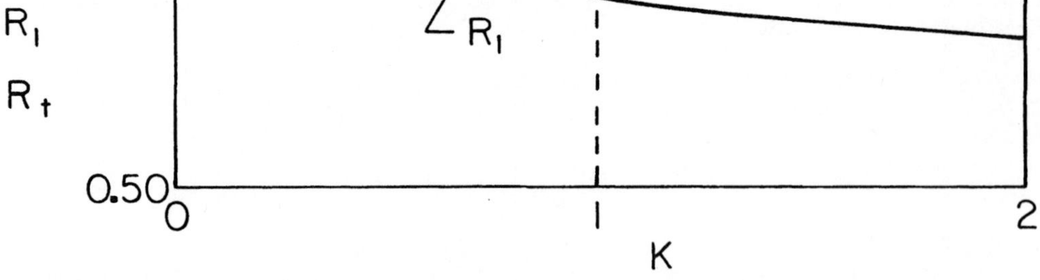

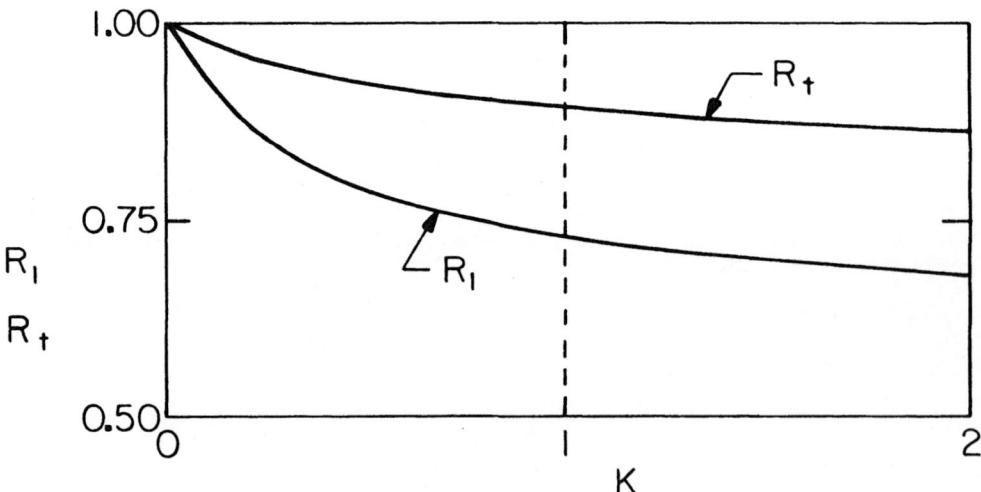

Figure 4. R_ℓ and R_t for general k are shown. K=1 corresponds to the weak coupling limit.

TABLE 1

	$F/(\sigma A\ \xi_\perp^{-1})$	R_ℓ	R_t
$\hat{d}$-soliton	$8(1-\frac{1}{2}k_2^2)^{1/2}$	0	0
$\hat{\ell}$-soliton	6.05 (bending)	0.829	0.341
composite soliton	8/ 5 (pure twist)	0.722	0.8944

ACKNOWLEDGMENT

Results reported here are obtained mostly by close collaboration with Pradeep Kumar at U.S.C. I am also grateful to Chris Gould and David Lee for sending me a copy of their experiment prior to publication and to David Mermin for a number of invaluable correspondences.

References

(1) P. G. de Gennes, Phys. Lett. 44A 271 (1973); and in "Proc. of the Nobel Symp. 24", 1973 (Academic Press, New York, 1974).

(2) K, Maki and T. Tsuneto, Phys. Rev. B 11 2539 (1975).

(3) K. Maki and H. Ebisawa, J. Low Temp. Phys. 23 351 (1976).

(4) K. Maki and P. Kumar, Phys. Rev. B 14 118 (1976) and Phys. Rev. B 14 3920 (1976).

(5) K. Maki, Phys. Rev. B 11 4264 (1975).

(6) V. Ambegaokar, P. G. de Gennes, and D. Rainer, Phys. Rev. A 9 2676 (1974).

(7) A. J. Leggett, Rev. Mod. Phys. 47 331 (1975).

(8) K. Maki and P. Kumar, in preparation.

(9) K. Maki and P. Kumar, preprint.

(10) W. F. Brinkman, H. Smith, D. D. Osheroff, and E.I. Blount, Phys. Rev. Lett. 33 624 (1974).

(11) K. Maki, "Proc. of the Sussex Symp. on Superfluid ^{3}He", (1976).

(12) M. J. Ablowitz, D. J. Kaup, A. C. Newell, and H. Segur, Phys. Rev. Lett. 30 1262 (1973).

(13) The analogous quantization condition is derived for Kortweg de Vries equation in G. B. Whitman, "Linear and Nonlinear Waves", (Wiley, New York, 1974).

(14) M. C. Cross and P. W. Anderson, "Proc. 14th Int. Conf. Low Temp. Phy., Otaniemi, Finland, 1975", edited by M. Krusius and M. Vuorio (North Holland, Amsterdam, 1975).

(15) D. N. Paulson, M. Krusius, and J. C. Wheatley, Phys. Rev. Lett. 22 1322 (1976).

(16) I am indebted to David Mermin for pointing out this fact. In the present formulation, this follows naturally from the Lagrangian.

(17) O. Avenel, M. E. Bernier, E. J. Varoquaux, and C. Vebet in "Proc. 14th Int. Conf. on Low Temp. Phy., Otaniemi, Finland, 1975, edited by M. Krusius and M. Vuorio. (North Holland, Amsterdam, 1975) Vol. 5, p. 429.

(18) C. M. Gould and D. M. Lee, Phys. Rev. Lett. 37 1223 (1976).

(19) M. C. Cross, J. Low Temp. Phys. 21 525 (1975).

EXPERIMENTAL ORBITAL DYNAMICS IN SUPERFLUID ^3He*

John C. Wheatley

Department of Physics, University of California at San Diego
La Jolla, California

Abstract: The principles and methods of producing dynamic orbital response in ^3He-A are discussed. Experiments and their interpretation are also presented for sudden magnetic field rotations and persistent orbital motions.

This paper represents a distillation of concepts and methods regarding ^3He-A which have been the subject of several reviews.[1,2,3] Orbital order in ^3He-A is expressed in part in terms of the orientation in space of an orthogonal triad of unit vectors: $\hat{e}_1$, $\hat{e}_2$, and $\hat{\ell} = \hat{e}_1 \times \hat{e}_2$, where $\hat{\ell}$ is the local direction of relative pair momentum along which no pairing takes place. Irrotational flow corresponds to a uniform $\hat{\ell}$ with a rotation of $\hat{e}_1$ and $\hat{e}_2$ about $\hat{\ell}$ as one moves in the direction of flow. Spin order is expressed in part by the direction of a unit vector $\hat{d}$, which is the local direction in space along which the spin of pairs is zero. The fields of $\hat{\ell}$ and $\hat{d}$ are coupled to one another by the coherent dipolar energy

$$\Delta F_D = -\tfrac{1}{2}\lambda_D(\hat{\ell} \cdot \hat{d})^2 , \qquad (1)$$

where $\lambda_D = \chi_N \Omega_A^2/\gamma^2$, where χ_N is the normal susceptibility, Ω_A is the longitudinal resonance (angular) frequency, and λ is the gyromagnetic ratio. The field $\hat{d}$ is coupled to the magnetic field $\vec{H}$ by the energy

$$\Delta F_H = + \tfrac{1}{2}\lambda_H(\hat{d} \cdot \vec{H})^2 , \qquad (2)$$

---

*Supported by the United States E.R.D.A. under contract E(04-3)-34, P.A. 143.

where $\lambda_H = (\chi_N - \chi_{\hat{d}})$ is the susceptibility anisotropy reflecting zero pair susceptibility along $\hat{d}$ ($\chi_{\hat{d}}$ is the susceptibility with $\vec{H}$ along $\hat{d}$). The field $\hat{\ell}$ is thought to be coupled to an electric field $\vec{E}$ by the energy

$$\Delta F_E = +\tfrac{1}{2}\lambda_E(\hat{\ell} \cdot \vec{E})^2 , \tag{3}$$

where the coefficient λ_E is currently in doubt. For a clamped normal fluid a uniform $\hat{\ell}$ field is coupled to an irrotational superfluid velocity field $\vec{v}_s$ by the energy

$$\Delta F_F = -\tfrac{1}{2}\lambda_F(\hat{\ell} \cdot \vec{v}_s)^2 , \tag{4}$$

where $\lambda_F = \rho_{s\perp} - \rho_{s/\!/} = \rho_{s/\!/}$, but this equation covers too specialized conditions and is probably only qualitatively useful. To get closer to reality one should consider general rotations of the $\hat{e}_1$, $\hat{e}_2$, $\hat{\ell}$ triad following Mermin and Ho,[4] Cross,[5] and Hall and Hook.[6] These correspond to both rotational and irrotational flows, reflecting gradient energies in the field of the $\hat{e}_1$, $\hat{e}_2$, $\hat{\ell}$ triad. There are also gradient energies in the field of $\hat{d}$ which then are coupled to the $\hat{\ell}$ field via the dipolar energy, an example of this being the problem solved recently by Maki and Kumar.[7]

Motion of the orbital triad can be induced by absorption of zero sound, as described by Wolfle[8] in a previous Sanibel symposium. Although Leggett and Takagi[9] have suggested how NMR may be used to detect the "flapping" motion of the triad, the absorption of high frequency zero sound has been used to date primarily as a tool for observing, via the $\hat{\ell} \cdot \vec{q}$ dependence of the absorption of sound of wave vector $\vec{q}$, the average orientation of the field of $\hat{\ell}$ for static or relatively slower motions. The sound attenuation method for probing the $\hat{\ell}$ field is both easy and effective. In Fig. 1 I show data of purely experimental significance from Paulson, Krusius, and Wheatley[10] (PKW) for the attenuation change relative to the attenuation at T_c for change of magnetic field direction from 35° to 15° with respect to $\vec{q}$ and for both 25 MHz zero sound at 33.5 Bar and 15 MHz sound at 24.1 Bar. There is a frequency dependent dead space along the $1-T/T_c$ axis, but otherwise the attenuation anisotropy is very substantial and useful to sufficiently large values of $1-T/T_c$.

Small angle relaxing rotational motion of the $\hat{\ell}$ field was induced by PKW[11] using a suddenly rotated, constant amplitude magnetic field. In this method the $\hat{d}$ field, rotated by susceptibility anisotropy, Eq. (2), acts to rotate the $\hat{\ell}$ field via the dipolar coupling, Eq. (1). Long-lived and persistent orbital motions in zero residual magnetic field were induced by PKW[12] in a sound cell with residual heat flow by subjecting the ^{3}He to a particular history of magnetic field magnitudes and directions. The magnetic field,

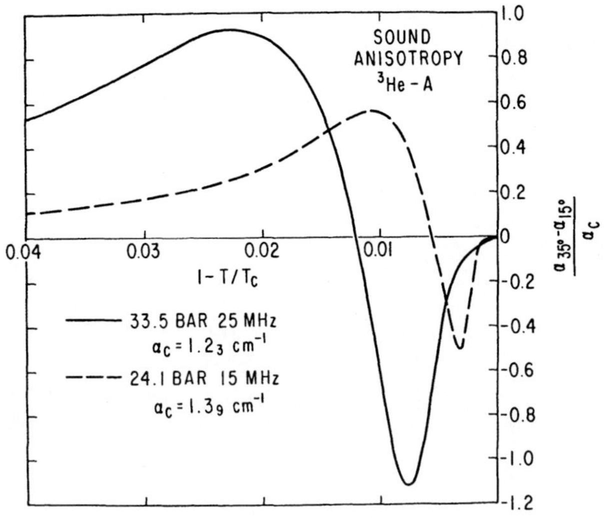

Figure 1. Experimental data from Ref. 10 showing the relative effect on zero sound attenuation at two pressures and frequencies of a change of magnetic field direction from 35° to 15° from the sound propagation direction.

via susceptibility anisotropy and dipolar coupling, forces the $\hat\ell$ field to have a largely two-dimensional character, while in zero field in presence of heat flow the $\hat\ell$ field is essentially three dimensional. When the final magnetic field is removed (and timing, direction, and field magnitude history are important) the transition to the zero field state may produce the orbital motions.

The experiments in which a field is suddenly rotated[11] through a small angle were interpreted by the following equation for the time rate of change of the orientation $\theta_{\hat\ell}$ of the $\hat\ell$ field in space:

$$\dot\theta_{\hat\ell} = - \theta_{\hat\ell} - \theta_{\hat\ell_o} / \tau_{\hat\ell} , \qquad (5)$$

where

$$\tau_{\hat\ell} = \mu [(\lambda_D)^{-1} + (\lambda_H H^2)^{-1}] \quad . \qquad (6)$$

Here μ is the Cross-Anderson[13] orbital viscosity coefficient:

$$\mu = (\pi^2/64)\ k_B^2 N(0)\ \tau(0) T_c^2\ (\Delta_o/kT_c)^3 , \qquad (7)$$

where $(\Delta_o/kT_c) = \pi^2(\Delta C/C_N)(1-T/T_c)$, N(0) is the density of states for spins of one sign; and $\tau(0)$, following Pethick and Smith,[14] is the lifetime of a quasiparticle at the Fermi surface at T_c. The experimental data were in excellent agreement with the forms of (5) and (6), even at a few tenths percent from T_c. The values of μ/λ_D were consistent with the predicted temperature dependence near to T_c [at 24 Bar $\mu/\lambda_D \cong 7$ msec $(1-T/T_c)^{\frac{1}{2}}$]; and when (7) was used to extract $\tau(0)T_c^2$, which we found to be about 0.3 μsec mK^2 in a range of pressure above the PCP, we obtained good consistency[3] with values of this quantity found via spin relaxation, viscosity, and ultrasonic attenuation. Thus the ideas regarding orbital viscosity of Cross-Anderson[13] and Pethick and Smith[14] are at least semi-quantitatively substantiated. The values of $\chi_{\hat{d}}/\chi_N$ deduced from λ_H via the field dependence of $\tau_{\hat{\ell}}$ are shown in Fig. 2 for two pressures. They are in quantitative disagreement with $\chi_{\hat{d}}/\chi_N$ determined from the equation

$$\chi_{\hat{d}}/\chi_N = (1+\tfrac{1}{4} Z_o)Y \;/\; 1+\tfrac{1}{4} Z_o Y \quad , \tag{8}$$

where Y is the Yosida function and the Landau parameter Z_o has been taken to be[3] - 3.05 at 24.1 Bar and -3.07 at 33.45 Bar. The nature of the disagreement is very similar to that found between static and dynamic measurements of the relative susceptibility χ_B/χ_N in the B phase: we show a compilation of data in Fig. 3 which includes the most recent NMR data from Helsinki,[15] some as yet unpublished dynamically calibrated static measurements of Sager, Warkentin, and Wheatley, and a theoretical curve for weak coupling and $Z_o = -3.05$. The discrepancy of $(\chi_N-\chi_{\hat{d}})/\chi_N$ in ^{3}He-A and the statically determined $(\chi_N\ \chi_B)/\chi_N$ in ^{3}He-B from theory is fractionally greatest near T_c, diminishing for the most accurate measurements at lower temperatures where thermodynamic measurements at melting pressure[16] suggest little difference between static and dynamic magnetism for $T/T_c < 0.8$.

The persistent orbital motions were discovered by PKW[12] quite accidentally in the course of routine measurements of sound attenuation at 33.9 Bar in which frequent (fractional minute) changes in magnetic field magnitude and direction were followed by an interval of zero ($< 10^{-2}$ gauss) field. Subsequent observations of 25 MHz sound attenuation by pulses at a few Hz rate disclosed a periodically varying transmitted sound signal. When an attempt was made to study these motions systematically we found that the usual condition of the superfluid ^{3}He in zero magnetic field was very stable, with no fluctuations. But when a special history of magnetic fields is provided then either long-lived or persistent oscillations of sound intensity were observed. The nature of these oscillations is shown in Fig. 4 and their temperature dependence in Fig. 5. They were observed in a cylindrical sound cell 8 mm in diameter and 3.8 mm across with flow connections to the outside via

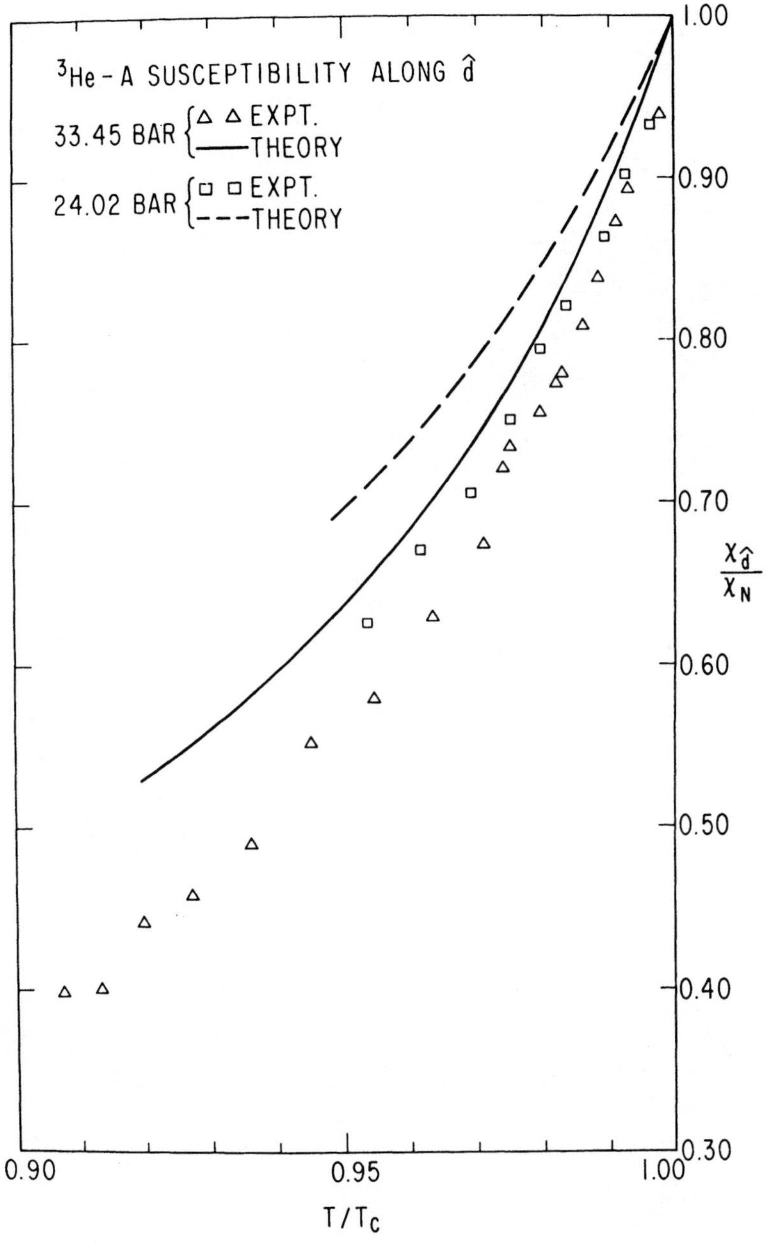

Figure 2. Magnetic susceptibility for field along $\hat{d}$ relative to its value in normal liquid as deduced from the sudden field rotation measurements of Ref. 11.

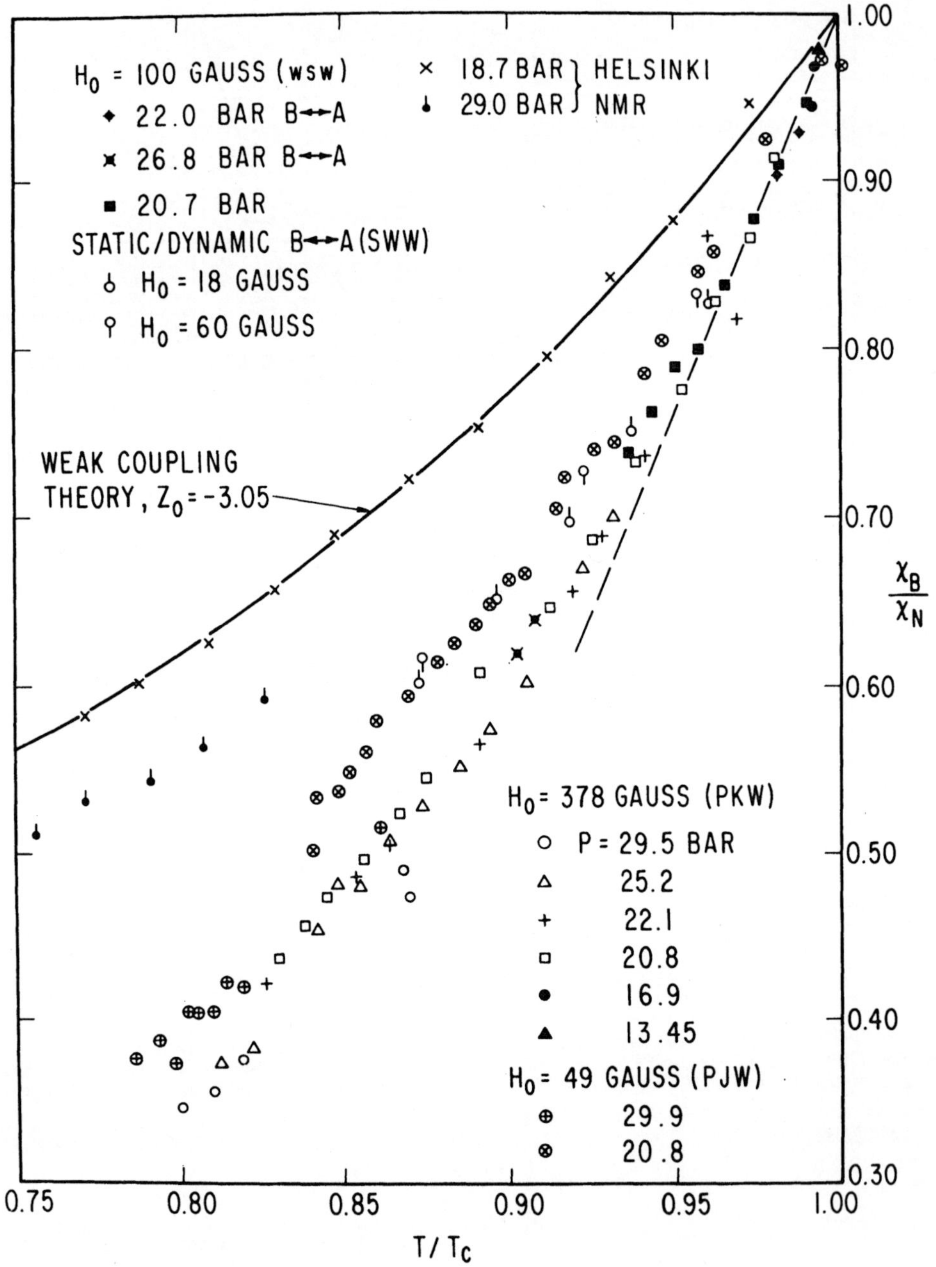

Figure 3. Compendium of the temperature dependence of the relative B-phase magnetic susceptibility.

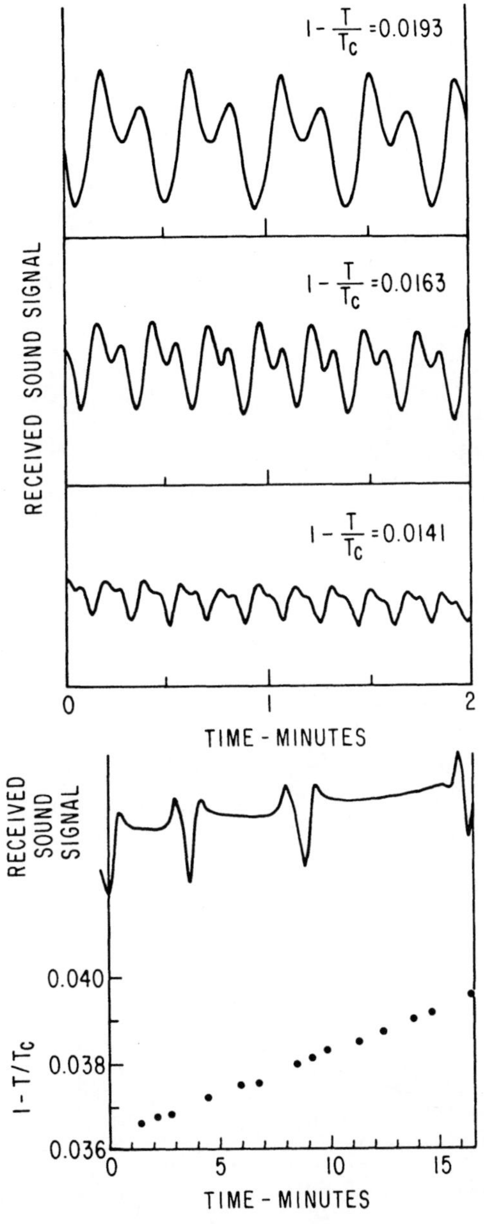

Figure 4. Sections of chart records showing how the received sound signal varied with time and temperature. The upper three 2-minute segments are cut from a continuous warming record of a persistent motion. The lower part of the figure shows part of a "low" temperature persistent motion and in addition the reduced temperature difference as a function of time. After Ref. 12.

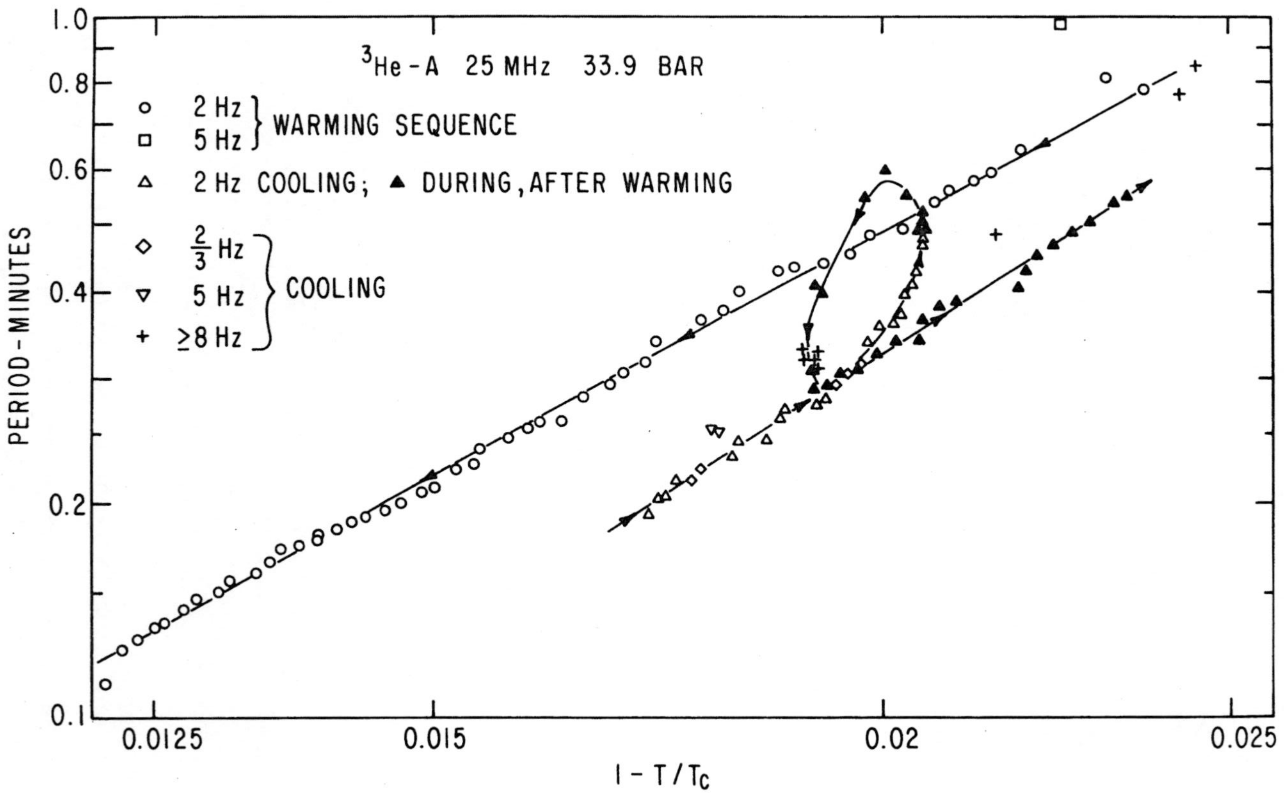

Figure 5. Dependence of observed period of persistent motion on 1-T/T_c showing the effect of temperature, temperature drift rate, and ultrasonic pulsing rate. After Ref. 12.

diametrically opposite ports in the cylindrical walls. They can persist for hours, reproducing minute details of the waveforms shown in Fig. 4. We concluded that they were probably driven in some way by the counterflow in the sound cell produced by heat flow, in view of the dependence of period on temperature drift rate: the counterflow is least on warming and most on cooling. If they were driven by flow then the time scale should be related to the orbital relaxation time

$$\tau_{\hat{\ell}} = \mu / \lambda_F v_s^2 \quad , \tag{9}$$

where μ is the orbital viscosity coefficient and $\lambda_F v_s^2$ is as in Eq. (4), for a uniform $\hat{\ell}$ field coming into equilibrium with a uniform irrotational superfluid velocity field. For a constant heat flow near T_c we expect $\lambda_F v_s$ to be constant and v_s to vary as $(1-T/T_c)^{-1}$. Then with the $(1-T/T_c)^{3/2}$ dependence of μ, Eq. (7), the time $\tau_{\hat{\ell}}$ would vary as $(1-T/T_c)^{5/2}$ near T_c for a constant heat flow. The periods shown on Fig. 5 varied as $(1-T/T_c)^n$ with n in the range 2.8 to 3.3. We obtained a crude estimate of $\lambda_F v_s^2$ at $(1-T/T_c) = .02$ by observing that a field H_c of 1/2 gauss changed the attenuation by about half the amount produced by a "strong" field of 2 gauss (or more). Then $\lambda_F v_s^2 \simeq \lambda_H H_c^2$, where λ_H is known from the sudden field rotation measurements. The value $v_s = 3 \times 10^{-3}$ cm/sec was obtained at $1-T/T_c = .02$, this velocity being well below critical velocities observed in the A phase. The corresponding period of persistent motions is about 20 sec. Using these barely quantitative numbers we then estimated from Eq. (9) that $\tau_{\hat{\ell}} = 3$ sec $[(1-T/T_c)/.02]^{5/2}$.

In the meantime Henry Hall's work[17] on hydrodynamics in the A phase led him to develop, with Hook, a proposed explanation[6] of the persistent motions as "orbital solitary waves." Although their theory appears complex, their results are conceptually simple enough to be reviewed here. In a spatially uniform counterflow along $\hat{z}$ and in a frame at rest with respect to the normal fluid they find the equation of motion for [Eq. (32) of Ref. 6 in ordinary units]

$$\begin{aligned} 2\,\tau_{\hat{\ell}} \frac{\partial \hat{\ell}}{\partial t} &= \left(\frac{\hbar}{2mv_s}\right)^2 \left[5\nabla^2\hat{\ell} - 5\hat{\ell}(\ell\cdot\nabla^2\hat{\ell}) \right] \\ &+ \frac{\hbar}{2mv_s}\left[5\hat{\ell} \times (\hat{v}_s\cdot\vec{\nabla})\hat{\ell} - \hat{v}_{s_\alpha}(\hat{\ell}\times\vec{\nabla})\hat{\ell}_\alpha - (\hat{\ell}\times\hat{v}_s)\,\mathrm{div}\hat{\ell} \right. \\ &\left. + 2(\hat{\ell}\cdot\hat{v}_s)(\hat{\ell}\times(\hat{\ell}\cdot\vec{\nabla})\hat{\ell}) \right] \\ &+ 2\left[\hat{v}_s(\hat{\ell}\cdot\hat{v}_s) - \hat{\ell}(\hat{\ell}\cdot\hat{v}_s)^2 \right] \quad , \end{aligned} \tag{10}$$

where $\tau_{\hat{\ell}}$ is as defined in (9) and $\vec{v}_s$ is the superfluid velocity field associated with rotations of the triad about $\hat{\ell}$. The characteristic time of the theory is $2\tau_{\hat{\ell}}$ and the characteristic distance $\hbar/2mv_s$: thus 6 sec $[(1-T/T_c)/.02]^{5/2}$ and 0.4 mm $[(1-T/T_c)/.02]$ for the conditions of our experiment. Hall and Hook solved for the motion of $\hat{\ell}$ in the case where $\hat{v}_s = \hat{z}$, $\vec{\nabla} = (\partial/\partial z)\hat{z}$, and $\hat{\ell}$ makes polar angles (θ,ϕ) with respect to the $\hat{z}$ axis. One sees immediately that the terms bilinear in $\hat{v}_s$ and gradients of $\hat{\ell}$, those coupling what might be called the irrotational and rotational parts of the velocity field, drive a precessional motion of the $\hat{\ell}$ field [e.g. $(\hat{\ell} \times \hat{v}_s)$ div $\hat{\ell} = \sin^2\theta(\partial\theta/\partial z)\hat{\phi}$]. Hall and Hook found a solution of (10) for which $\theta \to 0$ as $z \to -\infty$ and $\theta \to \pi$ as $z \to +\infty$ which was stable in space, precessed with a period $4\pi\tau_{\hat{\ell}}/1.557 \simeq 24$ sec $[(1-T/T_c/.02]^{5/2}$ for our experiment and had a spatial size of order $3(\hbar/2mv_s) \simeq 1$ mm$[(1-T/T_c)/.02]$ for our experiment. The agreement of the calculated period with the measured period is better than our limited knowledge of $\tau_{\hat{\ell}}$ would allow us to expect. The expected physical size of the predicted precessing object is not unreasonable with respect to the effect on the average attenuation through the cell which we observe, though analysis of the effective angular excursions of the average $\hat{\ell}$ field suggested them to be roughly temperature independent. As Hall and Hook suggest, the predicted precessional motion can be studied experimentally using two sets of sound transducers at right angles. It is a pity that we have no simple means to study the spatial dependence of the sound attenuation in the cell. Also it is interesting that only one persistently moving object appears to be formed via magnetic field history and that we have been unable to produce truly persistent motions at 24 Bar while there has been no real problem in doing this near melting pressure. Particularly with the development of a theory which may explain their existence, these persistent motions are one of the most fascinating new features of superfluid ^{3}He.

Other means of producing orbital motion are waiting to be tested using the method of ultrasound. We mention particularly the electric field orientation method,[18] which has yet to receive a really sensitive test and which, if effective, will have a major effect on the study of orbital dynamics as well as other orbital effects in ^{3}He. Also, during spin relaxation substantial effects on the $\hat{\ell}$ field probably occur, as for example in the metastable states discovered by Gianneta, Smith, and Lee.[19] Here I would think that the best defined experiments would be initiated either by turning off a uniform field or rotating the spin system uniformly through 180°, with controlled externally applied field gradients and defined geometries. And finally, of course, the study of heat or other types of flow in ^{3}He-A with the $\hat{\ell}$ field monitored ultrasonically will go a long way toward unraveling some of the present mysteries.

I wish to thank Doug Paulson, Matti Krusius, Bob Kleinberg, Ron Sager and Paul Warkentin, with whom the experiments in my laboratory at La Jolla have been done. Some of Sager and Warkentin's results on static magnetism are presented here for the first time. I am indebted to Tony Leggett and Shin Takagi for correspondence regarding the sudden field rotation experiments, to Henry Hall for correspondence during the development of his hydrodynamics of the A phase, and to Kazumi Maki and Pradeep Kumar for discussions of the properties of magnetic solitons.

References

(1) A. J. Leggett, Rev. Mod. Phys. 47, 331 (1975).

(2) J. C. Wheatley, Rev. Mod. Phys. 47, 415 (1975).

(3) J. C. Wheatley, "Further Experimental Properties of Superfluid ^{3}He," Progress in Low Temperature Physics, edited by D. F. Brewer (North-Holland, Amsterdam), in press.

(4) N. D. Mermin and T. L. Ho, Phys. Rev. Lett. 36, 594 (1976).

(5) M. C. Cross, J. Low Temp. Phys. 21, 525 (1975).

(6) H. E. Hall and J. R. Hook, "Orbital Solitary Waves in Superfluid ^{3}He-A," J. Phys. C. (in press).

(7) K. Maki and P. Kumar, "Composite Magnetic Solitons in Superfluid ^{3}He-A,"

(8) P. Wolfle, in Quantum Statistics and the Many Body Problem, ed. by S. B. Trickey, W. P. Kirk, and J. W. Dufty (Plenum Press, New York, 1975) p. 9.

(9) A. J. Leggett and S. Takagi, Phys. Rev. Lett. 36, 1379 (1976).

(10) D. N. Paulson, M. Krusius, and J. C. Wheatley, J. Low Temp. Phys. 26, 73 (1977).

(11) D. N. Paulson, M. Krusius, and J. C. Wheatley, Phys. Rev. Lett. 36, 1322 (1976).

(12) D. N. Paulson, M. Krusius, and J. C. Wheatley, Phys. Rev. Lett. 37, 599 (1976).

(13) M. C. Cross and P. W. Anderson, Proc. LT14, Vol. I, p. 29 (1975).

(14) C. J. Pethick and H. Smith, Phys. Rev. Lett. 37, 226 (1976).

(15) A. I. Ahonen, M. Krusius, and M. A. Paalanen, J. Low Temp. Phys. 25, 421 (1976).

(16) W. P. Halperin, C. N. Archie, F. B. Rasmussen, T. A. Alvesalo, and R. C. Richardson, Phys. Rev. B 13, 2124 (1976).

(17) H. E. Hall, J. Phys. C 9, L433 (1976).

(18) J. M. Delrieu, J. de Physique Lett. 35, L-189 (1974); ibid. 36, L-22 (1975).

(19) R. Gianneta, E. N. Smith, and D. M. Lee, "Evidence for a Metastable Mode of Superfluid ^{3}He-A."

ORBITAL HYDROSTATICS AND HYDRODYNAMICS OF ^{3}He-A*†

Tin-Lun Ho

Laboratory of Atomic and Solid State Physics and Materials Science Center, Cornell University
Ithaca, New York 14853

Abstract: The conditions for hydrostatic equilibrium in ^{3}He-A and the equations that describe the non-linear orbital hydrodynamics are given.

LOCAL EQUILIBRIUM AND TOTAL EQUILIBRIUM

The orbital part of the order parameter of the A phase is a complex vector $\hat{\phi}_1 + i\hat{\phi}_2$, $\hat{\phi}_1 \cdot \hat{\phi}_2 = 0$. In the London limit, if we neglect the dipolar coupling between the spin and orbital variables, the physical state is specified by the orientation of these two orthogonal unit vectors, or equivalently, by $\vec{v}_s \equiv \frac{\hbar}{2m} \underset{\sim}{\phi}_1 \cdot \vec{\nabla} \underset{\sim}{\phi}_2$, $\nabla_i \ell_j$, and $\hat{\ell}$.[1] The state of local equilibrium is described by the entropy density $s(\underset{\sim}{r})$, which is a function of the energy density $\varepsilon(\underset{\sim}{r})$, the mass density $\rho(\underset{\sim}{r})$, the momentum density $\vec{g}(\underset{\sim}{r})$, and the orientation of the order parameter $(\vec{v}_s, \nabla_i \ell_j, \hat{\ell})$.

$$s(\underset{\sim}{r}) = s(\varepsilon,\rho,\vec{g},\vec{v}_s,\nabla_i \ell_j,\hat{\ell}) \equiv s(\{x_i\}) \quad (1)$$

$$= s(\varepsilon+\vec{g}\cdot\vec{u} + \tfrac{1}{2}\rho u^2,\rho,\vec{g}+\rho\vec{u},\vec{v}_s+\vec{u},\nabla_i \ell_j,\hat{\ell}). \quad (2)$$

*Based on work supported in part by the National Science Foundation under Grant No. DMP 74-23494 and through the Materials Science Center of Cornell University, Technical Report No. 2776.
†Work based on a Ph.D. Thesis to be submitted to Cornell University.

Eq. (2) simply expresses the Galilean invariance of s. The conjugate variables $\partial s/\partial\varepsilon \equiv 1/T$, $\partial s/\partial\rho \equiv -\mu/T$, $\partial s/\partial\vec{g} \equiv -\vec{V}_n/T$ specify the equilibrium condition for exchange of energy, mass and momentum with a reservoir respectively: $T = T_r$, $\mu = \mu_r$, $\vec{v}_n = \vec{v}_r$.

Transformation properties of the conjugate variables $\partial s/\partial X_i$ can be obtained by differentiating (1) and (2) with respect to X_i and u. In particular, the chemical potential per unit mass μ defined as above (with $\vec{g}$ etc. held fixed) is not Galilean invariant transforming according to $\mu' = \mu - \vec{v}_n\cdot\vec{u} - 1/2u^2$, so that

$$\mu = \mu_o - \frac{1}{2}v_n^2 = \mu_1 - \vec{v}_n\cdot\vec{v}_s + \frac{1}{2}v_s^2 , \tag{3}$$

where μ_0 and μ_1 are the chemical potentials in the $\vec{v}_n = 0$ and $\vec{v}_s = 0$ frames. Galilean invariance also implies $-T\partial S/\partial \underset{\sim}{v}_s = \underset{\sim}{g} - \rho\underset{\sim}{v}_n \equiv \underset{\sim}{g}^o$, while rotational invariance of s provides another relation among various conjugate variables:

$$\underset{\sim}{\ell} \times \frac{T\partial S}{\partial\underset{\sim}{\ell}} + \underset{\sim}{\nabla}\ell_i \times \frac{T\partial S}{\partial\underset{\sim}{\nabla}\ell_i} + \nabla_i\underset{\sim}{\ell} \times \frac{T\partial S}{\partial\nabla_i\underset{\sim}{\ell}} + \underset{\sim}{g}^o \times \underset{\sim}{v}_s^o = 0 ,$$

$$\underset{\sim}{v}_s^o \equiv \underset{\sim}{v}_s - \underset{\sim}{v}_n . \tag{4}$$

The local equilibrium condition (1) can also be written as

$$Tds - d\varepsilon - \mu d\rho - \underset{\sim}{v}_n\cdot d\underset{\sim}{g} - \underset{\sim}{g}^o\cdot d\underset{\sim}{v}_s + \frac{T\partial S}{\partial\nabla_i\ell_j}\nabla_i d\ell_j + \frac{T\partial S}{\partial\underset{\sim}{\ell}}\cdot d\underset{\sim}{\ell} \tag{5}$$

or equivalently, as

$$\rho d\mu = -sdT + dP + \underset{\sim}{g}^o\cdot d\underset{\sim}{v}_s - \underset{\sim}{g}\cdot d\underset{\sim}{v}_n - \frac{T\partial S}{\partial\nabla_i\ell_j}\nabla_i d\ell_j - \frac{T\partial S}{\partial\underset{\sim}{\ell}}\cdot d\underset{\sim}{\ell} \tag{6}$$

Eq. (6) (the Gibb's-Duhem equation) comes from the thermodynamic relation

$$P = Ts - \varepsilon + \rho\mu + \vec{v}_n\cdot\vec{g} , \tag{7}$$

which follows in turn from the definition of the pressure $P(\underset{\sim}{r})$, $P \equiv T(\partial S^{tot}/\partial V)$ where S^{tot} is the total entropy in a sufficiently small volume V surrounding $\vec{r}$. The variation is done with the total mass, energy, and momentum, as well as the broken symmetries inside V held fixed.

The conditions of total equilibrium at temperature T can be obtained by minimizing the free energy

$$F = \int d^3x f \quad , \quad f = \varepsilon - Ts = f(T, \rho, \vec{g}, \vec{v}_s, \nabla_i \ell_j, \hat{\ell}), \tag{8}$$

subjected to the constraints $\hat{\ell}\cdot\hat{\ell} = 1$ and[2]

$$\nabla_i v_{sj} - \nabla_j v_{si} = \frac{\hbar}{2m} \underset{\sim}{\ell}\cdot\nabla_i \underset{\sim}{\ell} \times \nabla_j \underset{\sim}{\ell} \quad . \tag{9}$$

The conditions are, $\vec{v}_n$ = const, and

$$\underset{\sim}{\nabla}\cdot\underset{\sim}{g}^0 = 0, \tag{10}$$

$$\underset{\sim}{H} \times \underset{\sim}{\ell} = 0 \quad , \quad \underset{\sim}{H} = \frac{\partial f}{\partial \underset{\sim}{\ell}} - \nabla_\alpha \frac{\partial f}{\partial \nabla_\alpha \underset{\sim}{\ell}} + \frac{\hbar}{2m}(\vec{g}^o\cdot\vec{\nabla})\underset{\sim}{\ell} \times \underset{\sim}{\ell} \quad . \tag{11}$$

Eq. (11) is the generalization of the stationary condition for nematics. H will be referred to as the molecular field for ^{3}He-A. The expression for f is given by

$$f = \frac{1}{2}\vec{v}_s\cdot\bar{\bar{\rho}}_s\cdot\vec{v}_s + \vec{v}_s\cdot\bar{\bar{c}}\cdot\vec{\nabla}\times\hat{\ell} + \frac{1}{2}\vec{v}_n\cdot\bar{\bar{\rho}}_n\cdot\vec{v}_n + f_n \quad , \quad \bar{\bar{\rho}}_n + \bar{\bar{\rho}}_s = \bar{\bar{1}} \quad , \tag{12}$$

where $\bar{\bar{\rho}}_s = \rho_s\bar{\bar{1}} - \rho_o\hat{\ell}\hat{\ell}$, and $\bar{\bar{c}} = c\bar{\bar{1}} - c_o\hat{\ell}\hat{\ell}$; f_n is the deformation energy of ℓ only.[3] $\vec{v}_n$ is related to $\vec{g}$ by

$$\vec{g} = \bar{\bar{\rho}}_n\cdot\vec{v}_n + \bar{\bar{\rho}}_s\cdot\vec{v}_s + \bar{\bar{c}}\cdot\vec{\nabla}\times\hat{\ell} \quad . \tag{13}$$

$\vec{H}$ is then made up of (i) the nematic molecular field,[4] (ii) terms coming from the velocity dependent part of (13): $-(\rho_o\underset{\sim}{v}_s^o\cdot\underset{\sim}{\ell} + c_o\underset{\sim}{\ell}\cdot\underset{\sim}{\nabla}\times\underset{\sim}{\ell})\vec{v}_s^o$ $-c_o(\underset{\sim}{v}_s^o\underset{\sim}{\ell})\vec{\nabla}\times\hat{\ell} + c\vec{\nabla}\times\vec{v}_s^o - c_o\vec{\nabla}\times(\underset{\sim}{v}_s^o\cdot\hat{\ell}\hat{\ell})$, and (iii) a quantum contribution due to the $\vec{v}_s$ and $\hat{\ell}$ coupling, Eq. (9).

Note that for a rotating system with angular frequency $\vec{\omega}$, $\vec{v}_n = \vec{\omega}\times\vec{r}$, the equilibrium condition for $\hat{\ell}$ and $\vec{v}_s$ obtained by minimizing $\int(f - \vec{\omega}\times\vec{r}\cdot\vec{g})$ is still given by (10) and (11). However, unlike ^{4}He-II, the net quadratic term in $\vec{v}_n$, $(-\frac{1}{2}\vec{v}_n\cdot\bar{\bar{\rho}}_n\cdot\vec{v}_n)$, also plays a role in determining the texture because of the anisotropy of $\bar{\bar{\rho}}_n$.[5]

ORBITAL HYDRODYNAMICS

The relaxation from local equilibrium to total equilibrium is controlled by the conservation laws and the equation of motion of the order parameter. To preserve the orthonormality of the $\hat{\phi}^\alpha$, this must take the form $\partial/\partial t\, \hat{\phi}^\alpha = \vec{\omega}\times\hat{\phi}^\alpha$. In terms of $\hat{\ell}$ and $\vec{v}_s$,

$$\frac{\partial}{\partial t}\hat{\ell} = \vec{\omega}\times\hat{\ell}, \tag{14}$$

$$\frac{\partial}{\partial t}\vec{v}_s = -\frac{\hbar}{2m}\underset{\sim}{\ell}\cdot\vec{\nabla}\underset{\sim}{\omega} = \frac{\hbar}{2m}\vec{\nabla}(\underset{\sim}{\ell}\cdot\underset{\sim}{\omega}) + \frac{\hbar}{2m}\vec{\nabla}\underset{\sim}{\ell}\times\underset{\sim}{\ell}\cdot\frac{\partial}{\partial t}\ell\ . \tag{15}$$

The angular frequency $\vec{\omega}$ transforms under a Galilean boost as

$$\vec{\omega}' = \vec{\omega} + \frac{2m}{\hbar}\left(\underset{\sim}{v}_s\cdot\underset{\sim}{u} + \frac{u^2}{2}\right)\hat{\ell} + \underset{\sim}{u}\cdot\underset{\sim}{\nabla}\,\hat{\ell}\times\hat{\ell}\ . \tag{16}$$

The existence of a continuity equation for angular momentum requires the antisymmetric part of the stress tensor to be a total divergence $\vec{\varepsilon\pi} \equiv \varepsilon_{ijk}\pi_{jk} = -\nabla_\alpha B_{\alpha i}$.

The conservation laws and the equation of motion of the order parameter form a closed set of equations if the energy current $\vec{Q}$, the stress tensor $\bar{\bar{\pi}}$, and the angular frequency $\vec{\omega}$ can be expressed in terms of the original hydrodynamic variables. The form of these quantities can be determined from requiring the entropy production to be non-negative:

$$\frac{\partial}{\partial t}s = -\frac{1}{T}\underset{\sim}{\nabla}\cdot\underset{\sim}{Q} + \frac{\mu}{T}\underset{\sim}{\nabla}\cdot\underset{\sim}{g} + \frac{1}{T}v_n^j\nabla_i\pi_{ij} + \frac{\hbar}{2m}\frac{1}{T}\underset{\sim}{\ell}\,\vec{g}^{\,o}\cdot\vec{\nabla}\underset{\sim}{\omega} + \frac{\partial s}{\partial\nabla_i\ell_j}$$

$$\times\,\nabla_i(\omega\times\ell)_j + \frac{\partial s}{\partial\underset{\sim}{\ell}}\cdot\underset{\sim}{\omega}\times\underset{\sim}{\ell}\ . \tag{17}$$

The rotational symmetry (4), the $\vec{\nabla}\times\vec{v}_s$ constraint (9), and eq. (7) imply that (17) can be written as:

$$\frac{\partial s}{\partial t} = -\vec{\nabla}\cdot\left(s\vec{v}_n + \frac{\vec{q}}{T}\right) - \frac{1}{T}P\cdot A - \frac{1}{T}\vec{\xi}\cdot\frac{1}{2}\vec{\nabla}\times\vec{v}_n - \frac{1}{T}\vec{\Omega}^o\times\hat{\ell}\cdot\vec{H}^{\perp} - \frac{1}{T}\psi^o\underset{\sim}{\nabla}\cdot\underset{\sim}{g}^o$$

$$-\frac{\vec{q}}{T^2}\cdot\vec{\nabla}T \tag{18}$$

where

$$P_{ij} \equiv \pi^o_{ij} - P\delta_{ij} - g^o_i v^o_{sj} - \frac{T\partial S}{\partial \nabla_i \underset{\sim}{\ell}} \cdot \nabla_j \underset{\sim}{\ell} \quad , \tag{19}$$

$$\vec{\xi} \equiv \varepsilon\vec{\pi}\left(\frac{\hbar}{2m} \underset{\sim}{g}^o \hat{\ell} + \frac{\partial f}{\partial \underset{\sim}{\nabla}\hat{\ell}} \times \hat{\ell}\right) \quad , \tag{20}$$

$$\vec{\Omega}^o \equiv \vec{\omega}^o - \frac{1}{2}\vec{\nabla} \times \vec{v}_n \quad , \qquad \psi^o \equiv \frac{\hbar}{2m} \underset{\sim}{\ell} \cdot \underset{\sim}{\Omega}^o - \mu^o \quad , \tag{21}$$

$$\vec{q} \equiv Q^o - \frac{\hbar}{2m} g^o(\underset{\sim}{\ell} \cdot \underset{\sim}{\omega}^o) - \frac{T\partial S}{\partial \vec{\nabla} \underset{\sim}{\ell}} \cdot \underset{\sim}{\omega}^o \times \underset{\sim}{\ell} \,, \tag{22}$$

$$A_{ij} \equiv \frac{1}{2}(\nabla_i v^n_j + \nabla_j v^n_i) \quad , \qquad \vec{H}^{\perp} = \vec{H} - (\underset{\sim}{H} \cdot \ell)\hat{\ell} \; . \tag{23}$$

and the superscript "o" refers to quantities in the $\vec{v}_n = 0$ frame. $\partial s/\partial t$ is now of the form $1/T \sum_\alpha j_\alpha f_\alpha$ plus a total divergence, where the "fluxes" j_α are the quantities in (19) to (22), and $f_\alpha(\bar{A}, 1/2\vec{\nabla}\times\vec{v}_n, \vec{\nabla}\cdot g^o$, H and $\vec{\nabla}T)$ can be considered as forces that drive the system toward equilibrium. Considering j_α as responses to these forces, to the linear order of f_β, j_α(and hence $\bar{\bar{\pi}}^o$, Q^o, and $\vec{\omega}^o$) can be expressed as

$$\alpha = \sum_\beta \gamma_{\alpha\beta} f_\beta \equiv \sum_\beta (\gamma^r + \gamma^D)_{\alpha\beta} f_\beta \equiv j^r_\alpha + j^D_\alpha \quad . \tag{24}$$

The tensors γ^r and γ^D carry different number $\hat{\ell}$s, so that $j^r(j^d)$ has different (same) time reversal symmetry as f_α. The "reversible fluxes" j^r produce no entropy, while the dissipation is described by j^D. (j^D must be chosen in such a way to guarantee $\partial s/\partial t > 0$). The expression of the entropy production reduces to that of the ^{4}HeII and nematics by setting $\hat{\ell}$ and $\vec{v}_s$ equal to zero. The resulting hydrodynamic equations are

$$\frac{\partial}{\partial t}\rho + \underset{\sim}{\nabla} \cdot \underset{\sim}{g} = 0 \quad , \tag{25}$$

$$\frac{\partial}{\partial t} g_j = -\nabla_i \Big(P\delta_{ij} + \frac{\partial f}{\partial \nabla_i \underset{\sim}{\ell}} \cdot \nabla_j \underset{\sim}{\ell} + g^o_i v_{sj} + g_j v^n_i + \frac{1}{2}(1-\alpha)$$

$$\times\; \ell_v \cdot H^{\perp}_j - \frac{1}{2}(1+\alpha)\, H^{\perp}_i \ell_j + \zeta[A_{is}\ell_k \varepsilon_{skj}]^s$$

$$+ \zeta_1[(\hat{\ell}\times \vec{A}\hat{\ell})_i \ell_j]^s - \frac{1}{2}\varepsilon_{ijk}\ell_k \, \underset{\sim}{\nabla} \cdot \underset{\sim}{g}^o \frac{\hbar}{2m} + \pi^D_{ij}\Big) \quad , \tag{26}$$

$$\frac{\partial}{\partial t}s = -\vec{\nabla}\cdot(s\vec{v}_n + \frac{\gamma}{T}\hat{\ell}\times\vec{\nabla}T - \frac{\vec{K}}{T}\cdot\vec{\nabla}T) + R \quad , \tag{27}$$

$$\frac{\partial}{\partial t}\vec{v}_s = -\vec{\nabla}(\mu_1 + \frac{v_s}{2} + \frac{1}{2}\frac{\hbar}{2m}\underset{\sim}{\ell}\cdot\underset{\sim}{\nabla}\times\underset{\sim}{v}_n - [\nu_1\underset{\sim}{\nabla}\cdot\underset{\sim}{v}_n + \nu_2\,\underset{\sim}{\ell}\underset{\sim}{\mathcal{A}}\cdot\underset{\sim}{\ell} + \chi_1\,\underset{\sim}{\nabla}\cdot\underset{\sim}{g}^{o}])$$

$$+\frac{\hbar}{2m}\vec{\nabla}\underset{\sim}{\ell}\times\underset{\sim}{\ell}\cdot\frac{\partial}{\partial t}\underset{\sim}{\ell} \; , \tag{28}$$

$$\frac{\partial}{\partial t}\hat{\ell} + \underset{\sim}{v}_n\cdot\underset{\sim}{\nabla}\hat{\ell} = \frac{1}{2}(\vec{\nabla}\times\vec{v}_n)\times\hat{\ell} + (\beta\vec{H}\times\hat{\ell} + \alpha\overrightarrow{\ell\mathcal{A}}^{\perp}) + (\beta_1\vec{H}\times\hat{\ell} + \alpha_1\overrightarrow{\ell\mathcal{A}}^{\perp})\times\hat{\ell}; \tag{29}$$

where $\underline{\underline{\mathcal{A}}}$ is the shear flow $A-1/3\ \vec{\nabla}\cdot\vec{v}_n\ \underline{\underline{1}}$, $T^{s}_{ij} = 1/2(T_{ij} + T_{ji})$, and R is the entropy production that vanishes in the reversible limit. All the terms in (26) are reactive energy except π^{d}, which is symmetric. Its lengthy expression will be presented together with a detailed derivation of the above formulas elsewhere. Eq. (25) to (29) and the Gibbs-Duhem equation form a closed set. The following properties of the orbital hydrodynamics of ^{3}He-A should be noted:

(i) In normal fluids and ^{4}HeII, alternate orders in the gradient expansions of the fluxes correspond to dissipative and nondissipative processes; here, because of the time reversal symmetry of ℓ, both processes occur in all orders of the expansion. That both the shear flow and the temperature gradient contribute to the reactive part of fluxes is a special property of the axial state.[6] The $1/2\ \vec{\nabla}\times\vec{v}_n\cdot\hat{\ell}$ term in (28) represents the additional precession of $\hat{\phi}^{\alpha}$ in a non-inertia frame, as does the term $1/2(\vec{\nabla}\times\vec{v}_n)\times\hat{\ell}$ in (29).

(ii) The nonlinear coupling between $\hat{\ell}$ and $\vec{v}_s$ shows up in the last term of the equation (28). It describes the quantum fluctuation of vortices due to the motion of a non-uniform texture, and also implies the generation of a chemical potential gradient associated with the motion of $\hat{\ell}$ even in the case when $\vec{v}_s = 0$.

(iii) Besides those precessions stimulated by the normal fluid velocity gradient, the motion of $\hat{\ell}$ is controlled by the nonequilibrium molecular field $\vec{H}^{\perp}$. The dissipative term β_1 is related to the Cross's damping mechanism.[7]

(iv) The stress tensor has an antisymmetric part $\epsilon\vec{\pi} = -\nabla_{\alpha}B_{\alpha i}$, $B_{\alpha i} = \hbar/2m\ g^{o}_{\alpha}\ell_i + [(\partial f/\partial\nabla_{\alpha}\hat{\ell})\times\hat{\ell}]_i$. The total torque on the system $(\vec{L})$ is made up of (a) torques due to external stresses acting on the boundary $-\ \vec{r}\times d\vec{s}\cdot\overset{=}{\pi}$ and (b) $\epsilon\vec{\pi} = -\ d\vec{s}\underline{\underline{B}}$, interpreted as a torque on $\hat{\ell}$ at the boundary. According to the expression of f, eq. (13), and the boundary condition that at the surface $\hat{\ell}$ has to be parallel to the normal, the torque (b) is equal to $-\hbar/2m\int d\underset{\sim}{s}\cdot g^{o}\underset{\sim}{\ell}$

$-\int ds(c\vec{v}_s^{\,o} + k_3 \vec{\nabla}\times\hat{\ell})$.

(v) The momentum equation implies in equilibrium,

$$\nabla_i (P\delta_{ij} + \frac{\partial f}{\partial \nabla_i \ell} \cdot \nabla_j \ell + g_i^o v_{sj}) = 0 \tag{30}$$

Pressure is not a constant in the presence of non-uniform textures and superflow. The last two terms in (30) are the generalization of the "Ericksen stress" in nematics -- a stress produced by the deformation $\vec{r} \to \vec{r}' = \vec{r} + \vec{u}$, with $\hat{\ell}(r) = \hat{\ell}'(r')$. Eq. (30) can also be derived from the Gibbs-Duhem equation using the $\vec{\nabla}\times\vec{v}_s$ equation (9).

While this summary was being written, four preprints on the same subject[8] arrived at nearly the same time. The hydrodynamic equations derived here differ in many ways from the results of H. E. Hall and J. R. Hook, and that of J. M. Delrieu's. They agree with those of D. Lhuillier's and those of C. R. Hu and W. M. Saslow except that: (i) angular momentum conservation is not satisfied in Lhuillier's work because the term $-1/2\ \hbar/2m\ \varepsilon_{ijk}\ell_k \nabla\cdot g^o$ has not been included in the stress tensor, and (ii) the term $\gamma\vec{\nabla}T\ \hat{\ell}\times\vec{\nabla}T$ in the entropy current has been neglected by Hu and Saslow. We also believe the terms in their dissipative fluxes that contain the tensors $\bar{\bar{C}}$ (hence terms like $v_s^o \cdot \nabla\vec{v}_n$) in the entropy current) need not be included because they are of higher order in the velocities. If the free energy is considered as a power series of its arguments (T, ρ, $\vec{g}$, $\vec{v}_s$, $\nabla_i\ell_j$, ℓ), the term $\hat{\ell}\cdot\vec{\nabla}\times\vec{v}_n$ should not appear in their expression for f. If this term is included in f, terms like $\hat{\ell}\times\vec{v}_s^o\cdot\vec{\nabla}\mu$ and many more should also appear.

This work began after the author's attendance at N. D. Mermin's lectures on superfluid ^{4}He. He is grateful to him for discussion and criticism.

References

(1) This is equivalent to the tensor $\Omega_{ij} = -2m/\hbar\ v_{si}\ell_j + (\ell\times\nabla_i\ell)_j$ in Mermin and Ho, Phys. Rev. Lett. 36, 594 (1976).

(2) Equation (6) in Mermin and Ho, ibid.

(3) See de-Gennes, "The Physics of Liquid Crystals," (Clarendon Press, Oxford, 1974), p. 63. On symmetry grounds, there should also be a surface term $\nabla_i\ell_i\nabla_j\ell_j - \nabla_i\ell_j\nabla_j\ell_i$. It is neglected because it has no effects on determining the equilibrium texture and the dynamic molecular field.

(4) See de-Gennes, ibid. p. 68 and also footnote (3).

(5) The effect of $\bar{\bar{\rho}}_n$ has been overlooked by Buchholtz and Fetter, preprint, Stanford University ITP-542.

(6) The ζ terms in (27) are first discussed by Mario Liu, Phys. Rev. B 13, 4174 (1976).

(7) $\beta = L_o/\mu^2+L_o^2$, $\beta_1 = \mu/\mu^2+L_o^2$, where μ and L_o are the damping and intrinsic angular momentum coefficients introduced in Cross and Anderson, Proceedings of LT14, Vol. 1 (1975).

(8) H. E. Hall and J. R. Hook; J. M. Delrieu; D. Lhuillier; C.-R. Hu and W. M. Saslow, preprints and to be published.

GENERAL EQUILIBRIUM CONDITION AND HYDRODYNAMICS OF SUPERFLUID ^{3}He-A

Chia-Ren Hu

Department of Physics, Texas A&M University
College Station, Texas 77843

It is now generally accepted that the order parameter of superfluid ^{3}He-A is a complex tensor $A_{\mu i} = \Delta_0 \hat{d}_\mu (\hat{m}_i + i\hat{n}_i)$, where $\hat{d}$, $\hat{m}$ and $\hat{n}(\perp\hat{m})$ are three real unit vectors. A "texture" of ^{3}He-A refers to a slow spatial variation of the "orbital or anisotropy axis" $\hat{\ell} \equiv \hat{m} \times \hat{n}$, and the "spin axis" $\hat{d}$, possibly accompanied by a nonvanishing "superfluid velocity"[1] $\vec{V}^{(s)} \equiv (h/2M) m_j \vec{\nabla} n_j$ (where M is the mass of a ^{3}He atom), while the magnitude parameter Δ_o is usually fixed at the constant equilibrium value. Such textures are all quasi-degenerate in energy because the ^{3}He-A order parameter has spontaneously broken the gauge symmetry and the rotational symmetries in orbital and spin spaces. From the above definitions of $\vec{V}^{(s)}$ and $\hat{\ell}$, one may verify a relation first obtained by Mermin and Ho:[2]

$$\partial_i V_j^{(s)} - \partial_j V_i^{(s)} = (\hbar/2M)\hat{\ell}\cdot\partial_i\hat{\ell}\times\partial_j\hat{\ell} \quad , \tag{1}$$

which implies

$$\delta V^{(s)}_i = -(\hbar/2M)\hat{\ell}\cdot(\partial_i\hat{\ell}) \times \delta\hat{\ell} + (\hbar/2M)\partial_i(\delta\phi). \tag{2}$$

Thus if one minimizes the gradient free energy functional $f[\vec{V}^{(s)}, \hat{\ell}, \partial_j \ell_i, \hat{d}, \partial_j d_i]$, and invokes the boundary condition of Ref. 3 that $\hat{\ell}$ is anchored normal to a wall so that $\delta\hat{\ell} = 0$ there, one obtains the following general condition for the equilibrium textures of ^{3}He-A (in a stationary container):

$$\vec{\nabla}\cdot\vec{g}^{(s)} = 0, \quad \vec{g}^{(s)} \equiv \delta f/\delta\vec{V}^{(s)} \; ; \tag{3a}$$

$$\hat{\ell} \times \vec{\Psi} = 0, \quad \Psi_i \equiv \frac{2M}{\hbar}\left[\frac{\delta f}{\delta \ell_i} - \partial_j \frac{\delta f}{\delta(\partial_j \ell_i)}\right] - \varepsilon_{ijk}\ell_j(g^{(s)}\cdot\vec{\nabla})\ell_k ; \tag{3b}$$

$$\hat{d} \times \vec{\theta} = 0, \quad \theta_i \equiv \frac{2M}{\hbar}\left[\frac{\delta f}{\delta d_i} - \partial_j \frac{\delta f}{\delta(\partial_j d_i)}\right] . \tag{3c}$$

plus the additional boundary conditions that $\hat{N}\cdot\vec{g}^{(s)} = N_j[\delta f/\delta(\partial_j d_i)] = 0$ where $\hat{N}$ denotes the normal vector of a boundary surface.[4]

The many broken symmetries of ^{3}He-A, some of which are non-Abelian, also imply the existence of many low frequency dynamic modes, which may be studied in a model-independent way by a hydrodynamic approach. The general set of hydrodynamic equations, incorporating Eq. (1) plus the non-linear effects due to an inhomogeneous equilibrium texture and a steady-state flow (but neglecting coupling to the spin dynamics), has recently been derived by the author in collaboration with W. M. Saslow,[5] thus generalizing the hydrodynamics of Graham[6] (and Liu[7]) which is limited to a uniform texture with no steady-state flow. Omitting the detailed argument which will be published elsewhere,[5] this paper will be devoted to a summary and discussion of that work.

The fundamental thermodynamic relation for ^{3}He-A is:

$$d\varepsilon = Td(\rho s)+\mu d\rho + \vec{V}^{(n)}\cdot d\vec{g} + \vec{\lambda}^{(s)}\cdot d\vec{V}^{(s)} + \left(\frac{\hbar}{2M}\right)\psi_i d\ell_i + \left(\frac{\hbar}{2M}\right)\phi_{ij} \times d(\partial_j \ell_i) \tag{4}$$

with obvious notations, and $\vec{g} = \rho\vec{V}^{(n)} + \vec{\lambda}^{(s)}$ by Galilean invariance. The pressure is defined by $P = Ts\rho + \mu\rho + \vec{V}^{(n)}\cdot\vec{g} - \varepsilon$. Using Eq. (2) to eliminate $d\vec{V}^{(s)}$, and proceeding as in Ref. 6 but carefully handling the non-linear terms in $\vec{V}^{(s)}$, $\vec{V}^{(n)}$ and $\partial_i \ell_j$, we obtain the following complete set of hydrodynamic equations (except for the coupling to spin motion):

$$\partial\rho/\partial t + \partial_i g_i = 0, \quad \partial g_i/\partial t + \partial_m \sigma_{ij} = 0; \tag{5a}$$

$$\partial(\rho s)/\partial t + \partial_i(\rho s V_i^{(n)} + q_i^{D}/T) = R/T; \tag{5b}$$

$$(\frac{\hbar}{2M})\frac{\partial\phi}{\partial t} + \zeta_\phi = 0, \quad (\frac{\hbar}{2M})\frac{\partial \ell_i}{\partial t} + X_i = 0, \quad [\ell_i X_i = 0]; \tag{5c}$$

where[8]

$$\sigma_{ij} = P\delta_{ij} + g_i V_j^{(n)} + V_i^{(s)}\lambda_j^{(s)} + \partial_i \ell_k \phi_{kj} + \Sigma_{ij}^R + \Sigma_{ij}^D ; \tag{6a}$$

$$\Sigma_{ij}^R = \frac{1}{2}(\frac{\hbar}{2M})[(\ell_i\Psi_j - \ell_j\Psi_i) + \varepsilon_{ijk}\ell_k(\vec{\nabla}\cdot\vec{\lambda}^{(s)})] + \{[-\alpha\ell_i\delta^T{}_{jk}\Psi_k - 2(\phi_1\delta^T{}_{i\ell} + \phi_2\ell_i\ell_\ell)\varepsilon_{jpq}\ell_p A_{q\ell}] + (i\leftrightarrow j) ; \tag{6b}$$

$$\zeta_\phi = \vec{\mu} + \vec{V}^{(n)}\cdot\vec{V}^{(s)} + \frac{1}{2}(\frac{\hbar}{2M})\hat{\ell}\cdot\vec{\nabla}\times\vec{V}^{(n)} + \zeta_\phi^D ; \tag{6c}$$

$$\vec{X} = (\frac{\hbar}{2M})[(\vec{V}^{(n)}\cdot\vec{\nabla})\hat{\ell} - \vec{\Omega}\times\hat{\ell}] - \beta\vec{\Psi}\times\hat{\ell} - 2\alpha\overleftrightarrow{\delta}^T\cdot\overleftrightarrow{A}\cdot\hat{\ell} + \vec{X}^D ; \tag{6d}$$

$$R = -(\vec{q}^D/T)\cdot\vec{\nabla}T - \overleftrightarrow{\Sigma}^D:\overleftrightarrow{A} - \zeta_\phi^D\vec{\nabla}\cdot\vec{\lambda}^{(s)} + \vec{X}^D\cdot\vec{\Psi} ; \tag{6e}$$

In the above equations, $A_{ij} \equiv [\partial_i V_j^{(n)} + \partial_j V_i^{(n)}]/2$, $\Psi_i \equiv \Psi_i - \partial_j\phi_{ij} - \varepsilon_{ijk}\ell_j(\lambda^{(s)}\cdot\vec{\nabla})\ell_k$, $\delta^T{}_{ij} \equiv \delta_{ij} - \ell_i\ell_j$, and $\vec{\Omega} \equiv \frac{1}{2}\nabla \times \vec{V}^{(n)}$. In Eqs. (5) and (6), the dissipative currents $\vec{q}^D$, $\overleftrightarrow{\Sigma}^D$, ζ_ϕ^D and X^D are (in the lowest order) linear combinations of the four independent thermodynamic forces (i.e., affinities) $\vec{\nabla}T$, $\overleftrightarrow{A}$, $\vec{\nabla}\cdot\vec{\lambda}^{(s)}$ and $\hat{\ell}\times\vec{\Psi}$, with tensorial transport coefficients constructed with $\hat{\ell}$ and minimum powers of $(\vec{V}^{(s)} - \vec{V}^{(n)}$ and $\partial_j\ell_i)$ consistent with space- and time-inversion symmetries, so that R is even with respect to both symmetry properties.

While our σ_{ij} is not explicitly symmetric, local conservation of angular momentum is satisfied by ensuring that $\varepsilon_{ijk}\sigma_{ij}$ has a form $\partial_\ell f_{k\ell}$.[9,5] Energy conservation follows from Eqs. (4-6), and is not an independent equation.

For a non-dissipative equilibrium state all four affinities must vanish, from which we deduce T=constant, $\vec{V}^{(n)} = \vec{V}_o + \vec{\Omega} \times \vec{r}$ with constant $\vec{V}_o$ and $\vec{\Omega}$, and Eqs. (3a, b) generalized to allow $\vec{V}^{(n)} \neq 0$.

From Eqs. (5c), (6c), (6d) and $\partial\vec{\Delta}/\partial t = i(\partial\phi/\partial t)\vec{\Delta} - i\vec{\Delta} \times (\partial\hat{\ell}/\partial t)$, one may deduce the equation of motion for $\vec{\Delta} \equiv \Delta_o(\hat{m} + i\hat{n})$:

$$\partial\vec{\Delta}/\partial t + (\vec{V}^{n}\cdot\vec{\nabla})\vec{\Delta} - \vec{\Omega} \times \vec{\Delta} + (2Mi/\hbar)\ [\mu\vec{\Delta}$$

$$- \beta\hat{\ell}(\vec{\Delta}\cdot\vec{\Psi}) - 2\alpha i\hat{\ell}(\vec{\Delta}\cdot\overset{\leftrightarrow}{X}\cdot\hat{\ell}) \tag{7}$$

$$+ \zeta_{\phi}{}^{D}\vec{\Delta} + \vec{X}^{D} \times \vec{\Delta}] = 0$$

which allows a solution free of dissipation, corresponding to a $\vec{\Delta}$ (not just $\hat{\ell}$) rotating rigidly with a container (aside from the phase-winding due to μ). When $\hat{\ell}\cdot\vec{\Omega} \neq 0$, this rotation amounts to an additional phase winding, which explains the third term in Eq. (6c), and predicts and AC Josephson effect between rotating and stationary ^{3}He-A weakly linked to each other.

Finally we remark that Delrieu,[10] that Lhuillier,[11] Hall and Hook,[12] and Ho[13] have followed somewhat different approaches to study the hydrodynamics of ^{3}He-A and have independently reached very similar conclusions. While all these approaches (including Ref. 5) in fact complement each other in concepts, disagreements in conclusions do appear to exist at the present stage which must yet be removed to reach the final complete picture about the general hydrodynamics of ^{3}He-A.

References

(1) S. Blaha, Phys. Rev. Lett. 36, 874 (1976).

(2) N. D. Mermin and T.-L. Ho, Phys. Rev. Lett. 36, 594 (1976).

(3) V. Ambegaokar, P. G. de Gennes and D. Rainer, Phys. Rev. A 9, 2676 (1974).

(4) Near T_c, when the free energy f may be written in the general form $\gamma_1\partial_i A_{\mu i}{}^*\partial_j A_{\mu j} + \gamma_2\partial_i A_{\mu j}^*\partial_i A_{\mu j} + \gamma_3\partial_i A_{\mu j}{}^*\partial_j A_{\mu i}$, Eq. (3) may be shown to depend on $(\gamma_1+\gamma_3)$ and γ_2 only. [If one neglects the dipole coupling to the spin axis, then only the ratio$(\gamma_1+\gamma_3)/\gamma_2$ will appear.] The boundary condition

(4) (cont.) $0 = \hat{N}\cdot\vec{g}^{(s)} \propto 2\gamma_2 N\cdot\vec{V}^{(s)} - \gamma_3(\hbar/2M)\ \hat{N}\cdot\vec{\nabla}\times\hat{\ell}$ appears to depend on γ_3, but the second term may be shown to vanish on all smooth surfaces due to the boundary condition $\hat{N} \times \hat{\ell} = 0$. Thus to measure γ_1 and γ_2 independently, it appears that one must resort to rotating ^{3}He-A (see the third to last paragraph of this paper), or to a local measurement of $\vec{g}^{(s)}$.

(5) C.-R. Hu and W. M. Saslow, to be published in Phys. Rev. Lett.

(6) R. Graham, Phys. Rev. Lett. 33, 1431 (1974).

(7) M. Liu, Phys. Rev. B 13, 4174 (1976).

(8) In order to simplify the equations of Ref. 5, we have introduced the following change of notations: $\alpha_1+\alpha_2\to 2\alpha$, $\gamma_\perp^{(1)}+\gamma_\perp^{(3)}\to -2\phi_1$, and $\gamma^{(3)}_{\parallel} \to -\phi_2$.

(9) P. C. Martin, O. Parodi, P. S. Pershan, Phys. Rev. A 6, 2401 (1972).

(10) J. M. Delrieu, preprint.

(11). D. Lhuillier, preprint.

(12) H. E. Hall and J. R. Hook, preprint.

(13) T.-L. Ho, preprint.

THE MAGNETIC SUSCEPTIBILITY OF SUPERFLUID ^{3}He

J. W. Serene

Department of Physics - State University of New York
Stony Brook, New York 11794

D. Rainer

Low Temperature Laboratory
Helsinki University of Technology
SF-02150 Otaniemi, Finland

Measurements of the static magnetic susceptibility of superfluid ^{3}He played an important role in establishing the identity of the A and B phases. The B-phase susceptibility has continued to be of interest, both because of the unresolved discrepancy between static and NMR measurements, and also because, based on the existing (weak-coupling) theory, it appeared that measurements of $\chi_B(T)$ could be inverted to obtain both the energy gap $\Delta(T)$ and the unknown Landau parameter F_2^a. In this paper we will report the results of a strong-coupling calculation of the magnetic susceptibility of superfluid ^{3}He, including all corrections through first order in T_c/T_F. In particular, we will discuss the implications of our strong-coupling theory for the possibility of extracting $\Delta(T)$ and F_2^a from the measured susceptibility.

Measurements of the B-phase susceptibility have most frequently been interpreted in terms of an expression first derived by Leggett,[1]

$$\frac{\chi_B(T)}{\chi_N} = \frac{(1 + F_0^a)[\frac{2}{3} + \frac{1}{3}\,Y(T;\,\Delta(T))]}{1 + F_0^2[\frac{2}{3} + \frac{1}{3}\,Y(T;\,\Delta(T))]} \tag{1}$$

where χ_N is the normal Fermi liquid susceptibility, and $Y(T;\,\Delta(T))$ is the Yosida function. $Y(T;\,\Delta(T))$ depends on the temperature explicitly, and also depends on T implicitly through $\Delta(T)$. Leggett arrived at (1) by combining weak-coupling BCS theory and Landau's Fermi liquid theory. Leggett's derivation of (1) does not allow for strong-coupling effects, and in particular, the assumptions on which the derivation is based imply that $\Delta(T)$ should be the weak-coupling BCS energy gap $\Delta_{BCS}(T)$. An obvious scheme for including some strong-coupling effects in (1) is to replace $\Delta_{BCS}(T)$ by the true energy gap, which differs from $\Delta_{BCS}(T)$ both in its overall magnitude and in its temperature dependence. We will call this a trivial strong-coupling correction. If there were no non-trivial strong-coupling corrections to the susceptibility, then Eq. (1) could be inverted to give $\Delta(T)$ from the measured $\chi_B(T)$, since F_0^a can be obtained from the properties of normal ^{3}He. Unfortunately, even in the weak-coupling limit this procedure is questionable, because one additional Landau parameter, F_2^a cannot be obtained from normal-state measurements, so one cannot, in general, distinguish trivial strong-coupling corrections to the weak-coupling susceptibility from corrections due to F_2^a. Within the framework of the trivially renormalized weak-coupling theory, this difficulty can be avoided in two special limits. For $T \to T_c$, F_2^a does not affect $\chi_B(T)$ to order Δ^2, and the weak-coupling susceptibility is

$$\chi_B^{WC}(T) - \chi_N = -\chi_N \frac{7\zeta(3)}{12\pi^2 T_c^{\,2}} \frac{\Delta(T)^2}{1 + F_0^a} \quad . \tag{2}$$

Hence, in weak-coupling theory the slope of $\chi_B(T)$ just below T_c yields $\Delta(T)$, which is equivalent to the specific heat discontinuity, $\Delta C_B/C_N$.

The other special limit is $T \to 0$; here the susceptibility is independent of Δ, and given by

$$\frac{\chi_B(0)}{\chi_N} = \frac{(2/3)\ (1 + F_0^a)}{1 + (2/3)F_0^a + (1/15)F_2^a} \quad , \tag{3}$$

a result first obtained by Czerwonko.[2] Using (3), F_2^a can be found from the zero-temperature limit of the susceptibility, and once F_2^a is known, $\Delta(T)$ can be extracted from $\chi_B(T)$ for $0 < T < T_c$.

All of these schemes require that there be no significant non-trivial strong-coupling corrections to the susceptibility. Our strong-coupling calculation shows that there are non-trivial strong-coupling corrections, of order T_c/T_F, at all temperatures, and hence weak-coupling theory cannot be used to obtain F_2^a and $\Delta(T)$ from the measured susceptibility. The detailed derivation of these results will be published elsewhere. Here we will give a very brief description of the derivation, and then discuss the results for the magnetic susceptibilities.

Our calculation is based on a generalization of the microscopic free-energy functional formalism of Ref. 3 to include space and time dependent external perturbations, and includes all corrections of order T_c/T_F. The strong-coupling corrections fall into three classes: (1) feedback corrections to the irreducible interactions, of the type originally proposed by Anderson and Brinkman[4] to account for the stability of ^{3}He-A; (2) corrections of order T_c/T_F from the frequency-dependence of the normal-state irreducible interactions; (3) corrections of order T_c/T_F to the diagonal and off-diagonal self-energies. Corrections of types (1) and (2) are the basis of strong-coupling theories of the fourth order Ginzburg-Landau free energy.[3,5] Although the feedback corrections are intrinsically of order Δ^2 and hence first enter the superfluid response functions in order Δ^4, all corrections of type (2) and some corrections of type (3) are of zeroth order in Δ, and hence can modify the response functions to order Δ^2.

We will first summarize our results for the susceptibility near T_c. To order Δ^2, the weak-coupling susceptibility for an arbitrary $\ell = 1$ state has the form

$$\chi_{ij}(T)-\chi_N = -\chi_N \frac{1}{1+F_0^a} \frac{B_{WC}}{T_c^2} \int \frac{d\Omega}{4\pi}[\Delta_i(\hat{k})\Delta_j^*(\hat{k})+\Delta_j(\hat{k})\Delta_i^*(\hat{k})]. \quad (4)$$

Our notation is as in Ref. 3; the constant B_{WC} is $7\zeta(3)/8\pi^2$. We find an additional non-trivial strong-coupling correction to $\chi_{ij}(T)-\chi_N$ of exactly the same form as Eq. (4), with B_{WC} replaced by a strong-coupling coefficient B_{SC}. B_{SC} contains an overall factor of T_c/T_F multiplying a sum of weighted angular integrals over two factors of the quasiparticle scattering amplitude, with additional coefficients of order 1 given by convergent frequency sums. Using the s-p approximation for the quasiparticle scattering amplitude we find $(B_{SC}/B_{WC}) \simeq -0.1$ at the polycritical pressure.

Because B_{SC} is negative, the effect of the non-trivial strong-coupling effects in the Ginzburg-Landau region is to make the susceptibility closer to the weak-coupling susceptibility with the BCS energy gap. In other words, if the weak-coupling theory is used to obtain $\Delta(T)$ (or equivalently $\Delta C_B/C_N$) from the measured $\chi_B(T)$, the effect will be to underestimate the strong-coupling enhancement of $\Delta(T)$. For example, if B_{SC} were given correctly by the s-p approximation, then an analysis based on the trivially renormalized weak-coupling theory would give $\Delta C_B/C_N = 1.43$ when the true value was $\Delta C_B/C_N = 1.59$; if $|B_{SC}|$ were half the s-p value, the weak-coupling analysis would give $\Delta C_B/C_N = 1.43$ when the true value was $\Delta C_B/C_N = 1.51$.

Because the strong-coupling corrections of order Δ^2 have the same $\vec{\Delta}(\hat{k})$-dependence as the weak-coupling susceptibility, our strong-coupling theory preserves the weak-coupling result that the maximum A-phase susceptibility is equal to the normal susceptibility. We find that this result holds at all temperatures: $\chi_A^{max} = \chi_N$ through order T_c/T_F for $0 \leq T \leq T_c$. This provides a check on our calculation, since experimentally[6] $(\chi_A^{max} - \chi_N)/\chi_N \approx 5 \times 10^{-3}$. The observed small correction to χ_A^{max}, which is related to the A_1-A_2 splitting, is of higher order in T_c/T_F and hence outside the scope of our theory.

The strong-coupling magnetic free energy function for ^{3}He-B is given by

$$\Omega_B[\vec{m}_o,\vec{m}_2;\vec{H}] = \frac{N(0)}{2}\{-\frac{2|\vec{m}_0|^2}{A_0^a} - \frac{5|\vec{m}_2|^2}{A_2^a}$$

$$+ \frac{2}{3}\, y(T)\, |\vec{m}_0+\vec{m}_2-(1-A_0^a)\frac{\gamma\hbar}{2}\vec{H}|^2 - 2(1-A_0^a)\,|\frac{\gamma\hbar}{2}\vec{H}|^2\}. \tag{5}$$

Here A_0^a and A_2^a are Landau parameters,

$$A_0^a = \frac{F_0^a}{1 + F_0^a} ,$$

$$A_2^a = \frac{F_2^a}{1 + F_2^a/5} , \tag{6}$$

and $\vec{m}_0$ and $\vec{m}_2$ are $\ell = 0$ and $\ell = 2$ self-energies in the presence of the external magnetic field $\vec{H}$. The function $y(T)$ has a diagrammatic representation and contains all the strong-coupling corrections through order T_c/T_F. In the weak-coupling limit ($T_c/T_F = 0$) $y(T)$ is related to the Yosida function by

$$y_{WC}(T) = 1 - Y(T). \tag{7}$$

The strong-coupling contributions to y(T) are all given by angular averages of two powers of the quasiparticle scattering amplitude, weighted with frequency sums and multiplied by T_c/T_F. To obtain the magnetic free energy, and hence the susceptibility of ^{3}He-B, we must minimize $\Omega[\vec{m}_0,\vec{m}_2;\vec{H}]$ with respect to the self-energies $\vec{m}_0$ and $\vec{m}_2$. In this way we find

$$\Omega_B(\vec{H},T) = -\frac{\chi_N H^2}{2} \frac{(1+F_0^a)[\frac{2}{3} + \frac{1}{3}\tilde{Y}(T) + \frac{1}{5}F_2^a\tilde{Y}(T)]}{1+F_0^a(\frac{2}{3} + \frac{1}{3}\tilde{Y}(T)) + \frac{1}{5}F_2^a(\frac{1}{3} + \frac{2}{3}\tilde{Y}(T) + F_0^a\tilde{Y}(T))} \tag{8}$$

The quantity in curly brackets is the reduced B-phase susceptibility, $\chi_B(T)/\chi_N$; $\tilde{Y}(T)$ is a generalized Yosida function,

$$\tilde{Y}(T) = 1 - y(T), \tag{9}$$

which reduces to the original Yosida function in the weak-coupling limit (like the original Yosida function, y(T) and $\tilde{Y}(T)$ depend on temperature both explicitly and through $\Delta(T)$; to save space we have not shown this explicitly). From the s-p approximation we estimate that $\tilde{Y}(T=0)$ is negative and that $|\tilde{Y}(0)| \lesssim 0.05$. With $F_2^a = 0$ and $F_0^a = -0.75$, taking $\tilde{Y}(0) = -0.05$ changes $\chi_B(T=0)/\chi_N$ from 0.333 to 0.317. For comparison, in the weak-coupling limit ($\tilde{Y}(0) = 0$) a value of $F_2^a = 0.4$ would give $\chi_B(T=0)/\chi_N = 0.316$. Even for a larger F_2^a, with $\tilde{Y}(0) = -0.05$ the weak-coupling theory leads to significant errors. For $F_2^a = 1.0$ and $\tilde{Y}(0) = -0.05$, the strong-coupling theory gives $\chi_B(0)/\chi_N = 0.276$; using this susceptibility in the weak-coupling theory one finds $F_2^a \simeq 1.6$.

In summary, we have found non-trivial strong-coupling corrections to the B-phase susceptibility at all temperatures. Although these corrections are small relative to the weak-coupling susceptibility, they may be large enough to cause significant errors if the weak-coupling theory is used to obtain $\Delta(T)$ and F_2^a from measurements of the susceptibility.

References

(1) A. J. Leggett, Phys. Rev. 140, 1869 (1965).

(2) J. Czerwonko, Acta Phys. Pol. 32, 335 (1967).

(3) D. Rainer and J. W. Serene, Phys. Rev. B 13, 4745 (1976).

(4) P. W. Anderson and W. F. Brinkman, Phys. Rev. Lett. 30, 1108 (1973).

(5) W. F. Brinkman, J. W. Serene, and P. W. Anderson, Phys. Rev. A 10, 2386 (1974).

(6) D. N. Paulson, H. Kojima, and J. C. Wheatley, Phys. Lett. 47A, 457 (1974).

AMPLITUDE DEPENDENT EFFECTS IN THE MEASUREMENT OF THE ^{3}He A-PHASE SUPERFLUID DENSITY TENSOR*

P. C. Main, W. T. Band, J. R. Hook, and H. E. Hall

Physics Department, Manchester University
Manchester, M13 9PL, England

D. J. Sandiford

Laboratory of Atomic and Solid State Physics
Cornell University, Ithaca, N.Y. 14853

Physics Department, Manchester University
Manchester, M13 9PL, England

Abstract: An oscillating annular channel containing superfluid ^{3}He-A has been used to measure average values of the superfluid density for different amplitudes of oscillation. The results have the same form as the prediction of theory, and are in agreement with a ratio of two for the perpendicular and parallel components of the superfluid density tensor near T_c. A value of $\rho_{s\perp} = 1.1\rho(1-T/T_c)$ is obtained at 29.7 bar.

I. INTRODUCTION

Current theories, backed by experimental evidence, of the A and B phases of superfluid ^{3}He show that the B phase is isotropic with respect to flow properties such as the superfluid density

---

*Research supported by the Science Research Council, both through Research Grants and through the award of a Research Studentship to one of us (PCM).

tensor, whereas the A phase is anisotropic with an axial symmetry determined by the direction of the unit vector $\hat{\ell}$, which describes the orbital motion of the superfluid.

In the A phase, the field of the axial direction $\hat{\ell}$ is determined by a number of different factors: boundary walls, velocity field, magnetic field, etc. Thus, any meaningful measurement of the superfluid density tensor should be made in a defined texture ($\hat{\ell}$ field) controlled in a known way by the experimental conditions.

For zero magnetic and electric fields and very low velocity fields, the current view appears to be that the texture is entirely determined by the geometry of the container. Ambegaokar, deGennes, and Rainer[1] have shown that $\hat{\ell}$ is perpendicular to a boundary wall for an infinite half-space. Thus one might expect a simple uniform texture between two infinite boundaries, as indicated in Fig. 1a. For a finite system, one than has the problem of closing the container and finding the texture of lowest energy, possibly still maintaining $\hat{\ell}$ normal to all the walls. Fig. 1b shows a conjectural texture in two dimensions that satisfies the condition at the walls, but contains bending energy and singularities. Exactly what modifications to the texture are produced by the side walls is an unsolved problem at the moment.

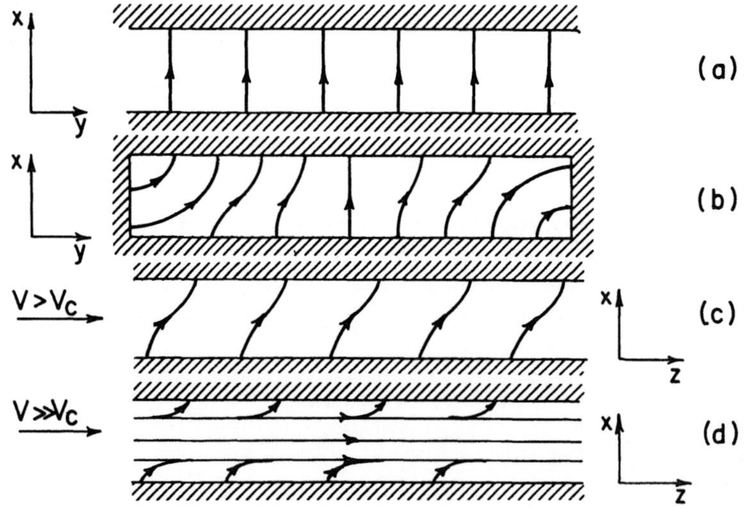

Figure 1 (a) Uniform texture between two infinite planes.
(b) Conjectural closed two-dimensional texture.
(c) Texture between two planes modified by a velocity field which is a little greater than critical $v > v_c$.
(d) Texture between two planes modified by a large velocity field, $v >> v_c$.

The effect of a velocity field on a simple texture has been tackled by deGennes and Rainer[2] and more recently by Fetter.[3] They showed that above a certain critical velocity the effect of flow is to send the $\hat{\ell}$ lines toward the direction of the velocity. Following Leggett,[4] near T_c, one can show that the free energy density for a homogeneous system is

$$f = \rho_{s||}(T)\, v_n^2 = \frac{1}{2}\rho_{s||}(T)\, (\hat{\ell}\cdot \underset{\sim}{v}_n)^2 \quad , \qquad (1)$$

where we assume $\underset{\sim}{v}_s = 0$ in the laboratory frame. Thus one sees that the free energy density is smallest for $\hat{\ell}$ parallel or anti-parallel to $\underset{\sim}{v}_n$.

A magnetic field influences directly the spin vector $\hat{d}$ and hence the configuration of $\hat{\ell}$ via the weak dipole interaction. Fetter[3] has been able to explain quantitatively the measurements of Berthold et al.[5] in a disc-shaped slab 50μm wide in terms of a superfluid density tensor whose principal values ρ_s and $\rho_{s||}$, are in the ratio of 2:1, with a texture controlled by a magnetic field. When the field is parallel to the plane of the slab, $\hat{d}$ and $\hat{\ell}$ can be parallel to one another and normal to the walls, for all values of the field, as in Fig. 1a. However, a perpendicular field of sufficient magnitude in a wide enough slab forces $\hat{d}$ and $\hat{\ell}$ towards directions parallel to the slab. The quantitative agreement between Berthold et al.'s results and Fetter's theory has very nicely supported the general ideas of the way texture of the $\hat{\ell}$ field may be controlled by boundaries and magnetic fields. These experiments were made at velociites small enough that a negligible effect on the texture was expected.

In an annular geometry (380 m wide) at higher velocities, a basically similar type of measurement by the Manchester group (Main et al.[6] gave average values of the superfluid density and viscosity. Within the scatter of the measurements, the results shows that the viscosity values were independent of amplitude of oscillation in both the A and B phases. However, a large variation in the derived values of ρ_s was apparent in the A phase, related to the amplitude of the velocity field. Measurements of ρ_s at low values of $\underset{\sim}{v}_n$ (to a first approximation $\underset{\sim}{v}_n$ is constant throughout the container, equal to the wall velocity) were interpreted as giving $\rho_{s\perp}$. Such an assignment is reasonable if Fig. 1b, for example, shows the texture for this case where the velocity field is normal to the figure and hence perpendicular to all the $\hat{\ell}$ vectors.

At large velocities one expects the texture to be altered, (Fetter[3]). Figs. 1c and 1d illustrate the way one might imagine the texture to change as a dc velocity field is imposed. An obvious question, however, is the relation of these figures to the experiment of Main et al. where the large velocity field oscillated at ∿ 60 Hz. Does the texture modification produced by a high velocity relax rapidly or slowly compared with the period of the oscillation? If the time constant is short, then the texture will vary considerably as the velocity field oscillates; for long time constants the variation will be small about a texture determined by the RMS value of the ac velocity. We will consider these problems in the next section.

2. SOLUTION OF THE TEXTURE EQUATION

Consider the following situation: ^{3}He-A is contained in a narrow annular channel between two concentric cylindrical surfaces. The container is rotated about its axis and the channel is moving with velocity $\underset{\sim}{v}$. We assume that the normal field rotates locked to the channel so that in the frame of reference moving with the channel $\underset{\sim}{v}_n = o$. If we have a channel width >> 6μm, so that $\hat{d}$ can be assumed to be parallel to $\hat{\ell}$ within the channel, then, according to Hall and Hook,[7] the equation of motion of $\hat{\ell}$ in the frame moving with the channel is

$$\frac{8\mu\, m^2}{\rho_{s||}\hbar^2}\frac{\partial\hat{\ell}}{\partial t} = 5\nabla^2\hat{\ell} - 5(\hat{\ell}\cdot\nabla^2\hat{\ell}) + \frac{8m^2}{\hbar^2}(\underset{\sim}{v}_s\cdot\hat{\ell})(\underset{\sim}{v}_s-(\underset{\sim}{v}_s\cdot\hat{\ell})\hat{\ell})$$

$$- \frac{2m}{\hbar}(\hat{\ell}\times\underset{\sim}{v}_s)\,\mathrm{div}\,\hat{\ell} - \frac{2m}{\hbar}\hat{\ell}\times(\hat{\ell}\cdot\underset{\sim}{\nabla})\underset{\sim}{v}_s - \frac{2m}{\hbar}\hat{\ell}\times\underset{\sim}{\nabla}(\hat{\ell}\cdot\underset{\sim}{v}_s)$$

$$+ 10\frac{m}{\hbar}\hat{\ell}\times(\underset{\sim}{v}_s\cdot\underset{\sim}{\nabla})\hat{\ell} + 10\frac{m}{\hbar}(\underset{\sim}{v}_s\cdot\hat{\ell})\hat{\ell}\times(\hat{\ell}\cdot\underset{\sim}{\nabla})\hat{\ell} \qquad (2)$$

Here, μ is the Cross-Anderson[8] viscosity coefficient. We assume, as have previous authors[1,3] that the circulation of $\underset{\sim}{v}_s$ is quantized, and that in the frame of reference moving with the channel the component of $\underset{\sim}{v}_s$ along the channel is $-\underset{\sim}{v}$.

Let us now consider the possibility of finding a time independent solution to (2) in which $\hat{\ell}$ is confined to the plane normal to the axis of the container. (In this situation the Mermin and Ho[9] formula for curl $\underset{\sim}{v}_s$ says that curl $\underset{\sim}{v}_s = o$.) Fetter[3] gives a method for calculating such a solution which amounts effectively to considering the first three terms on the RHS of (2). However, the other terms on the RHS of (2) are not zero for the solution obtained and are in the direction perpendicular to the plane. We deduce that the true solution for the uniformly rotating channel must be more complicated than those considered by Fetter.

There is, however, a way of 'stabilizing' the Fetter solution. To see this, we note that the terms in (2) that give the Fetter solution are all even in $\underset{\sim}{v}$ whereas those which create problems are odd in $\underset{\sim}{v}$. Suppose therefore we oscillate the channel so that

$$v = v_0 \cos \omega t.$$

and the frequency of oscillation ω is much bigger than the rate at which $\hat{\ell}$ can respond. The forces tending to take $\hat{\ell}$ out of the plane will then oscillate rapidly so that $\hat{\ell}$ will probably depart little from a planar configuration and within the plane we expect $\hat{\ell}$ to oscillate slightly around the stationary "Fetter" texture appropriate for the root mean square velocity $v_0/\sqrt{2}$. More detailed calculations confirm these results.

This situation corresponds to the Manchester experiment. The superfluid density we measure corresponds to the mean texture and to find this we solve (2) with $\partial\hat{\ell}/\partial t = 0$, the component of $\underline{v_s}$ parallel to the channel $= v_0/\sqrt{2}$, and ignore terms linear in $\underline{v_s}$. Equation (2) then becomes

$$\frac{5\ _{s||}\hbar^2}{8m^2}(\nabla^2\hat{\ell} - (\hat{\ell}\cdot\nabla^2\hat{\ell})\hat{\ell}) + \rho_{s||}(\underline{v_s}\cdot\hat{\ell})(\underline{v_s}-(\underline{v_s}\cdot\hat{\ell})\hat{\ell}) = 0 \quad (3)$$

The physical interpretation of this equation is that it is a statement that the total torque on $\hat{\ell}$ vanishes. The first two terms are the elastic torque and the final term the flow alignment torque. The boundary conditions at the walls of the channel are that $\hat{\ell}$ should be normal to the walls and the perpendicular component of the particle current should vanish.

We consider a superfluid order parameter of the same form as that considered by Fetter and like him we introduce an angle Θ which is the angle between $\hat{\ell}$ and the radial direction. Θ must be zero at the walls of the channel. In terms of Θ, (3) becomes

$$\frac{d^2\Theta}{dr^2} = \frac{-16\ _m^2}{5\hbar^2}\frac{\sin\Theta\cos\Theta}{(1+\sin^2\Theta)^2}\frac{v_o^2}{2} \quad (4)$$

where r is the radial coordinate.

For a channel of width W and for $v_0/\sqrt{2} \leq v_c = \sqrt{5}/4\ \pi\hbar/mW$, the appropriate solution of (4) is $=0$, i.e. $\hat{\ell}$ is everywhere radial. The measured value of ρ_s for this situation is $\rho_{s\perp}$. For velocity amplitudes greater than the above critical value Θ differs from zero. For $v_o >> v_c$, Θ is close to $\pi/2$ over most of the channel and the measured value of ρ_s tends to $\rho_{s||}$. At intermediate velocities a calculation of the particle current (see Fetter's equation (16)) provides the following result for $\bar{\rho}_s/\rho_{s\perp}$

$$\frac{\bar{\rho}_s}{\rho_{s\perp}} = \frac{2}{\pi}\sqrt{2}\,\frac{V_c}{V_o}\,(1+\sin^2\Theta_o)^{1/2}\int_0^{\Theta_o}\frac{d\Theta}{(\sin^2\Theta_o-\sin^2\Theta)^{1/2}(1+\sin^2\Theta)^{1/2}} \quad (5)$$

where $\bar{\rho}_s$ is the measured value and Θ_o, the value of Θ in the center of the channel is given by

$$(1+\sin^2\Theta_o)^{1/2}\int_0^{\Theta_o} d\Theta\,\frac{(1+\sin^2\Theta)^{1/2}}{(\sin^2\Theta_o-\sin^2\Theta)^{1/2}} = \frac{\pi}{2\sqrt{2}}\,\frac{v_o}{v_c}$$

It is assumed in deriving equation (5) that $\rho_{s\perp}/\rho_{s||} = 2$.

The amplitude of oscillation of $\hat{\ell}$ is indeed small since the condition for this for $v_o \sim v_c$ is easily shown to be[10]

$$\frac{\rho_{s||}v_c^2}{\omega\mu} << 1$$

which is well satisfied for the Manchester experiments. For $v_o >> v_c$ the condition for small oscillation is more difficult to deduce but rough calculations suggest that the assumption of small oscillation is still valid.

3. EXPERIMENTAL RESULTS

The experiments described by Main et al.[6] have been continued with improvements to the signal to noise ratio so that lower amplitude measurements have been possible, in addition to better measurements at higher amplitude. Further, the temperature scale (obtained by reference to the La Jolla T_c phase diagram given by Wheatley[11]) has been recalculated with an additional degree of freedom in the fitting function. Finally, the correction (previously estimated) for the effect of superfluid in the pores of the CMN has been measured at 29.7 bar, the pressure of the minimum in the melting curve. This measurement was obtained by injecting epoxy into the flow channel and measuring the effect of the pores alone. The result was essentially in agreement with our previous estimate and gave $n^2 \sim 15$, where n is the fourth-sound refractive index, assuming ρ_s in the pores is equal to $\rho_{s\perp}$.

Figure 2 shows $\bar{\rho}_s$ in the A phase at 29.7 bar for a low and a high amplitude oscillation as a function of temperature. It can be seen that the measured average ρ_s is linear in temperature with a slope that reduces with increasing amplitude of oscillation. The dashed line is the theoretical prediction calculated as in the last section. The critical velocity of 0.1mms^{-1} is that predicted for the size of flow channel used, but there is a fairly large (30%) undertainty in the amplitude calibration and it is possible that all the experimental points ought to be shifted from right to left, giving better agreement. The points themselves were calculated by evaluating $(\bar{\rho}_s/\rho)/(1-T/T_c)$ for every point between $0.05<1-T/T_c<0.15$ and taking an average. The lines describe the range of values and are not strictly error bars because some of the spread is due to systematic curvature in the plot.

The value of 1.1 for the slope at $v < v_c$ gives our best value for $\rho_{s\perp}$, measuring temperature on the La Jolla scale.[11] To compare with Berthold *et al.*'s value of 0.7 on the Helsinki scale,[12] one needs to correct by about 20% in this range of temperature, bringing our value down to about 0.9. Both experiments give values larger than the fourth sound results[13,14] and smaller than the pioneering experiment of Alvesalo *et al.*[15] on the melting curve.

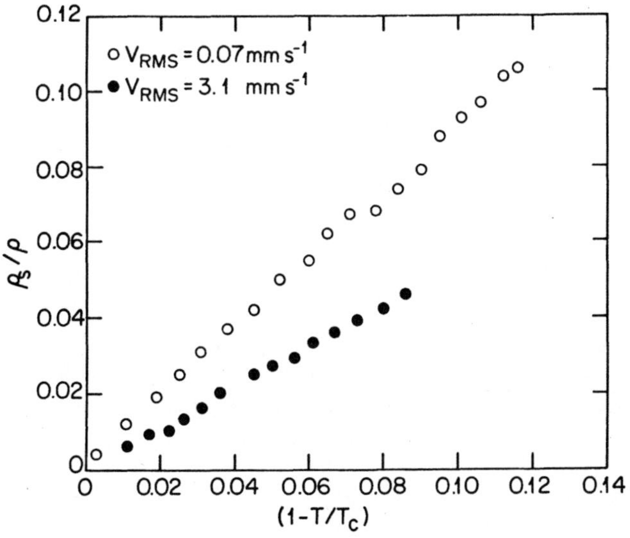

Figure 2. Superfluid density at 29.7 bar for two values (0.7mms^{-1} and 3.1mms^{-1}) of the RMS velocity field, V_{RMS}. v_c is expected to be 0.1mms^{-1}.

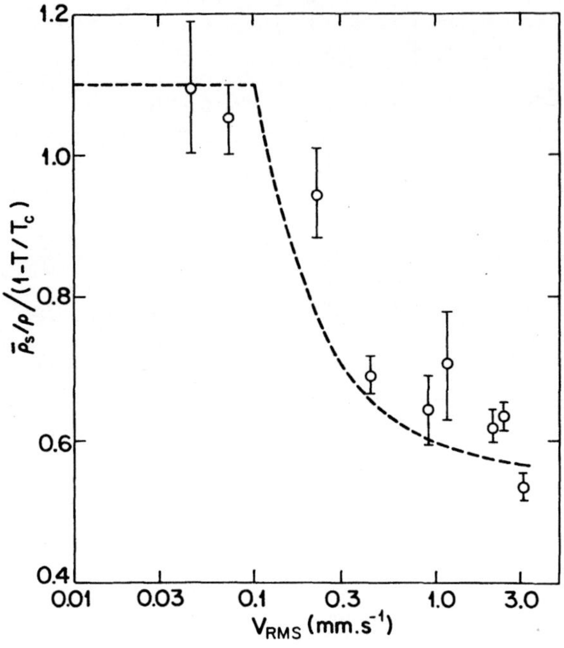

Figure 3. Slope of superfluid density as function of temperature $[(\rho_s/\rho)/(1-T/T_c)]$ plotted against the RMS velocity of the superfluid. The broken curve is the theoretical prediction.

References

(1) V. Ambegaokar, P. G. deGennes, and D. Rainer, Phys. Rev. A 9, 2676 (1974).

(2) P. G. deGennes and D. Rainer, Phys. Lett. 46 A, 429 (1974).

(3) A. L. Fetter, Phys. Rev. B 14, 2801 (1976).

(4) A. J. Leggett, Rev. Mod. Phys. 47, 331 (1975).

(5) J. E. Berthold, R. W. Gianetta, E. N. Smith, and J. D. Reppy, Phys. Rev. Lett. 37, 1138 (1976).

(6) P. C. Main, C. W. Kieweit, W. T. Band, J. R. Hook, D. J. Sandiford, and H. E. Hall, J. Phys. C., 9, L397 (1976).

(7) H. E. Hall and J. R. Hook, preprint submitted to J. Phys. C. (1976).

(8) M. C. Cross and P. W. Adnerson, "Proc. 14th Int. Conf. on Low Temperature Physics", edited by M. Krusius and M. Vuorio (North Holland, Amsterdam, 1975), Vol. 1, p. 29.

(9) N. D. Mermin and T-L Ho, Phys. Rev. Lett. 36, 594 (1976).

(10) P. C. Main, Ph.D. thesis, Manchester University (1976).

(11) J. C. Wheatley, Rev. Mod. Phys. 47, 415 (1975), Appendix A.

(12) For a discussion of temperature scales, see R. C. Richardson, Sussex Symposium on Superfluid ^3He, (1976), to be published.

(13) H. Kojima, D. N. Paulson and J. C. Wheatley, Phys. Rev. Lett. 32, 141 (1974) and J. Low Temp. Phys. 21, 283 (1975).

(14) A. W. Yanof and J. D. Reppy, Phys. Rev. Lett. 33, 631 and 1030 (Erratum) (1974).

(15) T. A. Alvesalo, Yu. D. Anufriyev, H. K. Collan, O. V. Lounasmaa and P. Wennerström, Phys. Rev. Lett. 30, 962 (1973). T. A. Alvesalo, H. K. Collan, M. T. Loponen, and M. C. Veuro, Phys. Rev. Lett. 32, 981 (1974). T. A. Alvesalo, H. K. Collan, M. T. Loponen, O. V. Lounasmaa and M. C. Veuro, J. Low Temp. Phys. 19, 1 (1974).

MEASUREMENTS OF THE STATIC AND DYNAMIC NMR SUSCEPTIBILITIES OF SUPERFLUID ^{3}He-B USING AN R-F BIASED SQUID*

R. A. Webb and Z. Sungaila

Argonne National Laboratory
Argonne, Illinois 60439

Abstract: Using an R-F biased SQUID, the temperature dependent susceptibility of superfluid ^{3}He-B has been measured both statically and via a pulse technique in a field of 309 gauss and in a pressure range 26.5 to 18 bar. In ^{3}He-B the dynamic susceptibility as determined using pulsed SQUID NMR agrees with the theoretical predictions for the BW state. However, the static susceptibility measured, using the same R-F biased SQUID and pick up coil, is significantly smaller than the dynamic susceptibility.

One of the major experimental discrepancies that still exists in superfluid ^{3}He is that of the difference between the susceptibility of ^{3}He-B as measured statically by SQUID techniques and dynamically using NMR techniques. Dynamic cw NMR measurements both on and off the melting curve[1,2] suggest that the susceptibility of ^{3}He-B is qualitatively the same as that predicted by the BW theory, while measurements of the static susceptibility using an R-F biased SQUID[3,4] suggest that the total temperature dependent susceptibility of ^{3}He-B is significantly smaller than the theoretical predictions. We have developed the techniques of SQUID NMR into a usable measuring tool in the millikelvin temperature range, and in this work have attempted to shed new light on this discrepancy by using the same detection system for the dynamic NMR measurements as for the static work.

A schematic drawing of the adiabatic demagnetization cell used in this work is shown in Fig. 1. The ^{3}He sample was contained in

*Work supported by the U. S. Energy Research and Development Administration.

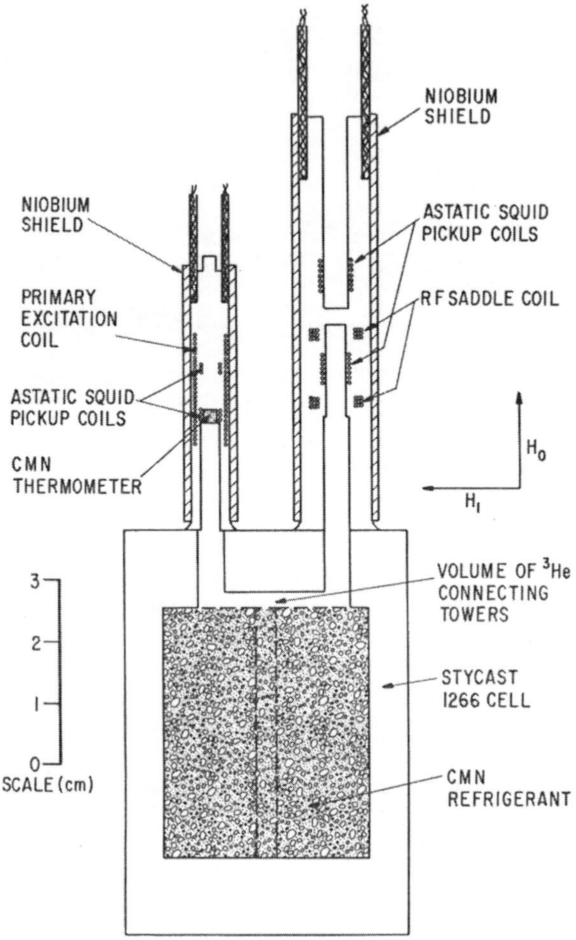

Figure 1. Schematic drawing of the adiabatic demagnetization cell used for the experiments on static and dynamic susceptibility of ^{3}He-B.

a 3mm I.D. tube. The detection system consisted of a 32 turn pick up coil on a 4.3mm I.D. wound astatically with .076mm Nb wire and was connected to the input coil of a two hole SQUID operated at 19 MHz in the flux loop configuration.[5] The static field H_o was provided by the 10.9mm I.D. Nb shield that surrounded the tower, 309 gauss in these measurements. A transverse H_1 field was produced by a 42 turn RF saddle coil wound using .076mm Nb wire on a diameter of 5mm. All temperatures were determined from 17 Hz mutual inductance measurements on 10mg of CMN located in the second tower of the cell.

The calibration of the magnetometer was performed in the conventional way[6] by holding the temperature of the cell constant and allowing liquid to slowly fill the towers while continually monitoring the output of the SQUID. The calibration constant determined from many such fills in the temperature range .33 K to 1.1 K was then checked in the same temperature range by using both adiabatic fast passage and 180° pulsed NMR. We find agreement between the calibration constant determined statically and dynamically to be better than ± 2%. Further, at 15mK in the pressure range 18 - 26.5 bar, the value of the paramagnetic susceptibility of ^{3}He determined using cw NMR, pulsed NMR and adiabatic fast passage all agree with the static calibration to ± 1%. The details of SQUID NMR as performed in this experiment will be given elsewhere.[7] However, it is important to point out here that one of the main advantages of SQUID NMR is that during the course of a measurement the change in the component of magnetization parallel to H_o is continuously monitored.

The static susceptibility of superfluid ^{3}He-B was determined using two techniques. The first was to measure, as a function of pressure, the size of the jump discontinuity upon slowly cooling from the A phase into the B phase. This proved to be the "reversible" transition.[6] The second technique employed to measure the static susceptibility was to cool deep into the B phase, establish temperature equilibrium and then rapidly, within a few seconds, increase the magnetic field on the main CMN and measure the change in magnetometer output as the ^{3}He in the magnetization tower warmed from the initial temperature T_o through the A phase into the normal phase. The change in magnetometer output is $\Delta\phi/\phi_o = \lambda\Delta\chi_B = \lambda(\chi_N(P)-\chi_B(P,T_o))$ where λ is the calibration constant. Using the known value of $\chi_N(P)$ and the measured value of $\Delta\chi_B$ the value of $\chi_B(P,T_o)$ can be determined at any temperature. In the field of 309 gauss the background magnetization was temperature independent.

The dynamic susceptibility of superfluid ^{3}He B was determined using pulsed SQUID NMR. We first held the temperature constant by slowly demagnetizing the main CMN and recording on film the recovery of the magnetization following an RF pulse. Then by stepping the frequency of the RF pulse off resonance by ± 50 KHz the magnitude of both the magnetic background and electronic time constant could be measured. The SQUID electronics had a total recovery time of approximately 50 μsec. The magnetic background was generally never larger than 3 - 4% of the total magnetization change of ^{3}He-B and was completely recovered within 80 to 180 μsec after the end of the RF pulse. For all of the work reported here, the width of the RF pulse was held constant at 25 μsec and its magnitude was varied so as to rotate the spins between 20° and 100°. A typical recovery of the magnetization of superfluid ^{3}He following a 89° RF pulse is shown in Fig. 2. For this data the background had completely re-

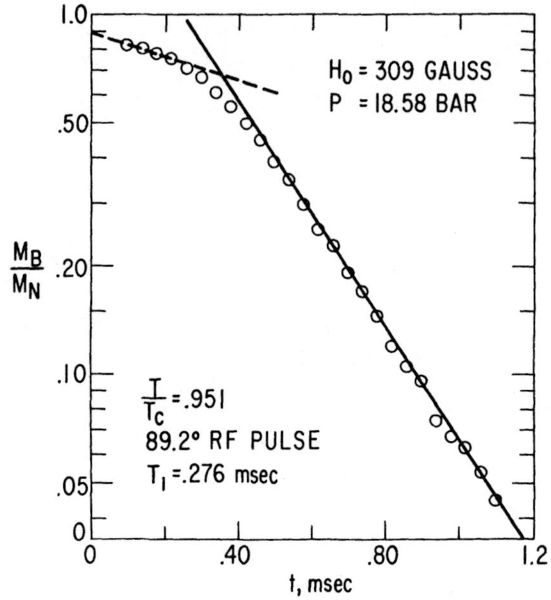

Figure 2. Typical data on the recovery of the magnetization of superfluid ^3He-B in a field of 309 gauss, following an 89° RF pulse at a reduced temperature T/T_c = .951 and at a pressure of 18.58 bar.

covered in less than 100 μsec and the data displayed were taken directly drom the photographic record with no background corrections made. The recovery of the magnetization after the first 0.5 msec is exponential with a time constant of .276 msec. This time constant was nearly temperature independent being ca. .24 msec at T/T_c = .84 to ca. 30 msec at T/T_c = .98 with a slight dependence on the magnitude of the RF pulse, T_1, being longer for a larger pulse. To extract the magnetization, the initial part of the recovery has to be extrapolated back to t = 0. Two techniques were used for this extrapolation. The first was to make a straight line extrapolation on a lnM vs. t plot, shown as a dashed line in Fig. 2. The second technique was based on the theoretical work of Leggett et al.[8] Briefly, the idea is that following a large angle RF pulse the energy of the spin system relaxes back to equilibrium at a rate linearly proportional to time. This implies that $d(\Delta M_z)^2/dt$ = constant, at least initially. We made an M^2 vs. t extrapolation for all our data and found generally that only the first 120 - 160 μsec of data following the recovery of the background could be used as compared with the exponential extrapolation where the first 160 - 240 μsec could be used. Further, the scatter in all the data was much greater when the M^2 extrapolation was used.

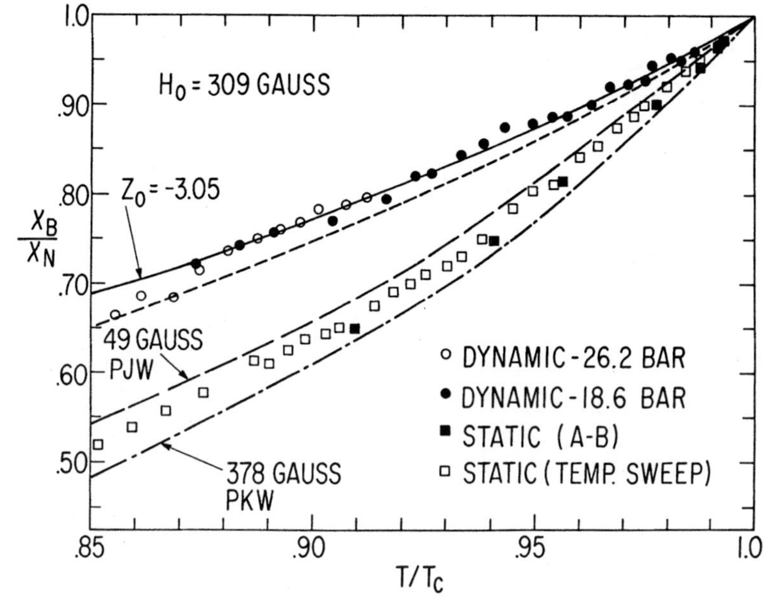

Figure 3. The normalized susceptibility of ^{3}He-B as a function of reduced temperature in a field of 309 gauss. The solid line is the theoretical BW susceptibility of ^{3}He-B, assuming Z_o = 3.05. The dashed lines labeled PKW and PJW are the smoothed static susceptibility data of references 4 and 5.

The results of all our static and dynamic data in a field of 309 gauss are shown in Fig. 3. The dynamic data displayed were obtained using the exponential extrapolation techniqe. The solid line is the BW theory using a value of - 3.05 for Z_o. The dotted line is a fit to all our χ_B/χ_N data determined using the M^2 vs. t extrapolation. The smooth static data of Paulson, Kojima and Wheatley[3] and Paulson, Johnson and Wheatley[4] are shown as dashed lines. The static susceptibility data determined from the A - B transition are systematically lower than the temperature sweep data. We believe that due to the reversibility of the A - B transition and the existence of a small temperature gradient over the sample region in the A phase, the change in magnetization at this transition will tend to be larger and hence give a smaller value for χ_B/χ_N. Near T_c our dynamic data can be described by $\chi_B/\chi_N = 1 + 2.6\ (T - T_c)/T_c$ while the static data follow $\chi_B/\chi_N = 1 + 3.9\ (T - T_c)/T_c$.

Clearly, there is a discrepancy between the static and dynamic susceptibility of ^{3}He-B as measured in this experiment. This discrepancy can no longer be attributed to the calibration of the magnetometer and is no doubt a real property of the superfluid as yet not understood. The magnitude of the discrepancy in the temperature coefficient was once thought to be almost a factor of 2 is now only 50% and even smaller if one assumes the M^2 vs. t extrapolation to be correct. The details of this experiment will be published elsewhere.[9]

References

(1) L. R. Corruccini and D. D. Osheroff, Phys. Rev. Lett. 34, 695 (1975).

(2) A. I. Ahonen, T. A. Alvesalo, M. T. Haikala, M. Krusius and M. A. Paalanen, Phys. Lett. A 51, 279 (1975).

(3) D. N. Paulson, H. Kojima and J. C. Wheatley, Phys. Rev. Lett. 32, 1098 (1974).

(4) D. N. Paulson, R. T. Johnson and J. C. Wheatley, Phys. Rev. Lett. 31, 746 (1973).

(5) R. P. Giffard, R. A. Webb and J. C. Wheatley, J. Low Temp. Phys. 6, 533 (1971).

(6) J. C. Wheatley, Reviews of Mod. Phys. 47, 415 (1975).

(7) R. A. Webb, to be submitted to Rev. Sci. Inst.

(8) A. J. Leggett and S. Takagi, Phys. Rev. Lett. 34, 1424 (1975).

(9) R. A. Webb and Z. Sungaila, to be submitted to J. Low Temp. Phys.

SOME NEW NMR EFFECTS IN SUPERFLUID ^{3}He-A*

R. W. Giannetta, C. M. Gould, E. N. Smith, and D. M. Lee

Laboratory of Atomic and Solid State Physics

and

Cornell Materials Science Center
Cornell University
Ithaca, N. Y. 14853

Abstract: Longitudinal and transverse satellite lines have been observed by CW techniques in superfluid ^{3}He-A under a variety of experimental conditions. A new metastable mode has been observed in superfluid ^{3}He-A using pulsed NMR techniques. The frequency shift of this mode is related to the frequency shift of the transverse satellite line.

I. INTRODUCTION

The earliest observations[1] of nuclear magnetic resonance in superfluid ^{3}He-A revealed a dramatic shift of the NMR frequency, ω, described by the equation $\omega^2-\omega_0^2 = \Omega_A^2(T)$ where ω_o is the Larmor frequency and $\Omega_A(T)$ is a field independent function of temperature. The large size of the observed frequency shift precluded any explanation in terms of the classical interaction between nearest neighbor nuclear dipoles. Leggett[2] was able to show that frequency shifts of the observed size could result from triplet pairing (S=1).

---

*Work supported in part by the National Science Foundation under grants No. DMR 76-21669 and No. DMR 75-15933, and through the Cornell Materials Science Center under grant No. DMR 76-01281.

He then went on to formulate an elegant theory[3] of the spin dynamics of triplet superfluids which has been most successful in explaining a great variety of magnetic phenomena in both the superfluid A phase and the superfluid B phase of liquid ^{3}He. The Leggett theory predicted the existence of a longitudinal resonance in both superfluid phases in addition to the usual transverse nuclear magnetic resonance. To observe longitudinal resonance the r.f. coil must be parallel to the applied steady magnetic field, in contrast to the transverse NMR configuration in which the r.f. coil is perpendicular to the field. The observation of longitudinal resonance in both the A and B phases by conventional NMR methods,[4,5] as well as by a novel transient method known as longitudinal ringing,[6] has provided strong evidence for the Leggett theory.

According to this theory, the large frequency shift and the other unexpected magnetic effects observed in superfluid ^{3}He are consequences of the coherent dipolar interactions of the pairs. This interaction is much larger than anything possible in classical theory. The general expression given by Leggett[3] for the dipolar Hamiltonian in superfluid ^{3}He is

$$H_D \propto \int \frac{d\Omega}{4\pi} \left\{ 3 \, |\hat{n}\cdot\vec{d}(\hat{n})|^2 - |\vec{d}(\hat{n})|^2 \right\} , \tag{1}$$

where $\hat{n}$ is the direction on the Fermi surface and $\vec{d}(\hat{n})$ is the vector order parameter. There is a great deal of evidence[7] indicating that superfluid ^{3}He-A corresponds to the Anderson-Brinkman-Morel (ABM) or axial state of p-wave pairing. For this state, the dipolar restoring torque following a departure from equilibrium in the spin system is proportional to $(\hat{d}\cdot\hat{l})(\hat{d}\mathrm{x}\hat{l})$. It is this dipolar torque which is responsible for the large frequency shift in superfluid ^{3}He-A. The spin dynamics in this phase is completely described by the Leggett equations of motion as written by Brinkman and Smith,[8]

$$\dot{S} = \omega_o \, \vec{S} \times \hat{H} + \frac{\Omega_A^2}{\omega_o} \, (\hat{d}\cdot\hat{l})(\hat{d}\mathrm{x}\hat{l}) \tag{2}$$

and

$$\dot{\hat{d}} = \omega_o \, \hat{d} \times (\hat{H} - \vec{S}) \tag{3}$$

where the units are chosen so that $\vec{S}$ is dimensionless and of unit length in equilibrium. The Larmor frequency is $\omega_o=\gamma H$, and $\hat{H}$ is a vector proportional to the vector sum of the static field and the r.f. field, which in equilibrium is a unit vector in the direction of the applied static field.

These equations have been solved by Brinkman and Smith[8] using a perturbation method to determine the frequency response of the ^{3}He spin system following an applied radio-frequency tipping pulse, with the tipping angle ϕ defined by $\phi = \gamma H_1 t$ where H_1 is the radio-frequency field strength and t is the pulse duration. The frequency shift $\Delta\omega(\phi)$ away from the Larmor frequency for the free induction decay tail following a pulse of tipping angle ω as compared with the frequency shift $\Delta\omega(0)$ for a vanishingly small tipping angle (which also corresponds to the CW resonance) is given by

$$\Delta\omega(\phi)/\Delta\omega(0) = \frac{1}{4} + \frac{3}{4}\cos\phi . \tag{4}$$

The above equation agrees well with data obtained previously by Osheroff and Corruccini[9] from pulsed NMR measurements at the melting pressure. We shall discuss experiments performed at Cornell which confirm the results of Osheroff and Corruccini[9] for pressures below the melting pressure. In the course of these experiments we have also observed a new metastable mode following a large angle tipping pulse. The NMR frequency response associated with this mode differs markedly from the response of the ordinary stable state of superfluid ^{3}He-A.

At the 14th International Low Temperature Conference, Avenel, Bernier, Varoquaux, and Vibet[10] of Orsay presented an interesting new phenomenon during the course of CW longitudinal NMR measurements on superfluid ^{3}He-A in a Pomeranchuk cell. In addition to the ordinary longitudinal line, with frequency $\Omega_A(T)$, these investigators also observed a small and rather broad satellite resonance whose frequency was given approximately by $\Omega_A(T)/2$. In recent experiments at Cornell[11] we have studied this longitudinal satellite in a variety of different geometries and magnetic fields. We have also found a previously unobserved transverse satellite resonance.

II. CW MEASUREMENTS

The CW nuclear magnetic resonance experiments were performed in a Pomeranchuk cell, so that the ^{3}He sample was always maintained at the melting pressure. The cell used in these experiments is shown in Fig. 1. The radio-frequency coil, a solenoid 1 cm. in diameter and 1.3 cm. long, was placed vertically in the cell. A second rf coil, consisting of a Helmholtz pair with axis perpendicular to the axis of the cell is also shown in Fig. 1. This latter coil was not used for CW NMR during these measurements.

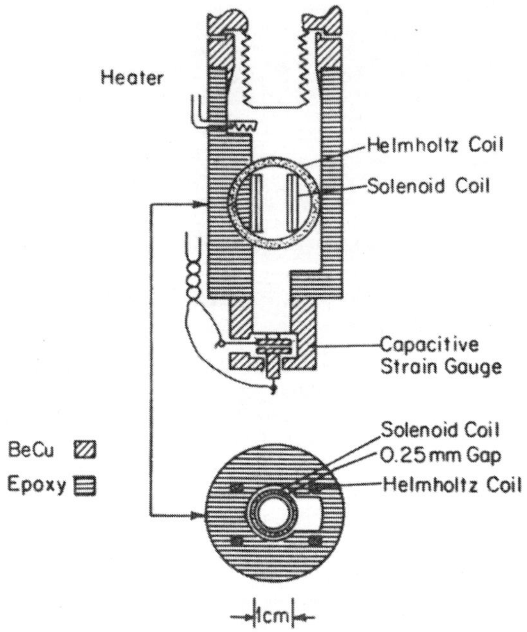

Figure 1. Pomeranchuk cell used to investigate the CW longitudinal and transverse satellite resonances. The sample geometry was altered by placing various inserts into the solenoid. The static magnetic field was provided by a large Helmholtz pair external to the cryostat.

The steady magnetic field was provided by a large Helmholtz pair external to the cryostat which was oriented with its axis in the vertical direction for longitudinal resonance experiments and in the horizontal direction for transverse resonance. The sample temperature was obtained from the melting pressure which was determined by means of a sensitive capacitance strain gauge at the bottom of the cell. Both longitudinal and transverse resonance in superfluid ^{3}He were detected by sweeping the temperature through the resonance line. The use of phase sensitive detection and high rf power levels greatly improved the signal to noise ratio, thereby allowing observation of very small NMR features. The high r.f. power levels did not saturate the signal in superfluid ^{3}He-A.

Chart recorder traces of a set of longitudinal and transverse resonance curves showing the main lines and the satellite structure are displayed in Fig. 2. The height of the longitudinal satellite line is about 20% of the height of the main longitudinal line, and the width of the longitudinal satellite is almost twice that of

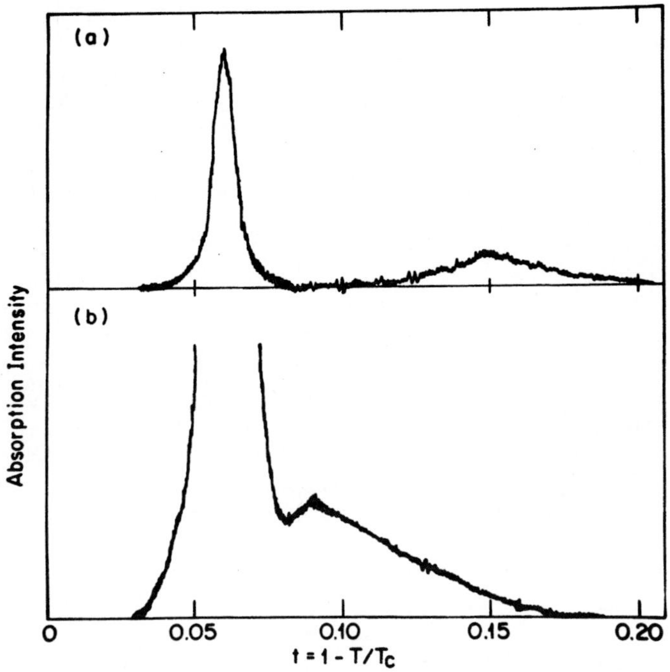

Figure 2. Temperature sweeps through (a) longitudinal resonance and (b) transverse resonance peaks. The vertical axis is the absorption intensity obtained while the frequency and steady magnetic field are held constant. The frequencies and fields in (a) and (b) have been chosen so that the main lines appear at the same temperature. Thus the satellite positions can be easily compared.

the main longitudinal line.

The main transverse line has a rapidly increasing linewidth and rapidly decreasing height as the temperature is reduced, and the transverse satellite follows a similar pattern. The transverse satellite is typically between 10 and 100 times shorter than the main line.

By observing the transverse satellite at a variety of different applied fields, we found that the difference between the square of the transverse satellite frequency $\omega_{t.s.}$ and the square of the Larmor frequency, ω_o, was always equal to a temperature dependent constant which, for convenience, we shall call the effective longitudinal frequency $\Omega^2_{t.s.}(T)$. Explicitly writing down the above statement as an equation, we have $\omega^2_{t.s.} - \omega^2_o = \Omega^2_{t.s.}(T)$. The longitudinal frequency of the main longitudinal line, Ω_{main}, the longitudinal satellite frequency, Ω_{sat}, and the transverse satellite shifting frequency, $\Omega_{t.s.}$ are all plotted against the reduced temperature $t = 1 - T/T_c$ in Fig. 3. Near T_c, these frequencies vary as $t^{1/2}$ as shown in Fig. 3. A more detailed comparison of the temperature dependences of the satellite lines with the main longitudinal line is obtained by studying the ratios of the longitudinal and transverse satellite frequencies to the frequency of the main longitudinal line. These ratios, defined by $R_L(t) \equiv \Omega_{sat}(t)/\Omega_{main}(t)$ and $R_T(t) \equiv \Omega_{t.s.}(t)/\Omega_{main}(t)$, respectively, are plotted against the reduced temperature in Fig. 4. The ratio $R_L(t)$ relating the longitudinal satellite to the main longitudinal line varies from 0.74 at T_c to 0.67 at the B→A transition (B´) rather than being equal to $.707 = 1/\sqrt{2}$ as suggested in the earlier work of Avenel et al.[10] On the other hand, the ratio $R_T(t)$ appears from Fig. 4 to be temperature independent with a value of $\Omega_{t.s.}/\Omega_{main} = .835$. The difference, both in magnitude and temperature dependence of $R_T(t)$ and $R_L(t)$ suggests the possibility that the longitudinal and transverse satellites stem from different underlying causes.

The measurements were conducted under a wide variety of experimental conditions with the hope of gaining further insight into the nature of the satellite resonances. Static magnetic fields ranging from zero to 1 kilogauss were applied during studies of the longitudinal satellite. For fields above 20 gauss, variation of the steady field produced no change in the longitudinal satellite

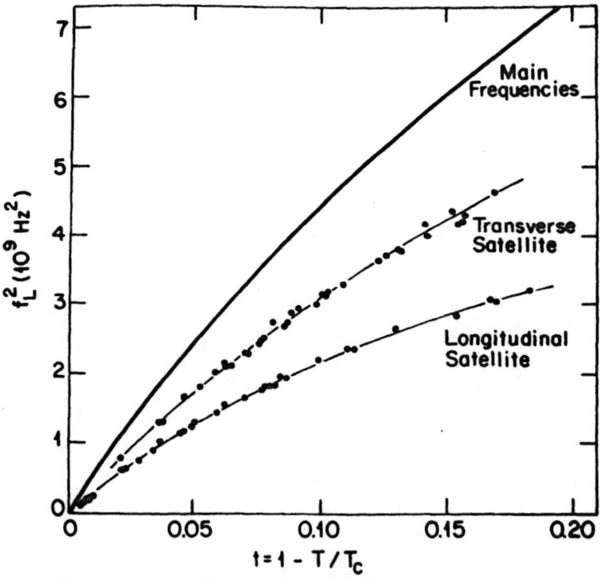

Figure 3. Longitudinal satellite frequency and trasverse satellite shift frequency as a function of temperature. The main, longitudinal and transverse shift frequencies are represented by the heavy line.

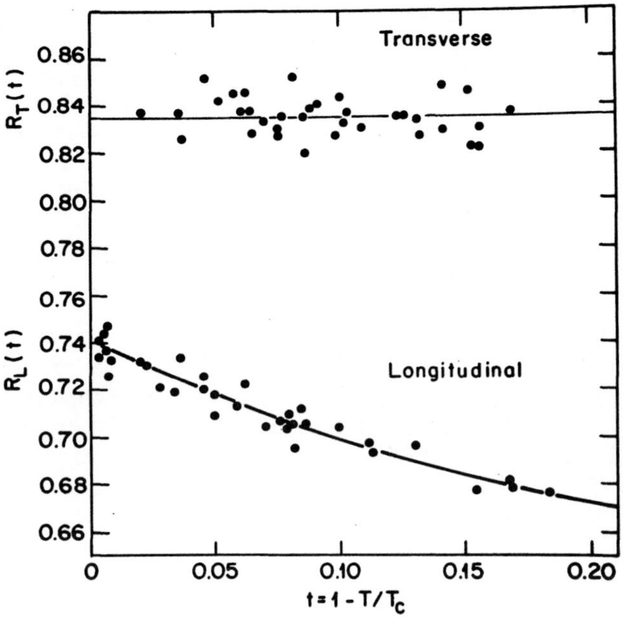

Figure 4. Ratio of the longitudinal and transverse satellite frequencies to the main longitudinal frequency. This figure shows the different temperature dependences of the two types of satellites.

or the main longitudinal line. Below 20 gauss both lines decreased in magnitude, probably as a result of the anisotropic order parameter becoming randomly distributed as a consequence of the reduction in the orienting strength of the applied magnetic field.

Varying the rf power level over three orders of magnitude also had no effect on either the main longitudinal line or its satellite, nor did the application of up to 300 nW of heat in the region just above the rf coil. Finally, a variation of the amount of the solid in the cell did not lead to any change in either the longitudinal or the transverse satellite.

It was possible to induce changes in the longitudinal and transverse satellites by changing the sample geometry and by varying the rate of compression or decompression of the Pomeranchuk cell. The geometry of the sample region was altered by placing different inserts in the space surrounded by the r.f. coil. Geometries employed included: (1) no insert, (2) parallel plates with 1 mm plate separation oriented vertically in the cell, allowing vertical flow to occur, (3) vertical 1 mm diameter cylindrical holes which also permitted vertical flow, (4) a horizontal cap which blocked the top of the coil region, and (5) a stack of horizontal plates 0.9 mm apart which were contained in the cylindrical tube comprising the r.f. coil form, with only miniscule gaps at the edges of the plates to provide pressure equilibrium. In this latter geometry the longitudinal and transverse satellite lines were unobservable for low compression rates, and only became evident when the pressure in the cell changed rapidly, whereas in the first four geometries the longitudinal and transverse satellites were always present.

The presence of satellites appears to be related to flow in the cell. The horizontal parallel plate geometry severely restricted vertical flow while the other four geometries permitted more or less unrestricted vertical flow over distances much greater than 1 mm. Vertical supercurrents in a Pomeranchuk cell can be driven by the thermal gradient of 3μK/cm which is thought to be present in Pomeranchuk cells as a result of the hydrostatic pressure which leads to a vertical distribution of melting pressures in the cell.

Further evidence indicating that the satellite lines are affected by flow comes from the observation that the satellite lines increased and the main lines decreased in size during rapid compression or decompression (corresponding to rapid temperature changes ∿ 20μK/sec taking place in the cell). Under these conditions both transverse and longitudinal satellites could even be observed in the horizontal parallel plate geometry, despite the fact that they were unobservable in this geometry for slow compressions.

Short pulses of compression or decompression also produced temporary increases in the transverse and longitudinal NMR signal over a wide range of temperatures below the temperature corresponding to the satellite absorption line, provided that the sample remained in the A phase. Thus, as a result of transient pressure changes, the liquid responds to the r.f. for frequencies other than the equilibrium main line and satellite frequencies.

III. PULSED NMR EXPERIMENTS

Pulsed nuclear magnetic resonance experiments were performed in the cell shown in Fig. 5. This cell was cooled into the superfluid region by a separate Pomeranchuk cell thermally connected to the sample cell by copper wires and specially designed heat exchangers. By this indirect Pomeranchuk cooling technique[12] it was possible to study superfluid ^{3}He at arbitrary pressures below the melting pressure and at temperatures as low as 1.7 mK. The temperature was measured by means of a small NMR coil surrounding some finely divided platinum powder immersed in the ^{3}He sample contained in the cylindrical appendage shown in Fig. 5. The nuclear susceptibility and the spin-lattice relaxation time of the platinum as obtained by a pulsed NMR technique, were used to determine the temperature. The ^{3}He pulsed NMR measurements were performed in a rectangular slab shown in Fig. 5. The sample cavity in this slab had dimensions of 1 cm x 0.08 cm x 0.4 cm, and thus the cavity closely approximated a parallel plate geometry. An iron electromagnet external to the cryostat provided a horizontal steady magnetic field, whose direction could be varied by rotating the magnet about the vertical axis. Most of the pulsed NMR measurements were performed at a field of 284 gauss. The ^{3}He NMR coil was carefully designed to provide a uniform field over the sample region and to cut off sharply at the ends of the coil to prevent eddy current heating, by minimizing the r.f. field in the neighborhood of a sintered copper heat exchanger at the top of the cell. The magnitude of the r.f. field strength, H_1, was typically 2 gauss for most of these experiments.

Measurements of the frequency shifts were made by mixing the amplified r.f. NMR signal with the output of a stable frequency

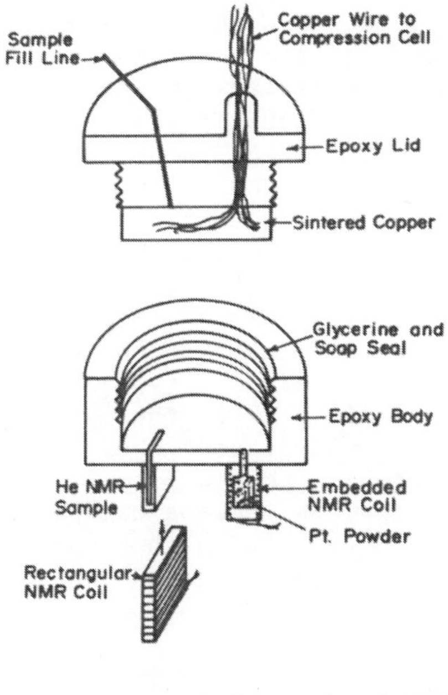

Figure 5. Experimental cell used for pulsed NMR measurements. The ^{3}He NMR tail and the Pt NMR tail are shown. The ^{3}He coil configuration is also shown in this figure. The external static field could be rotated in the plane perpendicular to the cell axis. Rotating the field produced no observable effect on the NMR response of superfluid ^{3}He-A.

synthesizer set at a frequency 1 kHz below the Larmor frequency. The resulting audio frequency signal was stored in a transient recorder and then redisplayed on a computer graphics terminal. The beat frequency could be measured to within $\pm$ 20 Hz.

Measurements of the frequency shift vs. tip angle were performed in these experiments at pressures of 23 bar and 27 bar in superfluid ^{3}He-A. The $1/4 + 3/4 \cos \phi$ dependence of frequency shift on tip angle ϕ observed by Osheroff and Corruccini[9] on the melting curve was confirmed in these experiments. A typical set of data illustrating the agreement between theory and experiment is shown in Fig. 6. Equally good agreement was obtained for a variety of different temperatures at both 23 bar and 27 bar.

A new metastable mode of superfluid ^{3}He-A induced by large angle tipping pulses ($\phi \gtrsim 55°$) has been observed in these experiments. This mode is characterized by a considerably different response to small angle tipping pulses than that observed in ordinary superfluid ^{3}He A. Fig. 7 portrays a series of observations for which the square of the effective longitudinal frequency, defined by the equation $\Omega^2_{eff} = \omega^2_{observed} - \omega^2_o$, is plotted against $1 - T/T_c$, where the observed frequency is obtained from the free induction decay tail of a small (11.5°) tipping pulse. The data at 23 bar, shown in Fig. 7, were obtained while the sample was being cooled. The downward arrow at $T/T_c = .955$ corresponds to the time at which a 180° pulse was applied. At this point, Ω^2_{eff}, obtained from the free induction decay of small angle tipping pulses, changes from the square of the ordinary longitudinal frequency Ω^2_{main}, to a lower frequency given by $\Omega_{eff}/\Omega_{main} = 0.83 \pm 0.06$. As the sample is cooled further, this ratio maintains its new value, which should be compared with the corresponding ratio for the CW transverse satellite, $\Omega_{t.s.}/\Omega_{main} = 0.835$. The close agreement between these ratios suggests that the metastable mode and the transverse satellite may be related to one another. At $T/T_c = 0.94$, another large angle tipping pulse was applied, causing the small angle tip frequency to shift back to the upper curve of Fig. 7. The newly observed mode could be destroyed or created by the application of large angle tipping pulses ($\phi \gtrsim 55°$). A large number of similar observation of this phenomenon were made at a pressure of 23 bar, with steady field oriented either parallel or perpendicular to the slab face. The results did not depend on which of the two field orientations was chosen.

The stability of the new mode exhibited a temperature dependence. In the temperature region between 0.90 T_c and 0.99 T_c, the mode was quite stable, usually persisting for five minutes or more. In this temperature range a large angle pulse ($\phi \gtrsim 55°$) was generally required to destroy the metastable mode. Below 0.9 T_c the frequency relaxed to the ordinary ^{3}He-A transverse frequency in several seconds, indicating a much shorter lifetime for the meta-

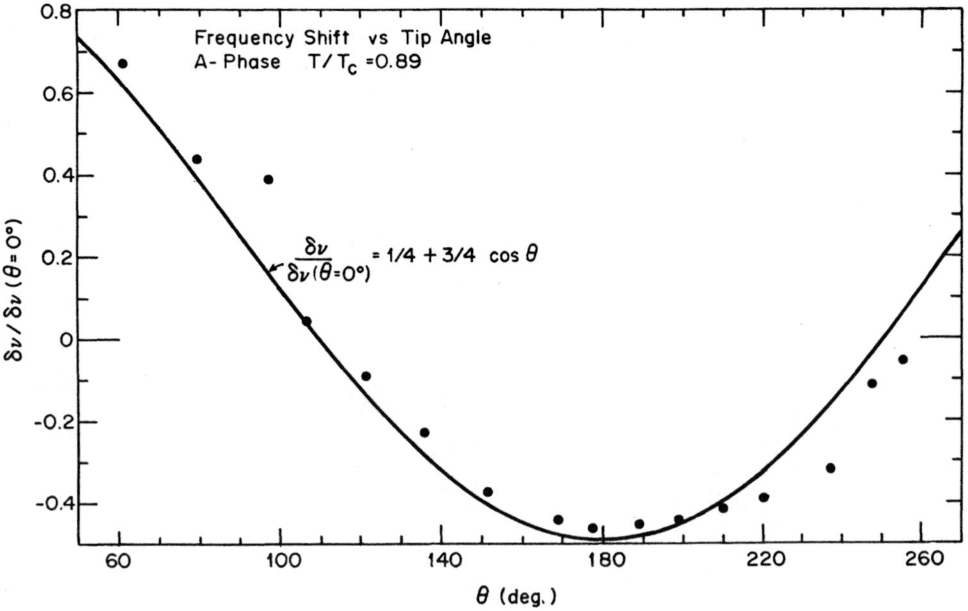

Figure 6. Spin precession frequency vs. tipping angle. The figure shows the change in the superfluid ^{3}He A transverse frequency following various initial tipping angles. $\delta\nu(0) = \nu - \gamma H_0/2\pi$ is the transverse shift for CW resonance or very small tipping angles. The solid curve is the theoretical prediction.

stable mode at lower temperatures. In the immediate vicinity of T_c, the metastable mode was not observed.

In addition to exhibiting different frequency shift behavior, the free induction decay tail following a small tipping pulse was usually shorter for the metastable mode than for the ordinary superfluid ^{3}He-A signal. A comparison of the two decay tails is shown in Fig. 8. Both signals have the same initial amplitude, indicating that the whole liquid seems to be responding at the metastable frequency. This is in marked contrast to the CW case where only a small fraction of the sample appears to respond at the transverse satellite frequency (see Fig. 2). Furthermore, the free induction decay tail for the metastable mode does not exhibit beats as might be expected for the Fourier transform of the doubly peaked spectrum observed in the CW case.

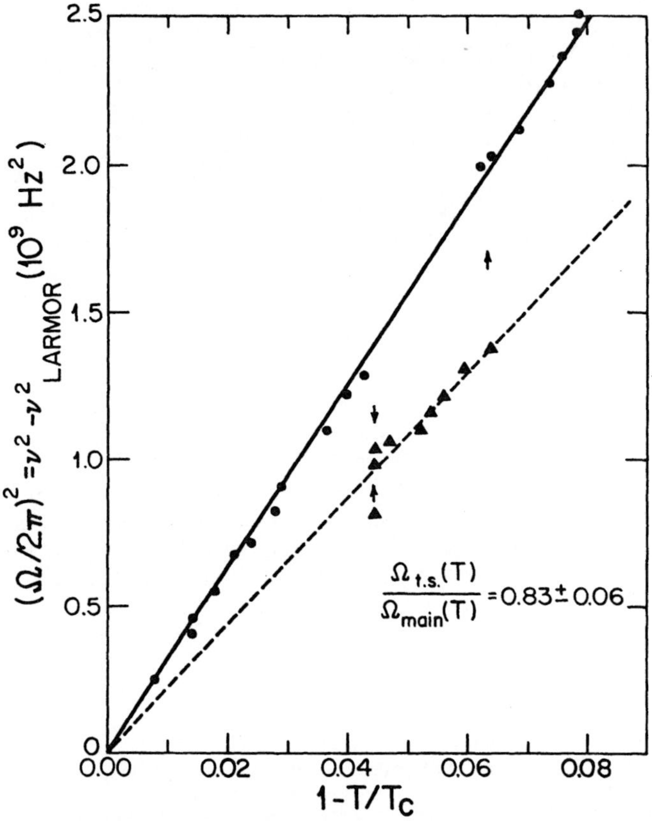

Figure 7. Effective longitudinal frequency squared versus reduced temperature. Application of a 180° pulse at $T/T_C = .955$ causes the frequency to shift to the lower curve. At a later time, a 90° pulse returns the system to the upper curve.

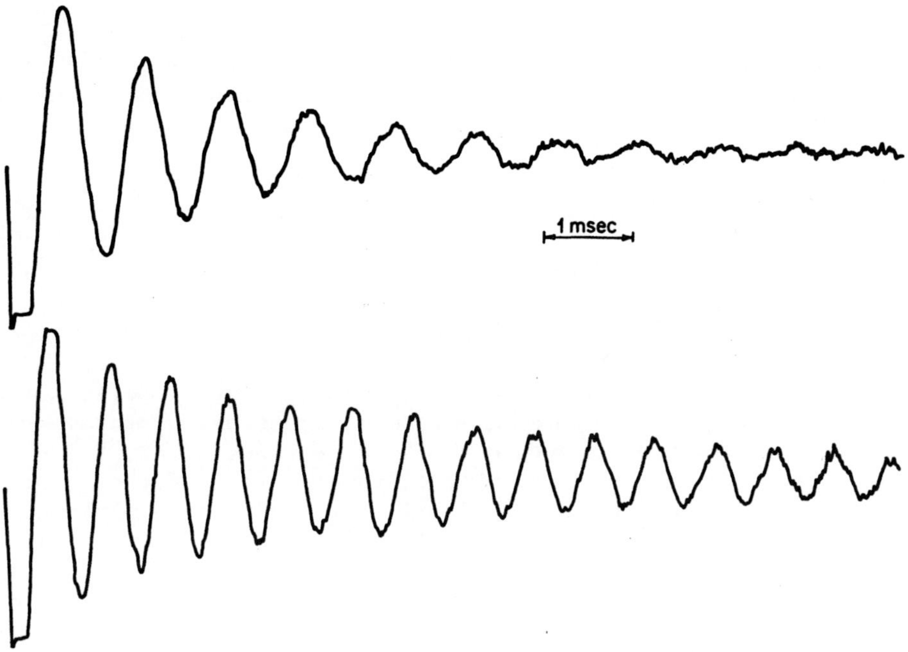

Figure 8. Traces of the free induction decay signals of two 11.5° tipping pulses at $T/T_c = .93$. The upper trace corresponds to the metastable mode. By applying a 90° pulse and a subsequent 11.5° pulse, we obtain the lower trace, characteristic of the usual A phase signal. Oscillations correspond to the beat frequency between the actual NMR signal and a local oscillator set at a frequency 1 kHz below the Larmor frequency. Note the shorter decay time and smaller frequency shift of the metastable mode (upper trace).

IV. CONCLUSION

We have not yet found an explanation for the phenomena observed in these experiments. It is possible that large tipping pulses or flow can introduce in superfluid ^{3}He-A distortions of textures in which both $\hat{d}$ and $\hat{l}$ are non-uniform and are twisted with respect to one another. An example of such a structure is the magnetic soliton introduced by Maki[12] who suggests that the frequency shift should be altered in the region of a soliton. The twisting of $\hat{d}$ relative to $\hat{l}$ should give rise to a change in the dipolar torque ($\propto (\hat{d}\cdot\hat{l})(\hat{d}\times\hat{l})$) and hence a change in the frequency shift. Large r.f. field gradients such as those found at the ends of our coil favor the formation of solitons during a large angle tipping pulse. It is possible that other macroscopic structures induced by flow or by large angle tipping pulses might also lead to similar effects.

Finally, a simple method for flow induced alignment of $\hat{l}$ in a magnetic field can be made to yield a double peaked CW NMR spectrum but one which is not in quantitative agreement with the data. Experiments are presently underway to further characterize these phenomena.

ACKNOWLEDGEMENTS

We express our deep gratitude to R. C. Richardson who introcued us to pulsed NMR and provided much valuable advice. In addition, we thank H. M. Bozler, W. J. Gully and E. K. Zeise for their participation in the early phases of these experiments. We are also grateful to D. D. Osheroff, L. R. Corruccini, W. F. Brinkman and K. Maki for discussing various aspects of their work with us. Finally we thank N. D. Mermin for patiently explaining and clarifying a number of theoretical points.

References

(1) D. D. Osheroff, W. J. Gully, R. C. Richardson, and D. M. Lee, Phys. Rev. Lett. 29, 920 (1972).

(2) A. J. Leggett, Phys. Rev. Lett. 29, 1227 (1972).

(3) A. J. Leggett, Ann. Phys. (NY) 85, 11 (1974).

(4) D. D. Osheroff and W. F. Brinkman, Phys. Rev. Lett. 32, 584 (1974).

(5) H. M. Bozler, M. E. R. Bernier, W. J. Gully, R. C. Richardson, and D. M. Lee, Phys. Rev. Lett. 32, 875 (1974).

(6) R. A. Webb, R. L. Kleinberg, and J. C. Wheatley, Phys. Rev. Lett. 33, 145 (1974).

(7) J. C. Wheatley, Rev. Mod. Phys. 47, 415 (1975).

(8) W. F. Brinkman and H. Smith, Phys. Lett. 51A, 449 (1975).

(9) D. D. Osheroff and L. R. Corruccini, Phys. Lett. 51A, 447 (1975).

(10) O. Avenel, M. E. Bernier, E. J. Varoquaux, and C. Vibet, "Proc. Fourteenth Int. Conf. Low Temp. Physics", edited by M. Krusius and M. Vuorio (North Holland, Amsterdam 1975), Vol. 5, p. 429.

(11) C. M. Gould and D. M. Lee, Phys. Rev. Lett. 37, 1223 (1976). For a detailed discussion of some of the experimental methods, see also W. J. Gully, C. M. Gould, R. C. Richardson, and D. M. Lee, J. Low Temp. Phys. 24, 563 (1976).

(12) E. N. Smith, H. M. Bozler, W. S. Truscott, R. C. Richardson and D. M. Lee, "Proc. Fourteenth Int. Conf. Low Temp. Physics", edited by M. Krusius and M. Vuorio (North Holland, Amsterdam, 1975), Vol. 4, p. 9.

(13) K. Maki, Proc. of the Sussex Symposium, 1976 (to be published in Physica).

PULSED NMR FREQUENCY SHIFT IN SUPERFLUID ^{3}He-B*

C. M. Gould

Laboratory of Atomic and Solid State Physics and the Materials Science Center - Cornell University
Ithaca, New York 14853

Abstract: A significant discrepancy exists between theory and measurements of the magnetization precession frequency following large angle tipping pulses in superfluid ^{3}He-B. Here it is shown that the resolution of this discrepancy lies in a more complete treatment of the equations of spin dynamics.

Immediately following the discovery of the superfluid phases of ^{3}He, Leggett[1] developed a theory of spin dynamics which has provided a wealth of information concerning these phases. Brinkman and Smith[2] considered the specific problem of pulsed NMR experiments in the B phase. Osheroff and Corruccini[3] performed these experiments finding excellent agreement with the prediction for tipping angles less than about 140°, but significant deviations at higher angles. The present work proposes a resolution of this discrepancy.

The order parameter of superfluid ^{3}He-B can be conveniently parameterized by a unit vector $\underset{\sim}{n}$ and an associated rotation angle ρ. Specializing Leggett's theory to the B phase, Brinkman[4] first derived the equations of spin dynamics under the above parameterization in dimensionless form as

$$\frac{d\underset{\sim}{S}}{dt} = \omega \underset{\sim}{S} \times \underset{\sim}{H} + \underset{\sim}{n} \frac{4}{15} \frac{\Omega_B^2}{\omega} \sin \rho \, (1 + 4\cos \rho) \tag{1}$$

*Work supported by the National Science Foundation through Grant DMR76-21669 and DMR76-01281 (Materials Science Center, Technical Report #2774).

$$\frac{d\rho}{dt} = \omega\, \underset{\sim}{n} \cdot (\underset{\sim}{S} - \underset{\sim}{H}) \tag{2}$$

$$\frac{d\underset{\sim}{n}}{dt} = -1/2\ \omega\ \underset{\sim}{n} \times (\underset{\sim}{S} - \underset{\sim}{H}) - 1/2\ \omega \frac{\sin\rho}{1-\cos\rho}\left\{\underset{\sim}{n}[\underset{\sim}{n}\cdot(\underset{\sim}{S}-\underset{\sim}{H})]-(\underset{\sim}{S}-\underset{\sim}{H})\right\} \tag{3}$$

where $\omega = \gamma H_o$, Ω_B is the longitudinal resonance frequency, and $\underset{\sim}{H}$ includes both the static field $\underset{\sim}{H}_o$ as well as any rf field $\underset{\sim}{H}_1$ applied to the fluid. Relaxation effects are ignored here.

The simplest pulsed NMR experiment which can be done is to apply a short intense rf pulse at the Larmor frequency at right angles to $\underset{\sim}{H}_o$, of amplitude H_1, and observe the precession of the magnetization following the pulse. Brinkman and Smith[2] concluded that for tipping angles Φ less than the equilibrium rotation angle $\rho_o = \cos^{-1}(-1/4) \simeq 104°$ the resulting precession should be exactly as in the normal Fermi liquid. However, for $\Phi \gtrsim 104°$ they predicted that the magnetization precession frequency would be greater than the Larmor frequency by an amount

$$\frac{\delta\omega}{\omega} = -\frac{4}{15}\left(\frac{\Omega_B}{\omega}\right)^2 (1 + 4\cos\Phi)\ . \tag{4}$$

This is shown as the lighter line in the figure. As mentioned above, Osheroff and Corruccini[3] found for $\Phi \gtrsim 140°$ significant disagreement with this prediction. Further experiments by Corruccini and Osheroff[5] as well as experiments performed in smaller fields at Cornell find similar deviations from the prediction.

The central result of this work is an explicit calculation demonstrating that at least one source of the discrepancy lies in the connection between theory and experiment, namely, the evaluation of the quantity Φ. In the theoretical treatment Φ is simply identified as the angle the magnetization forms with the $\underset{\sim}{H}_o$ field. The experimentalist, however, only controls the amplitude of the rf field H_1 and the length of the pulse, t. (Of course, here I am ignoring complications arising from non-ideal H_1 field profiles.) In the normal liquid this is no problem since then $\Phi = \gamma H_1 t$. The B phase presents the problem that for convenient field strengths this simple constitutive relation does not hold.

To illustrate this, consider the limiting case $H_1 \gg \Omega_B^2/\gamma^2 H_o$. In this limit we simply ignore the latter term in Eq. (1). Despite the fact that the remaining set of equations is desperately nonlinear, there is an analytic solution describing the motion of $\underset{\sim}{S}$, ρ, and $\underset{\sim}{n}$ during the tipping pulse. As usual, it is easiest to describe the motion in a frame rotating at the Larmor frequency. For $\underset{\sim}{H}_1$ along the y-axis, the magnetization precesses in the x-z plane at a constant rate so that $\Phi = \gamma H_1 t$ (only) in this case. The $\underset{\sim}{n}$ vector precesses in a plane whose normal forms an angle of about 38° with the $\underset{\sim}{H}_1$ field in the x-y plane. The precession is anharmonic but periodic, the average rate of precession being one half that of the magnetization's rate. The rotation angle also varies irregularly but periodically between 104° and 256°. When the magnetization has gone 360°, the $\underset{\sim}{n}$ vector is along $-\underset{\sim}{z}$, and $\rho \cong 256°$.

Unfortunately, the limit $H_1 >> \Omega_B^2/\gamma^2 H_o$ is not easily reached in practice. In order to be able to resolve the frequency shift accurately, the right hand side of the inequality must not be made too small. The H_1 field cannot be made too large either owing to heating effects. Hence we are left with the full set of equations which have so far defied a similar analytic solution. Qualitatively, however, we can argue that at the beginning of the tipping pulse the motion must be similar since initially $1 + 4 \cos \rho = 0$. As the tipping continues, ρ gets larger, exerting a small but continuously growing push on the magnetization, forcing it out of the x-z plane. By this means the magnetization is never tipped all the way to 180°. Instead, it moves to the side at a maximum tipped angle which is a complicated function of H_1, H_o, and Ω_B.

To study the motion quantitatively, I have numerically integrated Eqs. (1) - (3) to compare directly with experimental results. Tipping and precession are studied separately. Generally during a tipping pulse the magnetization stays fairly close to the x-z plane until it is tipped about 104° at which point it begins moving more to the side than down. After the rf field is removed the magnetization bobs up and down and rotates at a rate which itself varies irregularly. Since in order to resolve a small frequency shift a long time interval must pass, the experiment measures only the average rate of this complicated rotation. Results for the precession frequencies immediately following the rf pulse are compared with experiments done at Bell and Cornell in Fig. 1. Qualitative features such as the abrupt change in frequency in the upper graph and the reduced height and width of the lower one appear in both the calculated and experimental results, though there remains a quantitative disagreement.

Recently Fomin[6] has developed an approximate solution to this problem which describes the average motion of the magnetization quite accurately. All qualitative results developed in this work and by Fomin are in agreement. A complete description of the present work including relaxation effects and a comparison with Fomin's work will be presented elsewhere.

I thank D. D. Osheroff and L. R. Corruccini for useful discussions concerning their experiments, H. Smith for a discussion of his theory, R. Giannetta, E. N. Smith, and E. K. Zeise for permission to cite the results of their experiments prior to publication, and

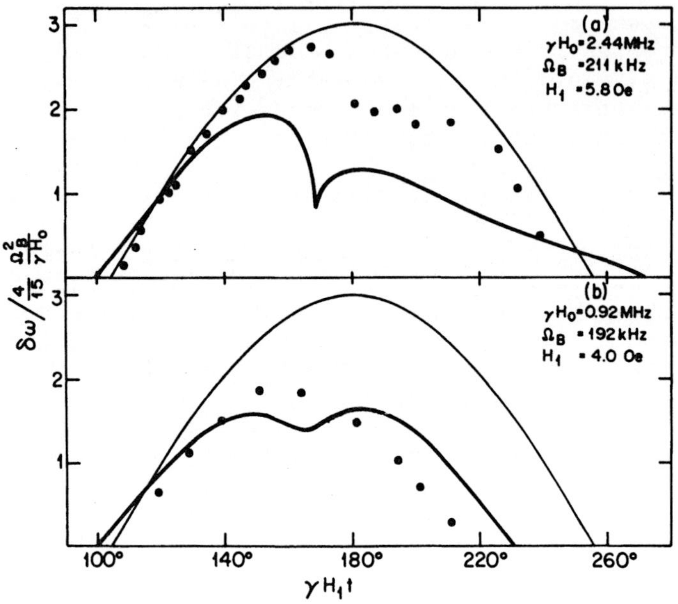

Figure 1. Initial free precession frequencies. The heavier line shows the results of numerical solutions for parameters appropriate to experiments performed at (a) Bell and (b) Cornell. The lighter line is the function $-(1 + 4 \cos \Phi)$ from Eq. (4) assuming $\Phi = \gamma H_1 t$.

especially D. M. Lee for his support and encouragement on this project.

References

(1) A. J. Leggett, Ann. Phys. 85, 11 (1974).

(2) W. F. Brinkman and H. Smith, Phys. Lett. 53A, 43 (1975).

(3) D. D. Osheroff and L. R. Corruccini, Proc. LT14, V. 1, p. 100 (1975).

(4) W. F. Brinkman, Phys. Lett. 49A, 411 (1974).

(5) L. R. Corruccini and D. D. Osheroff, Phys. Rev. B (to be published).

(6) I. A. Fomin, ZhETF 71, 791 (1976) (in Russian).

NMR IN FLOWING SUPERFLUID ^{3}He IN CONFINED GEOMETRIES

E. B. Flint, E. D. Adams and R. M. Mueller

Department of Physics - University of Florida
Gainesville, Florida 32611

Abstract: The transverse NMR of superfluid ^{3}He A and B has been studied as a function of flow velocity in two different channel geometries. In the A phase, for the parallel plate geometry, a fraction or all of the signal was found to shift to lower frequencies with flow. In addition, the NMR signal of the flowing liquid developed a time dependent structure under certain conditions. The signal of the static liquid confined between parallel plates was found to split for a field angle of 45° relative to the plates and for temperatures near the A-B transition. None of the above effects was observed in the B phase.

I. INTRODUCTION

This paper discusses measurements of the transverse NMR in ^{3}He A and B in the presence of flow. These studies were motivated by works of A. L. Fetter[1] and S. Takagi[2] which indicate that one expects to see a shift in the frequency of the transverse absorption in the presence of flow under certain conditions.

We describe first the flow apparatus and the two flow channel geometries where the cw NMR absorption took place. Next, we discuss three flow induced features of the transverse NMR that have been observed in the A phase liquid. These include the development of structure in the signal which we shall refer to as satellite peaks, height reductions of the signal, and a shift in the frequency of a fraction of the NMR absorption in the slab geometry. Finally, we discuss measurements which show a splitting of the static liquid signal in the A phase near the A-B transition under certain conditions.

II. APPARATUS

The flow was produced in a channel which connected two Pomeranchuk cells by compressing one cell while maintaining constant ^3He pressure in the other cell. The pressure regulation was accomplished through the use of a feedback system which sensed the ^3He pressure in the regulated cell and used the error signal to control the ^4He pressure used to compress this cell. The details of the Pomeranchuk cells have been described by Kummer, Mueller, and Adams,[3] and a more complete description of the technique used to generate the flow can be found in a paper by Mueller, Flint, and Adams.[4] The pressure regulation system was able to maintain a temperature stability of 1μK in the regulated cell during the largest flows. Continuous flow rates between the two cells were limited to 5.3×10^{-5} cm^3/sec by the room temperature plumbing; however, for short periods, flow rates a factor of 5 larger could be produced.

The experiments that are to be described below used two different channel geometries. The first was a 1.5 mm diameter, horizontal, cylindrical channel, and the second consisted of a stack of vertical glass plates which formed ten parallel, slab type flow geometries with a wall spacing of 135μm. The flow through the parallel plates was also horizontal. The $\vec{H}_o$ field was rotatable through an angle ψ relative to the flow in the horizontal plane, and the H_1 field was vertical. The transverse NMR circuitry was a frequency swept Rollin circuit. The absorption was observed in a static field of 150 gauss for the cylindrical channel, and 267 gauss for the parallel plate geometry.

III. RESULTS

Satellite peaks and structure - An effect which was observed frequently in the cylindrical geometry for $\psi = 45°$ and occasionally in the parallel plate geometry for $\psi = 45°$ and $60°$ is shown in Fig. 1a. Each trace is a frequency sweep through the resonance in the liquid. The uppermost sweep was taken without flow in the channel and lower sweeps were taken at successively larger flow rates. The absorption at the low frequency end of the sweep was due to the small amounts of solid that were present in the flow channel. Relatively small flow velocities often produced structure in the main peak and small satellite peaks which were shifted to lower frequencies. Increasing the flow increased the number of satellites and the size of the shift.

These satellite peaks were not constant in time. In Fig. 1b the first trace was a sweep through resonance with flow in the channel and the second and third traces were 3 second time sweeps

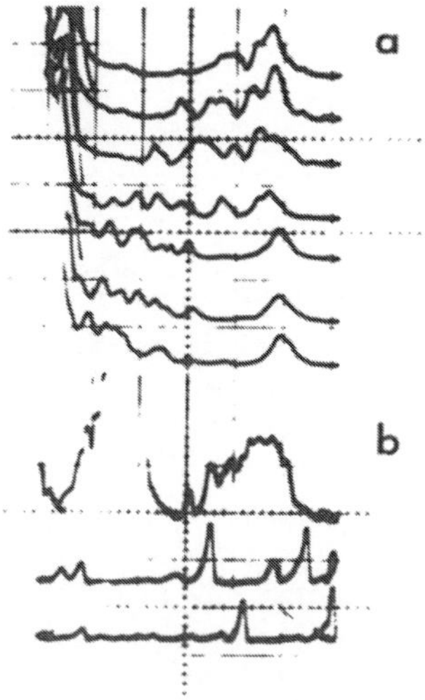

Figure 1. Flow induced structure on the NMR signal. (a) $\psi = 60°$ and T = 2.58 mK; for the upper trace $v_s \approx 0$ and increases to $v_s \approx 0.33$ mm/sec for the lower trace. (b) The upper trace is a frequency sweep through the resonant frequency and the lower traces are 3 sec observations at the center frequency of the upper trace showing the time dependence of the flow induced structure.

with the frequency held at the center frequency of the first trace. This demonstrates that the satellite peaks are a time varying phenomena. In the absence of flow, such satellites do not appear. The same procedure was performed for a frequency greater than the frequency of the static liquid absorption and no satellites were observed. It should be noted that the calculations of Fetter[1] and Takagi[2] indicate a shift towards lower frequencies with flow. These satellite peaks may therefore be due to such a texture related effect.

<u>Signal height reductions due to flow</u> - A second effect that we have observed for both channel geometries and for the various field angles studied was that the peak height was reduced in the presence of flow. An example of this is shown in Fig. 2a where the sweeps through the absorption frequency were at successively higher flows and have been displaced horizontally. These data were taken in the cylindrical channel. Several step-like reductions in signal height are seen here with a final decrease in height after which there is relatively little change in the signal with further increases in the flow velocity. By assuming that the normal fluid is at rest,[5] and by assuming a specific form for the superfluid fraction, it is possible to determine the superfluid

velocities, v_s, at which the steps occur. The functional form that we have chosen to use is $\rho_s/\rho = 0.6(1-T/T_c)$.[6,7] At the lower velocities it was difficult to determine the velocity at which a given step occurs due to signal instabilities such as those produced by the structure already discussed. The velocity at which the final step occurs, however, has been obtained as a function of temperature. Fig. 2b shows the superfluid velocities as a function of temperature for field angles of $\psi = 90°$ and $45°$ again for data taken in the cylindrical channel. It can be seen that the temperature range spanned by this plot may be divided into three regions in which the critical velocity is relatively constant. The critical velocity was highest near T_c, $v_s = 0.64$ mm/sec, and was only $v_s = 0.14$ mm/sec at the lower temperatures. Takagi[2] has calculated a characteristic velocity of ~ 0.1 cm/sec for the flow induced frequency shift of the NMR absorption near T_c which is near the corresponding velocity obtained in this work. In addition, these critical velocities are similar to those found by Johnson et al.[8] and Wheatley[6] in their heat flow measurements, and may be related.

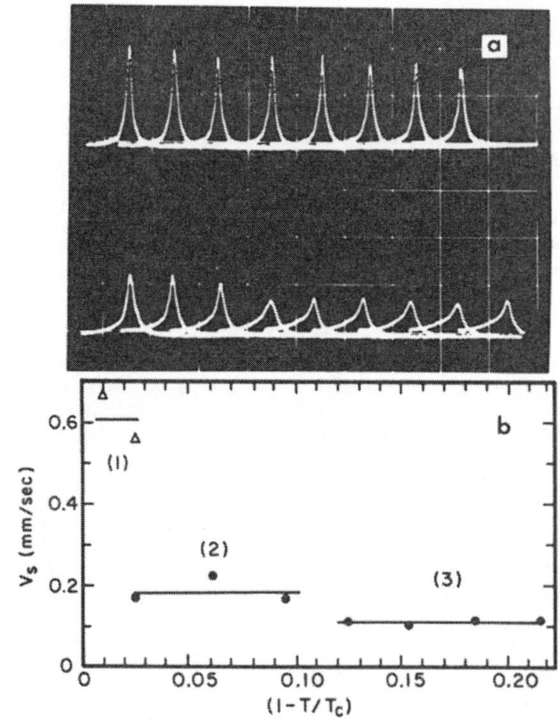

Figure 2. Signal height reductions. (a) Successive sweeps displaced from upper left to right with flow increasing in approximately equal increments. Steps occur at arrows. (b) The velocity at which the final step occurs versus $(1-T/T_c)$. Δ, $\psi = 45°$; 0, $\psi = 90°$.

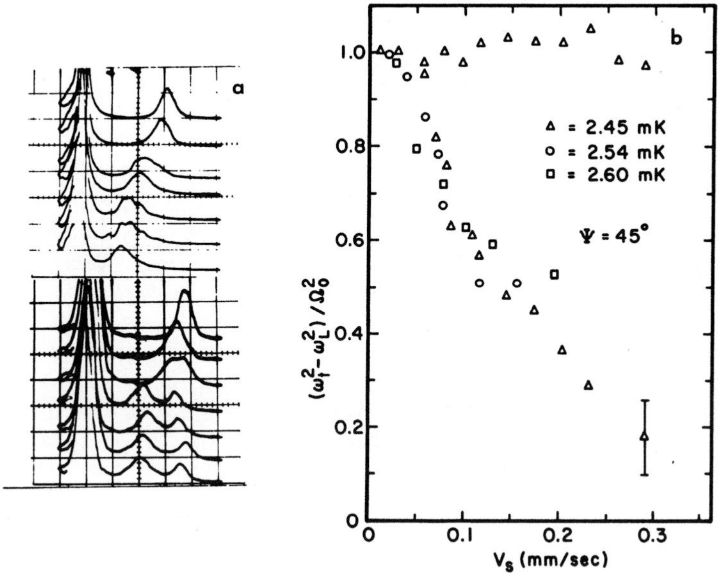

Figure 3. Splitting and shift of the signal for flow between parallel plates. (a) The upper set of traces were for $\psi = 45°$, T = 2.60 mK and the lower set were for $\psi = 45°$, T = 2.25 mK. The liquid signal shifts towards the Larmor frequency with increasing v_s. (b) The relative frequency shift versus v_s for $\psi = 45°$.

Splitting and shifting of the signal - A third effect which was observed in the parallel plate geometry, and is probably related to the two already discussed is shown in Fig. 3a. The sweeps through the resonant frequency with $\psi = 45°$ were taken at successively higher flow velocities and were displaced downward. At 2.6 mK the entire signal was shifted towards the lower Larmor frequency in the presence of flow, but when the system was cooled the signal split into two peaks with increasing flow velocity; one of which shifted to lower frequency and another which was unshifted. Fig. 3b shows this frequency shift as a function of velocity for several temperatures at $\psi = 45°$. The shift has been plotted as the difference in the squares of the liquid signal and the Larmor frequencies normalized by the square of the parallel ringing frequency, Ω_0. It can be seen that the shift for various temperatures falls along the same curve. Extrapolating the data, the intercept with the v_s axis is at $v_s \sim 0.4$ mm/sec. This intercept appears to change with angle; however, it is not possible to make a precise statement. Two qualitative remarks regarding these data should be made. The first is that the splitting of the signal into two distinct peaks takes place at higher temperatures for field angles of $\psi = 22.5°$ and $45°$ relative to the flow and at lower temperatures for field angles of $\psi = 60°$ and $90°$, so that the splitting appears to depend on the angle of the H_o field relative to the flow. The second is that the portion of the signal which shifts to lower frequency with flow was broader for the smaller field angles and for the lower temperatures.

Several comparisons between these data and the flow induced frequency shifts predicted by Takagi[2] can be made. First, the observed shifts are towards lower frequency as expected. In addition, this shift when treated as the difference between the squares of the liquid frequency and the Larmor frequency normalized by the square of the parallel ringing frequency, appears to be nearly independent of temperature again as expected. There is not, however, quantitative agreement. The observed shifts for each field angle and flow velocity are larger than the calculated shifts. For example, at 22.5° a shift back to the Larmor frequency was observed for $v_s \sim 0.7$ mm/sec, but it is expected[2] that a velocity in excess of 1 mm/sec is required to produce this shift. There are a number of possible explanations for this discrepancy. One is that solid formed in the flow channel during the compression may constrict the channel causing the flow velocity to be higher than calculated from the volume flow rate and the cross-sectional area of the channel. A second possible explanation is that treating flow between plates spaced by 135μm as bulk flow may not be justified even though the effect of the walls is expected to extend only about 10μm into the liquid. A third possibility is that the characteristic velocity calculated by Takagi[2] for temperatures near T_c should not be used at lower temperatures where much of the data reported here were taken. It should also be noted that the observed splitting of the signal with flow is not expected in bulk liquid. This again may indicate that the sample studied here cannot be treated as a bulk sample. In spite of these difficulties, we believe that this effect, like the others, is probably a flow induced texture phenomenon.

The B phase signal was also examined in the presence of flow in the parallel plate geometry; however, no effects, such as those described for the A phase, were seen. In fact, flow rates a factor of five greater than any used in the A phase produced no noticeable effect at all on the NMR signal in the B phase. Higher characteristic velocities in the B phase may account for the absence of changes in the NMR signal in this experiment.

In addition to the flow induced phenomena described, one rather interesting effect in the static liquid was also observed in the parallel plate geometry. This was observed with a field angle of $\psi \sim 45°$ with respect to the plates. Starting at about 2.4 mK, the static liquid signal broadened with decreasing temperature, with two distinct peaks being in evidence at 2.26 mK. The spreading of the peaks continued with decreasing temperature so that the center frequencies were separated by about 1.7 KHz in the supercooled A phase at T = 1.98 mK. Cooling into the B phase restored the signal to a single peak. This splitting was not observed at field angles near 90° or 0° relative to the plates.

References

(1) A. L. Fetter, Phys. Lett. 54A, 63 (1975).

(2) S. Takagi, J. Phys. C. 8, 1507 (1975).

(3) R. B. Kummer and E. D. Adams, to be published in J. Low Temp. Phys.

(4) R. M. Mueller, E. B. Flint and E. D. Adams, Phys. Rev. Lett. 36, 1460 (1976).

(5) The time constant for the normal fluid to come to rest was calculated to be ∿ 100 msec. for the cylindrical channel and ∿ 0.1 msec. for the parallel plate geometry.

(6) J. C. Wheately, Rev. Mod. Phys. 47, 415 (1975), and references cited therein.

(7) H. Kojima, D. N. Paulson and J. C. Wheatley, Phys. Rev. Lett. 32, 141 (1974), and J. Low Temp. Phys. 21, 283 (1975); A. W. Yanof and J. D. Reppy, Phys. Rev. Lett. 33, 631 (1974); T. A. Alvesalo, H. K. Collan, M. T. Loponen, O. V. Lounasmaa and M. C. Veuro, J. Low Temp. Phys. 19, 1 (1975); D. D. Osheroff and L. R. Corruccini, Phys. Rev. Lett. 34, 695 (1975).

(8) R. T. Johnson, R. L. Kleinburg, R. A. Webb, and J. C. Wheatley, J. Low Temp. Phys. 18, 501 (1975).

SOME UNUSUAL CHARACTERISTICS OF ^{3}He-B

D. D. Osheroff

Bell Laboratories
Murray Hill, New Jersey 07974

Abstract: This paper analyzes three unusual characteristics of ^{3}He-B which have been observed in continuous wave nuclear magnetic resonance studies over the past three years. They include evidence for the existence of textural singularities in ^{3}He-B, unusual spin wave related behavior of B phase samples in highly homogeneous magnetic fields, and power dependent distortions of the B phase absorption lineshape.

INTRODUCTION

Since the discovery of superfluidity in liquid ^{3}He, the richness of the spin dynamical properties of both phases has been exploited extensively. During that interval, many seemingly inexplicable phenomena have been observed which have served to whet the curiosities of theorists and experimentalists alike before either being fully explained or reluctantly discarded in favor of more fruitful pursuits. Several observations have fallen into a third category: Those phenomena which we can almost understand, but about which lingering doubts persist. In the remainder of this article I attempt to describe and explain partially three such observations.

A. Textural Singularities in ^{3}He-B

In the original study of the NMR properties of superfluid ^{3}He at Cornell University,[1] and again in work at Bell Laboratories,[2] observations were reported which indicated that stable, localized domain effects could exist in bulk samples of ^{3}He-B contained within cylindrical sample chambers several millimeters in diameter and several centimeters long. Anderson and Brinkman[3] suggested that in the later experiment, where a static magnetic field, $\vec{H}_0$, was oriented vertically along the cylinder axis, the localized effect might be due to a point singularity in the B phase texture. Either a positive or negative monopole[4] could explain the observations. Such singularities would be stable provided that in the portion of the sample far from the chamber walls, $\hat{n}$ pointed parallel to $\vec{H}_0$ above the singularity and antiparallel to $\vec{H}_0$ below the singularity, or vice versa. In the Cornell experiments the static field and the cylinder axis were perpendicular to one another, and such point singularities would not be stable. Provided that $\hat{n}$ still changed its orientation from being parallel to $\vec{H}_0$ at one end of the cylinder to being antiparallel to $\vec{H}_0$ at the other end, however, a textural domain wall would develop in which $\hat{n}$ twisted by 180° over a relatively short distance along the cylinder axis in the Cornell configuration.

The spatial extent of both textural effects must be comparable to R_H, the field texture bending length. At melting pressures, $R_H \approx [36(1-T/T_c)/H_0]$ cm, when H_0 is expressed in Oersteds.[5,6] Since the healing of $\hat{n}$ back to its equilibrium orientation along $\vec{H}_0$ away from the disturbance is an exponential process, the region over which $\hat{n}$ is disturbed should be about $3R_H$ on either side of the singularity. The domain wall will, of course, have a comparable extent.

In both experiments, the effects were observed by transverse NMR absorption studies. A gradient in $\vec{H}_0$ was placed along the cylinder axis so that one could obtain information about the spatial distribution of the absorption signal by sweeping the frequency.[1,2] For the regions of the sample far from the misorienting textural disturbance, $\hat{n}$ is nearly parallel to $\vec{H}_0$, and the resonant frequency of each infinitesimal volume of sample, $V(\vec{r})$, is essentially $\gamma H_0(\vec{r})$. Near the disturbance, and in fields sufficiently large that the resonant frequency, ω_0, is much larger than the longitudinal resonant frequency, Ω_L, we have:

$$\omega_0 \approx \gamma H_0(\vec{r}) + \Omega_L^2/2\ \gamma H_0(\vec{r})\ \sin^2\chi$$

Here χ is the angle between $\hat{n}$ and $\vec{H}_0$, and γ is the gyromagnetic ratio. As we sweep the resonant frequency through the sample, from low to high frequency, we discover missing absorption at the frequency corresponding to the site of the disturbance. As we continue sweeping beyond this frequency, we will find additional absorption added to the normal B phase absorption as the sample near the textural disturbance comes into resonance along with other portions of the sample.

In Fig. 1 is shown a set of three NMR absorption profiles obtained in the Bell Labs experiment. The frequency is swept upward from the left, and the bottom of the NMR region corresponds to the right hand side. The dotted line was recorded while the liquid was in the A phase. Because of the large frequency shift of the A phase resonance, this profile shows only the solid ^{3}He absorption signal. The peak off to the extreme right is due to solid ^{3}He at least 1 cm below the bottom of the NMR region. After recording the A phase profile, the cell was cooled to $T/T_c = 0.671$, and the solid trace was recorded. Finally the cell was warmed just above T_{AB} and then cooled back to $T/T_c = 0.671$ to eliminate the singularity. The dashed trace which was then recorded shows a typical B phase absorption profile in which no singularity is present. By comparing the dahsed and solid curves, one can clearly see the additional absorption signal on the high frequency side of the singularity. The sharp peak at the right hand edge of the singularity comes from a sample region in which the gradient in χ just balances the gradient in $\vec{H}_0$ to give a uniform NMR freuqency over an extended region of the sample.

In Fig. 2 is shown a series of B phase absorption profiles like the solid trace in Fig. 1, taken upon warming from $T/T_c = 0.567$ to $T/T_c = 0.761$. They all show the same singularity, which seems to have moved downward in the cell between traces (a) and (b).

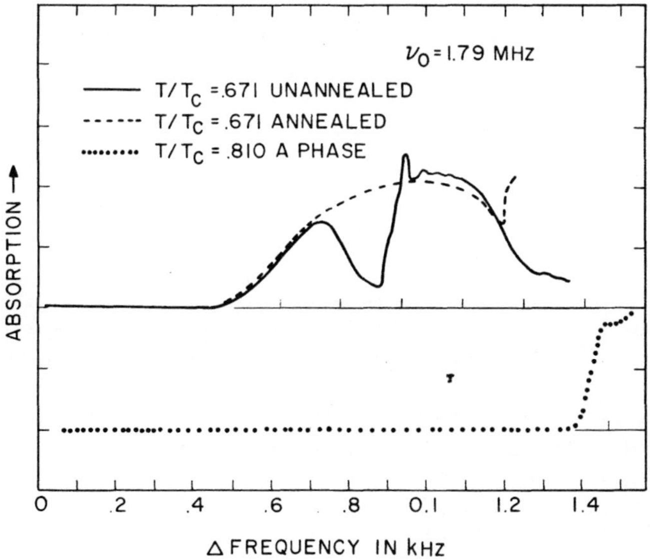

Figure 1. NMR absorption profiles showing a singularity in the B phase (solid trace). The dotted trace shows the solid distribution near the NMR region, and the dashed trace is a normal B phase profile with no singularity. The top is on the left and the bottom is on the right.

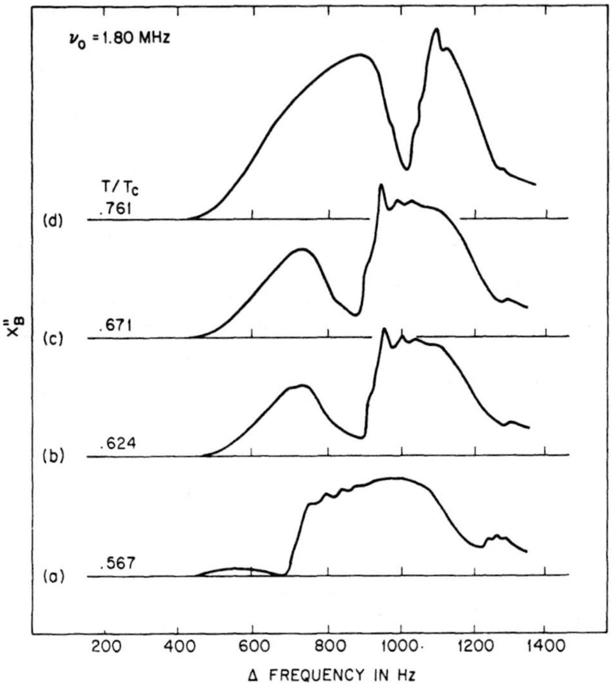

Figure 2. NMR absorption profiles similar to Fig. 1 (solid trace) showing a singularity at four different temperatures. Notice that as the temperature is raised, the width of the disturbance in the profile due to the singularity decreases. Each trace was taken at a different gain setting (y-axis).

As the sample is warmed, the width of the textural disturbance is clearly seen to decrease, as is expected from the temperature dependence of R_H. Since the width of the NMR region was about 1.2 cm, we can estimate that the characteristic width of the textural disturbance in trace (d), for example, was about 7.5 R_H.

Although at Cornell textural disturbances such as we have discussed were rare, at Bell they were observed frequently. It is interesting to speculate how $\hat{n}$ could reverse its direction across the sample chamber so easily. One possibility recently brought to light by M. C. Cross involves the boundary conditions on the anisotropy axes in both ^{3}He-A and ^{3}He-B at the A-B interface.[7] This interface moves across the sample during the A → B phase transition. Cross finds that the condition on $\hat{n}$ is that $|\hat{S}\cdot\overleftrightarrow{R}\cdot\hat{d}| = 1$, where $\hat{S}$ is the interface normal vector, $\hat{d}$ is the A phase spin anisotropy axis, and $\overleftrightarrow{R}$ is the B phase rotation matrix, which rotates the spin coordinates about $\hat{n}$ by $\cos^{-1}(-1/4) \approx 104^o$. Since $\vec{H}_0$ is in the vertical direction, $\hat{d}$ must be horizontal in the

A phase in the Bell experiments. As the interface moves upward (or downward) following B phase nucleation, it is likely that $\hat{S}$ will wobble about the vertical direction in some manner. If at some point $\hat{S}\cdot\hat{d} = 1/4$, which only requires that $\hat{S}$ be oriented 14° away from the vertical direction, then $\hat{n}$ will be horizontal. At this point it is easy for $\hat{n}$ to rotate back in the wrong direction as the interface continues to move upward through the sample, hence resulting in a reversal of $\hat{n}$.

B. Spin Wave Effects in Uniform Samples of ^{3}He-B

Recent experiments[6,8] at Bell laboratories have shown that textural spin wave modes can be excited in ^{3}He-B contained between closely spaced parallel surfaces using a uniform radio frequency field. The nature of these modes is well accounted for by existing theory.[9] Although these results represent the only instance so far where spin wave effects could be accounted for theoretically, other evidence of spin wave effects in ^{3}He-B have been observed which, although less well defined, are nonetheless quite interesting. One example of such an effect was reported by Webb _et al._,[10] who showed that a magnetic disturbance could propagate in ^{3}He-B over relatively large distances. In this section, I present another very different example of spin wave phenomena.

In a variety of high resolution NMR absorption studies at Bell Laboratories, it has been noted that a peculiar distortion of the B phase absorption lineshape may develop as the sample is cooled, provided the field gradient across the sample is less than several $x10^{-5}H_o$ and that the magnetic field is larger than about 500 Oe. These effects have been observed in cylindrical samples of various orientations with respect to $\vec{H}_0$ ranging from 3 mm to 6 mm in diameter. The distortions, which are manifested as a series of spikes which grow out of the absorption lineshape as the sample is cooled, are most pronounced at the lowest temperatures and in the highest magnetic fields.

In Fig. 3 are shown three absorption spectra which illustrate the changes mentioned above. The vertical scales are arbitrary, and different for each trace. Trace (a) shows the Fermi liquid lineshape in a 6 mm diameter sample at a field of 1.76 kOe just above T_c. Although the leading edge of the lineshape is made purposely sharp using externally applied field gradients, we have little control over the distribution of field gradients across the sample. The characteristic width of (a) is $\sim 2x10^{-6}\ \gamma H_0$, typical for this particular run. Upon cooling into the B phase, near to T_{AB} lineshapes similar to (a) were observed. Once the sample was cooled to $T/T_c \sim 0.39$, the lowest temperature obtainable in the compression apparatus, however, the lineshape distorted considerably. Trace (b) is the lineshape obtained at $T/T_c = 0.39$. Notice the

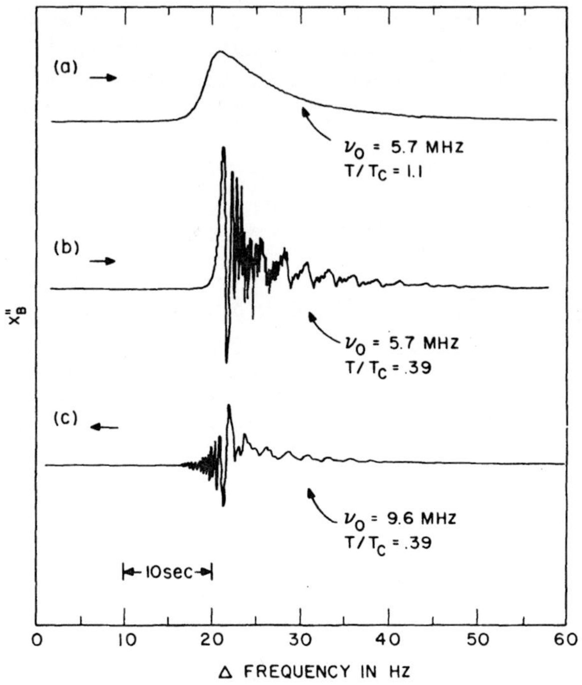

Figure 3. NMR absorption spectra showing spin wave effects in ^{3}He-B. The upper trace shows the Fermi liquid signal at 5.7 MHz. The middle trace shows a B phase spectrum at 5.7 MHz and T/T_c = 0.39, obtained by sweeping the frequency upward (left to right) at the rate of 1 Hz/sec. The lower trace shows a B phase spectrum at 9.6 MHz and T/T_c = 0.39 obtained by sweeping the frequency downward. Notice the free induction signal on the left hand edge of the absorption lineshape.

very uniform spacings between the sawtooth structure which has developed on the high frequency side of the line. In higher fields, as many as twelve such steps could be observed in the absorption spectra. The sharp hash on the left hand edge of (b) results mainly from spins which have been tipped by the rf field and continue to precess in phase for quite some time, beating against the local oscillator frequency. Even though we have swept through the line at the rate of 1 Hz/second, we have not swept slowly enough! Trace (c) is a similar absorption spectrum, but this time obtained by sweeping the frequency downward (arrow) in a field of 3.0 kOe. After we have finished sweeping through the line, we are left with a free induction decay signal caused by spins precessing together and beating against the local oscillator frequency. The characteristic dephasing (or decay) time for these spins is ~2 seconds, indicating that the resonance associated with this signal is probably between 0.2 Hz and 0.5 Hz broad.

We have studied the spacing of the sawtooth structure seen in trace (b) as a function of temperature. We find that from T/T_c = 0.39 to T/T_c = 0.64, the spacing of the peaks is proportional to the square of the spin wave velocity,[6,8] to within $\pm 4\%$. We are unable to make any meaningful statements concerning the field and/or field gradient dependence, however, because these two quantities are not independent variables in the experiments.

These spin wave effects are presumably closely related to the spin wave modes discussed recently by Maki and Tsuneto,[11] in which a magnetic field gradient is placed across a sample of ^{3}He-B in a parallel array and spin wave modes are excited with a uniform rf field. We cannot press this comparison too far, however, because: (a) Our spacings all appear to be uniform, unlike the Maki-Tsuneto results, and (b) we doubt any surface is acting to confine the spinwaves in our experiment. Even if a surface does play some part, it certainly is not a flat surface.

Two additional comments concerning these spin wave effects must be made: (A) The time scale of the free induction signal decay shown in trace (c) is long compared to the spin-lattice relaxation times, T_1, measured in ^{3}He-B by Corruccini and Osheroff.[12] At $T/T_c \sim 0.39$, Corruccini and Osheroff would claim that $T_1 \sim 100$ ms in ^{3}He-B, and that the spin recovery along the static field should be exponential. If that were indeed the case, all but 2×10^{-9} of the net magnetization should have recovered along $\vec{H}_0$ after two seconds, when we still observe a significant free induction signal. Even allowing for the quadratic relationship between M_x and M_z ($\vec{H}_0 || \hat{z}$), the measured T_1 appears to predict a much more rapid decay than is observed. Either we are seeing a signal precessing in the x-y plane long after the magnetization has totally recovered along the $\hat{z}$ direction, which seems unlikely, or the spin-lattice relaxation process is very different for uniformly precessing spins in a lowest spin wave mode (which we presumably have here) than it is in the limit in which it was measured by Corruccini and Osheroff.

(B) There appears to be absolutely no reason why the spin wave behavior observed in trace (b) should not also be observed in ^{3}He-A, at least to some degree. The only complication in ^{3}He-A is that the spin wave velocity is highly anisotropic, and that uniform d textures ($\hat{d}$ being the spin anisotropy axis in ^{3}He-A) may not exist over large regions of the sample as uniform $\hat{n}$ textures do in ^{3}He-B. Still, we find it interesting that absolutely no such lineshape distortions are observed in ^{3}He-A in identical fields, field gradients and temperatures. This absence may be an indication that spin waves are much more highly damped modes in ^{3}He-A than in ^{3}He-B.

C. Power Dependent Absorption Spectra in ^{3}He-B

In the NMR absorption studies in the 3 mm cylincrical sample by Corruccini and Osheroff[12] it was observed that, provided the absorption linewidth was less than several $x10^{-5}$ H_0, a very pronounced asymmetry in absorption spectra was obtained by sweeping the frequency in opposite directions when the rf level was sufficiently high. This effect occurred even in relatively low fields and high temperatures, where the spiked structure just discussed would not be observed. In Fig. 4, the upper trace shows a normal B phase absorption spectrum which would reproduce itself no matter which direction the frequency was swept. The spikes shown are just the onset of the phenomenon discussed in the previous section. It was obtained using an rf field (rotating component), h_1, equal to $2x10^{-5}$ Oe. When the rf field was increased to $4x10^{-4}$ Oe the spectra shown in the lower trace were obtained. With the exception of the oscillations at point B, the dashed spectrum, obtained by sweeping the NMR frequency downward, resembles the spectrum shown in (a). The spectrum obtained by sweeping the frequency upward, however, shows a severe distortion. At point B the absorption level rises very rapidly above the true absorption level associated with the line, and reaches a peak over three times as high as the dashed peak. Clearly the area under the solid curve is substantially greater than the area under either the dashed curve in (b) or the

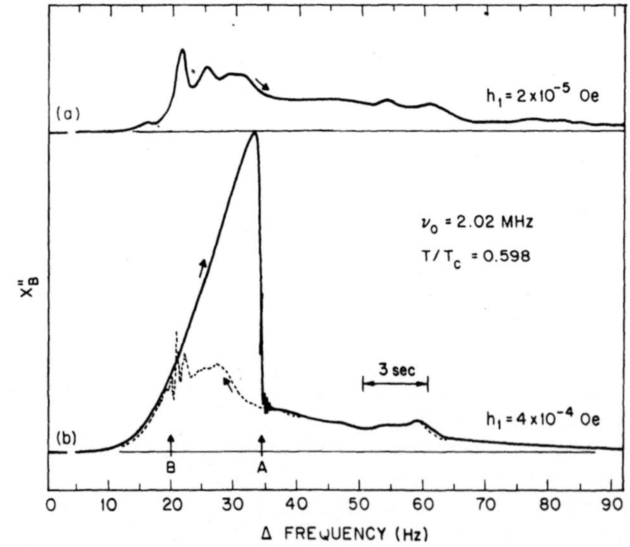

Figure 4. NMR absorption spectra showing power dependent effects in ^{3}He-B. All traces should be identical except for power dependent effects. Trace (a) reproduces itself precisely, independent of sweep sense. Trace (b) shows excessive absorption when frequency is swept upward.

curve shown in (a). At point A the absorption level observed upon sweeping the frequency upward dropped abruptly back onto the usual curve. Oscillations in the rf level across the NMR coil were observed at this point which persisted (at lower rf levels) for between one and two seconds.

By regulating the temperature and taking repated sweeps at successively higher rf levels, it was found that point A moved progressively to the right, and the maximum peak height decreased. For a value of $h_1 \approx 1.8\times10^{-3}$ Oe, point A shifted upward by about 50 Hz.

Presumably the mechanism which causes the above distortion involves a dependence of the resonance frequency upon tipping angle, such as is seen in ferromagnetic resonance.[13] As the spins are tipped by the rf field, their resonant frequency increases. This brings them back into resonance along with other portions of the sample. As the resonant frequency is swept upward, then, the rf field eventually pushes all or most of the sample, at resonance, along with it. Eventually, loss mechanisms prevent the spins from being tipped further, and the radio frequency rises above the resonant frequency of the sample. At this point the absorption level drops abruptly, as is seen at point A in Fig. 4. Spin wave effects such as those discussed above in Section B would certainly enhance this behavior.

The above analysis leaves us with two questions still unanswered: (1) What causes the frequency shift? and (2) How far are the spins actually being tipped to create that shift? Clearly these questions are interrelated. As long as the h_1 field is applied and the magnetization feels a torque resulting from it, there seems to be no question that a frequency shift will be produced. Unless we can answer the questions raised in the preceding section regarding spin relaxation, however, we cannot expect to fully understand the mechanisms responsible for the power dependence which we have just considered.

ACKNOWLEDGMENTS

I wish to thank W. F. Brinkman and M. C. Cross for their generous help in understanding the phenomena discussed in this manuscript. I also thank L. R. Corruccini for his useful comments concerning the spin-lattice relaxation problem.

References

(1) D. D. Osheroff, W. J. Gully, R. C. Richardson, and D. M. Lee, Phys. Rev. Lett., 29, 920 (1972).

(2) D. D. Osheroff and W. F. Brinkman, Phys. Rev. Lett. 32, 584 (1974).

(3) P. W. Anderson and W. F. Brinkman, in "The Helium Liquids" (Academic Press, London, 1975) p. 407.

(4) W. F. Brinkman and M. C. Cross, to be published in Progress in Low Temperature Physics.

(5) D. D. Osheroff, S. Engelsberg, W. F. Brinkman, and L. R. Corruccini, Phys. Rev. Lett. 34, 190 (1975).

(6) D. D. Osheroff, to be published as part of the proceedings of the Sussex Symposium, Physica B.

(7) M. C. Cross. These Proceedings.

(8) D. D. Osheroff, W. van Roosbroeck, H. Smith, and W. F. Brinkman, Phys. Rev. Lett., 38, 134, (1977).

(9) H. Smith, W. F. Brinkman, and S. Engelsberg. To be published in Phys. Rev. B.

(10) R. A. Webb, R. E. Sager, and J. C. Wheatley, Phys. Lett. 54A, 243 (1975).

(11) K. Maki and T. Tsuneto, preprint.

(12) L. R. Corruccini and D. D. Osheroff, Phys. Rev. Lett., 34, 564 (1975).

(13) P. W. Anderson, private communication.

EXTRAORDINARY MAGNETISM ON SURFACE LAYERS OF ^3He

A. I. Ahonen, J. Kokko, O. V. Lounasmaa, M. A. Paalanen, R. C. Richardson,* W. Schoepe,† and Y. Takano

Low Temperature Laboratory
Helsinki University of Technology
SF-02150 Espoo 15, Finland

Abstract: The nuclear magnetic susceptibility of ^3He, confined within the open space around 9 nm carbon particles at liquid pressures, grows very large at low temperatures. Below 2 mK the inverse susceptibility is proportional to $T - \Delta$; Δ has the value 0.75 mK at 6 bar and 0.85 mK at 28 bar pressure of the liquid.

We have observed extremely large changes in the low temperature susceptibility χ of ^3He mixed with fine carbon particles[1] at liquid pressures. For example, when the temperature doubles from 1.5 mK to 3.0 mK, the ^3He susceptibility typically decreases by a factor of 3. Apart from the superfluid modification, bulk liquid at these temperatures maintains the constant susceptibility of a fluid of fermions. Even solid ^3He is widely thought[2] to have a susceptibility varying less rapidly than the Curie law would indicate, because an antiferromagnetic transition is near at these tempeatures. Our results seem to show that a ferromagnetism occurs on the surface of liquid ^3He at very low temperatures. This effect appears to be unrelated to the superfluid transition which takes place in bulk liquid.

The experimental methods for cooling and for measuring the temperature and susceptibility of ^3He are the same as those we have recently described.[3] A small cylinder, 4.0 mm long and

---

*Guggenheim Fellow on leave from Cornell University, Ithaca, N.Y., USA.

†On leave from Regensburg University, Regensburg, West Germany; supported by the Deutsche Forschungsgemeinschaft.

3.2 mm in diameter, filled approximately to 10% with carbon particles having a characteristic diameter of 9 nm, was immersed in the ^3He volume of our nuclear demagnetization cryostat. The field of the NMR coil was shielded from the external bulk liquid by a cylindrical copper sleeve and by a pair of copper gauze caps at each end of the coil. The carbon grains were confined within the coil volume by membranes perforated with 5 μm holes. The apparatus is illustrated in Fig. 1.

The temperature was determined by measuring the magnetic susceptibility of powdered platinum confined in a similar cylinder and immersed in ^3He. The Curie law thermometer was calibrated with the superfluid transition temperature of the bulk liquid monitored with a NMR coil (Coil 2 in Fig. 1).

The results of our measurements at 28 bar and in a magnetic field of 28 mT are shown in Fig. 2. The arrows indicate the superfluid transition observed in the bulk liquid outside of the chamber filled with carbon particles. In Fig. 3 we have plotted $1/\chi$ for ^3He vs. temperature at two pressures, 6 bar and 28 bar, both at two magnetic fields, 28 mT and 14 mT.

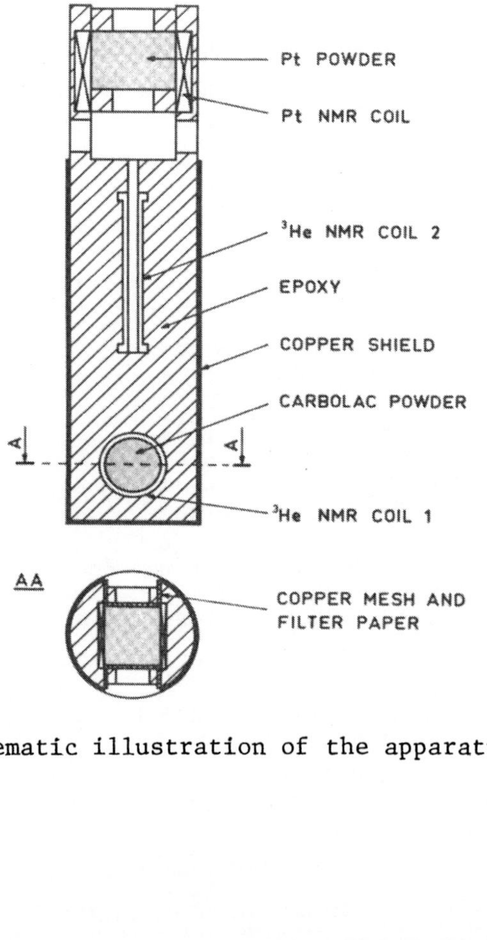

Figure 1. A schematic illustration of the apparatus immersed in liquid ^3He.

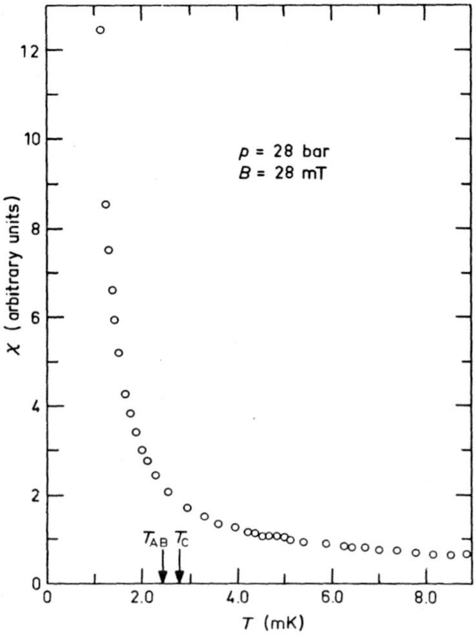

Figure 2. The susceptibility of ^{3}He intermixed with small carbon particles vs. temperature. The arrows indicate the superfluid transitions observed in the bulk liquid outside of the experimental cell.

The data obtained may be fitted to an equation of the form

$$\chi = \frac{A_s}{T - \Delta} + B_\ell \quad . \tag{1}$$

We regard the fact that $1/\chi$ vs. T has a positive intercept Δ on the temperature axis as evidence of a ferromagnetic enhancement of the ^{3}He susceptibility. Δ is independent of the magnetic fields we have used but depends upon the pressure; it has the value 0.85 ± 0.02 mK at 28 bar and 0.75 ± 0.02 mK at 6 bar. In similar plots the inverse susceptibility of bulk solid ^{3}He has been found to have a negative Δ.

The $1/\chi$ in Fig. 3 is in arbitrary units; the relative susceptibility between measurements at 28 bar and 6 bar was calibrated by scaling the data obtained at temperatures greater than 20 mK to the Fermi liquid coefficients in Wheatley's[4] Table V. There may be a magnetic field dependence in the absolute susceptibility which is masked by the changes in the decay time of the free induction signal when the field or temperature is changed.

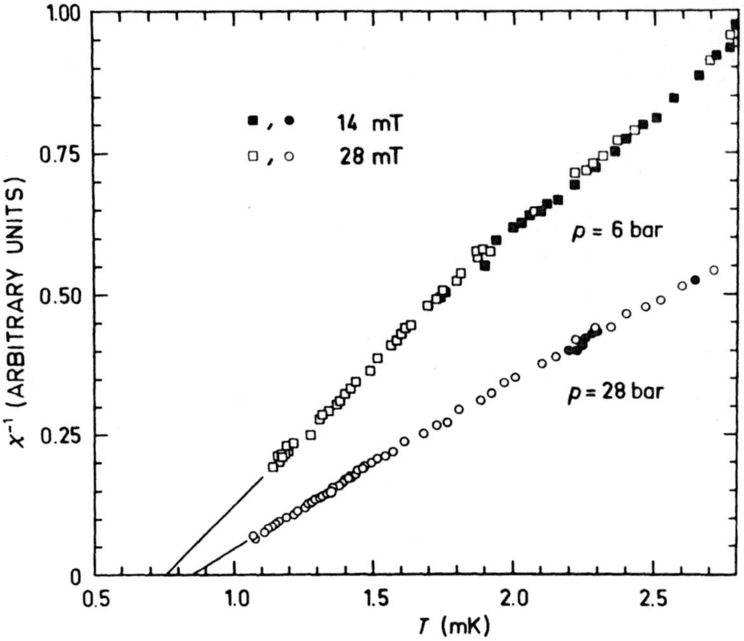

Figure 3. $1/\chi$ of ^{3}He intermixed with carbon particles vs. temperature.

These will be discussed later, but we find that the relative changes in χ, measured at a given pressure, are the same within 5% in both magnetic fields at all temperatures below 4 mK.

The ratio of constants A_s and B_ℓ in Eq. (1) also depends on the fluid pressure. At 6 bar, $A_s/B_\ell = 13 \pm 1$ mK; at 28 bar $A_s/B =$ 11 ± 1 mK. The magnitude of A_s/B_ℓ at 6 bar is consistent with a model which assumes that a monolayer of ^{3}He atoms on the carbon surface, with the typical density[5] of 1×10^{15} atoms cm^{-2}, behaves like a solid and produces the contribution $A_s/(T - \Delta)$ in Eq. (1) and that the remaining ^{3}He in 90% of the volume has the temperature independent bulk susceptibility of a Fermi liquid, as reflected in the magnitude of coefficient B_ℓ.

This model is similar to the ones used by the Sussex group[6] in their analyses of the susceptibility of ^{3}He contained in Vycor glass. At 28 bar this model predicts that A_s/B_ℓ should be 56% of the ratio at 6 bar because of the increase in the liquid density. Our larger ratio indicates that relatively more atoms are participating in the phenomenon responsible for the term A_s at higher pressures. This result is also reflected in the magnitude of $1/\chi$ for the two pressures shown in Fig. 3. At low temperatures the A_s term completely dominates the data.

Somewhat to our surprise, we found that the apparent decay time τ_s^* of the free induction signal, following an applied rf pulse, was temperature dependent. At 32 bar we studied τ_2^* in more detail and found that below 20 mK the signal envelope V(t) is

$$V(t) = V(0)\ \{[A_s/(T - \Delta)]\exp(-t/\tau_s)+B_\ell f(t/\tau_\ell)\}, \qquad (2)$$

where τ_s and τ_ℓ are temperature independent decay constants. V(0) is a constant related to detector sensitivity and $f(t/\tau_\ell)$ is a decaying function of time which crosses zero at $t = \tau_\ell$. The experimental values of the decay constants are: $\tau_s = 0.25 \pm 0.02$ ms, $\tau_\ell = 0.67 \pm 0.07$ ms at 28 mT; $\tau_s = 0.35 \pm 0.02$ ms, $\tau_\ell = 1.2 \pm 0.2$ ms at 14 mT.

Eq. (2) is an extension of the model used for Eq. (1): a surface layer with a decay constant τ_s and the bulk fluid with the constant τ_ℓ. The apparent decay time τ_2^* at 6 bar and at 28 bar can be accounted for by Eq. (2), without assuming a temperature dependence for τ_s or $f(t/\tau_\ell)$. Although we have not yet made detailed studies, τ_s and τ_ℓ appear to have only a weak pressure dependence if any. The intrinsic relaxation time τ_2 of ^{3}He in the same carbon powder has been measured by Kelly[7] using spin echo techniques. Since τ_ℓ is much shorter than τ_2 and varies inversely with the applied magnetic field, it is likely that τ_ℓ is due to inhomogeneous dephasing of the signal caused by the local gradient of our applied magnetic field. The shorter τ_s, as compared with τ_ℓ, implies an additional dephasing mechanism for the surface layer. The strong field dependence of τ_s in low magnetic fields makes it unlikely that τ_s is the characteristic time of the motionally narrowed ^{3}He signal on the surface layer. It seems more likely that local magnetic impurities cause a field dependent inhomogeneous broadening in the surface layer.

No change occurred in the magnetic behavior of ^{3}He in our carbon filled coil when the bulk fluid outside the chamber passed through the superfluid transition. There was neither a discontinuity in χ near T_c nor a shift in the resonance frequency. It may be that since the characteristic distance between particle surfaces, $\sim$ 10 nm, was of the same order as the coherence length in superfluid ^{3}He, T_c was strongly depressed. However, the time for thermal equilibrium was perhaps shorter when the outside liquid was superfluid; the scatter of the data in Fig. 3 is noticeably less for points measured below T_c.

We have recently[3] reported studies of the susceptibility of liquid ^{3}He contained between mylar sheets spaced 4 μm apart. In this case the susceptibility was much smaller than the one reported here for carbon particles. However, when the results of the two experiments are scaled with the relative surface areas, the susceptibility appears to have been about 10 times larger for mylar. In our mylar experiments the most detailed studies were performed at zero pressure; the parameter Δ was +0.5 mK. We also reported a field dependence of χ. It is possible that the field dependence of τ_2^* led to an error in our value of the Fermi liquid susceptibility, which was used when comparing changes in χ at 28 mT and at 7.7 mT. The field dependence of the susceptibility enhancement and the details of changes in τ_2^* suggested by Eq. (2) are both fertile subjects for future studies.

Three different explanations for the observed enhancement in χ have been suggested or occur to us as plausible; (1) Solid ^{3}He in the surface layer has a ferromagnetic transition near 1 mK; (2) There is a ferromagnetic transition in liquid ^{3}He near the surface, induced by the boundary condition that ^{3}He ends at the wall and that the normal liquid is already nearly ferromagnetic, a possibility suggested in a recent paper by Beal-Monod and Doniach[8]; and, (3) The large magnetic polarization is the result of a spin pumping of the ^{3}He atoms when they collide with polarized paramagnetic surface imperfections on the walls. This is similar to the mechanism invoked in theories for the anomalous heat transfer between liquid ^{3}He and paramagnetic salts (CMN).[9]

We conclude by observing that if explanation (1) is correct, a ferromagnetic surface transition should be included in interpretations of the thermodynamic processes occurring in a ^{3}He compressional cooling cell.[10] The entropy differences deduced from most latent heat measurements of compression cells are those between liquid ^{3}He and the solid in the surface sheath around the liquid. If the atoms on the surface of solid ^{3}He behave in the same way as those responsible for the effects we report here, the transition measured at 1.1 mK in compression experiments[10] is likely to be a ferromagnetic transition in the surface layer. The bulk solid may yet prove to have the expected antiferromagnetic transition if it can be adequately cooled.

We wish to thank L. Rehn for her assistance in the data analysis. We are very much in the debt to M. T. Beal-Monod for here interest in this work. We have had valuable discussions with W. F. Brinkman and S. Doniach. We also wish to acknowledge financial support by the Academy of Finland.

References

(1) The particles have the trade name "Carbola I" and were obtained from the John Cabot Corporation, Boston, Mass., USA.

(2) See, for instance, the review by S. B. Trickey, W. P. Kirk, and E. D. Adams, Rev. Mod. Phys. 44, 668 (1972).

(3) A. I. Ahonen, T. Kodama, M. Krusius, M. A. Paalanen, R. C. Richardson, W. Schoepe, and Y. Takano, J. Phys. C 9, 1665 (1976).

(4) J. C. Wheatley, Rev. Mod. Phys. 47, 415 (1975).

(5) See, for instance, J. G. Daunt, S. G. Hedge, and E. Lerner in Monolayer and Submonolayer Helium Films, edited by J. G. Daunt and E. Lerner (Plenum Press, New York, 1973), p. 19.

(6) J. Rolt and D. F. Brewer, Phys. Rev. Lett. 29, 1485 (1972).

(7) J. F. Kelly, Thesis (Cornell University, 1974, unpublished); and J. F. Kelly and R. C. Richardson, Proc. of 13th Internat. Conf. on Low Temp. Phys., Vol. I, edited by Timmerhaus *et al.* (Plenum Press, New York, 1974), p. 167.

(8) M. T. Beal-Monod and S. Doniach, preprint (1976) (unpublished).

(9) A. J. Leggett and M. Vuorio, J. Low Temp. Phys. 3, 359 (1970); and R. A. Guyer, J. Low Temp. Phys. 10, 157 (1973).

(10) W. P. Halperin, C. N. Archie, F. B. Rasmussen, R. A. Buhrman, and R. C. Richardson, Phys. Rev. Lett. 32, 927 (1974).

THERMAL TIME CONSTANT BETWEEN POWDERED CMN AND SUPERFLUID ^{3}He

T. Chainer and H. Kojima*

Department of Physics, Rutgers University
New Brunswick, N. J. 08903

We report measurement of the thermal relaxation time constant of a mixture of powdered cerium magnesium nitrate (CMN) and liquid ^{3}He in its superfluid phases. At the superfluid transition of liquid ^{3}He, there is a discontinuity in the thermal time constant. A thermal time constant discontinuity would be expected from a simple model for the CMN-^{3}He mixture as two coupled thermal baths. The discontinuity may be used to extract the ratio of heat capacities of ^{3}He below and above the superfluid transition temperature.

Our experimental arrangement is similar to that used by Abel, et al.,[1,2] who discovered the anamolously short thermal time constant of a CMN-^{3}He mixture. A cylindrical cavity made of an epoxy (Stycast 1266) in the form of right circular cylinder (diameter = height = 2.15 cm) is tightly packed with 12.25 grams of ground CMN powder which passed through an NBS 200 sieve (74 micron opening). The amount of CMN used in our cell is significantly greater than that in Ref. (1). The ^{3}He gas used in the experiment contained less than 4 ppm of ^{4}He impurity.[3] The cell is precooled by the mixing chamber of a dilution refrigerator to about 13 mK. The cell is cooled into the superfluid phase of ^{3}He by adiabatic demagnetization of the CMN. The CMN is also used as our magnetic thermometer by measuring its magnetic susceptibility using a 15 Hz mutual inductance bridge. A d.c. magnetic field (varied from 2 to 5 gauss depending on temperature) is applied (or removed) on the cell to heat (or cool) the CMN. The temperature

*Alfred P. Sloan Foundation Research Fellow

versus time relationship following a removal of an applied magnetic field is fitted to an exponential relation to extract the time constant. The time constant measurement was taken only upon removal of an applied field since otherwise there is a complicated transient due to the anisotropy of the CMN crystals.[2]

The CMN-^{3}He mixture in the present experiment may be regarded as two thermal baths of heat capacities C_{CMN} and C_3 coupled through a thermal resistance R.[1,2] The thermal relaxation time τ is then given by

$$\tau = \frac{R\, C_{CMN}\, C_3}{C_{CMN} + C_3} \quad . \tag{1}$$

If the C_{CMN} is much larger than the C_3 Eq. (1) simplifies to

$$\tau \approx R\, C_3 \quad (C_{CMN} >> C_3) . \tag{2}$$

The heat capacity jump of C_3 at the superfluid transition leads to a jump in the time constant.

Fig. 1 shows the results of the time constant measurements as a function of magnetic temperature with liquid ^{3}He pressurized at 29.95 bars. Different symbols refer to separate runs. The minimum temperature reached was $T^* = 1.60$ mK. The measurement was made as the cell warmed up by residual heat leak. The time constant increases from about 16 sec at the minimum temperature to a maximum of 26 sec at $T_c^* = 2.20$ mK. Then τ decreases sharply by a factor of 2.7 just above this temperature and continues to increase again as the temperature rises but at a smaller rate than below T_c^*. The superfluid transition temperature T_c^* of ^{3}He was determined in this experiment by monitoring the warm up rate and noting the temperature at which the warm up rate increase abruptly, as was done by Webb, et al.[5] The warm up rate and time constant jumps occur at the same temperature within $\sim$ 2%. We conclude that the time constant jump arises from the ^{3}He heat capacity jump as was expected from Eq. (2).

Measurements such as shown in Fig. 1 were carried out at several different pressures. In Fig. 2 we show the ratio $\tau_< / \tau_>$ as a function of pressure, where $\tau_<$ and $\tau_>$ are the time constants just below and just above T_c^*, respectively. The ratio $\tau_< / \tau_>$ varies from 2.7 at 30 bar to 2.4 at 17.4 bar. At 17.4 bars the transition temperature $T_c^* = 1.87$ mK. We estimate that the ratio

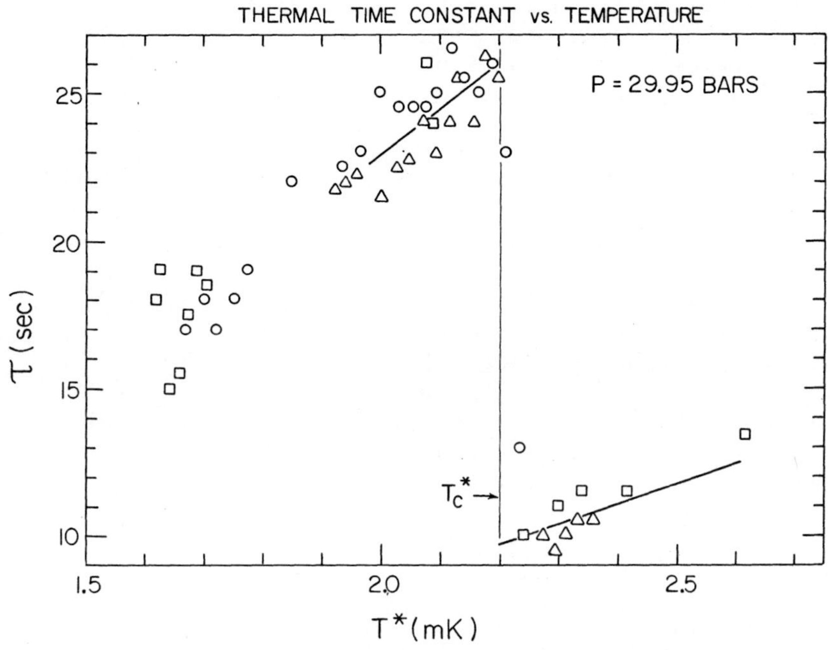

Figure 1

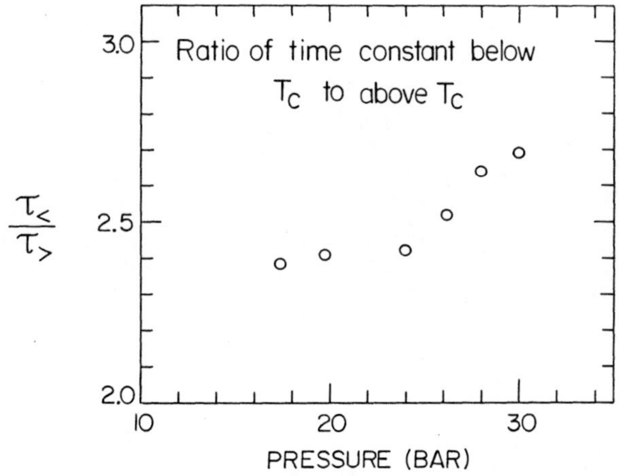

Figure 2

$C_{3>}/C_{CMN}$ in our cell is no more than 3 - 4% at $T_c^* \lesssim 2.2$mK, or at pressures less than 30 bar, so that Eq. (2) should hold true for our cell. If we assume that the thermal resistance R is continuous across the superfluid transition, then $\tau_</\tau_> = C_{3<}/C_{3>}$. Under this assumption we may compare the ratio $\tau_</\tau_>$ with other $C_{3<}/C_{3>}$ measurements. Within the scatter, our data in the pressure range from 30 bar to 24 bar appear to be consistent with the ratio $C_{3<}/C_{3>}$ obtained by Webb, et al.[5] who used conventional heat capacity measurement technique for a cell similar to ours. Our data indicate that below 20 bars $C_{3<}/C_{3>}$ levels off at about 2.4 which is close to the BCS value. The technique used in our experiment to obtain $C_{3<}/C_{3>}$ has an advantage when $C_3 \ll C_{CMN}$ over standard heat capacity measurement technique, since the large background heat capacity of CMN does not enter Eq. (2).

It is our pleasure to acknowledge extremely valuable discussions with Prof. John Wheatley. We also thank Prof. Bill Glaberson for numerous useful discussions.

References

(1) W. R. Abel, A. C. Anderson, W. C. Black, and J. C. Wheatley, Phys. Rev. Letters 16, 273 (1966).

(2) W. C. Black, A. C. Mota, J. C. Wheatley, J. H. Bishop, and P. M. Brewster, J. Low Temp. Phys. 4, 391 (1971).

(3) Monsanto Research Corporation, Mound Laboratory, Miamisburg, Ohio.

(4) J. C. Wheatley, Ann. Acad. Sci. Fenn. Ser. A. VI, 15 (1966).

(5) R. A. Webb, T. J. Greytak, R. T. Johnson, and J. C. Wheatley, Phys. Rev. Letters 30, 210 (1973).

CALCULATION OF SURFACE ENERGIES IN A AND B PHASES OF ^3He

M. C. Cross*

Bell Laboratories
Murray Hill, New Jersey 07974

IMPORTANCE OF SURFACE ENERGIES

In superfluid ^3He two rather different kinds of surfaces exist: firstly the interface between two different but co-existing phases - typically the A and B phases at T_{AB} - and secondly the interface between a liquid phase and a solid (or perhaps the vapour). The surface energies at these interfaces are important in phase nucleation, phase stability in small geometries, equilibrium textures and dynamical boundary conditions. Here we concentrate on the surface energies relevant to phase nucleation, and in particular to the problem of nucleating the B phase on cooling. Many authors[1,2,3] have considered the latter three applications, and we present a few additional remarks here.

CALCULATION OF A-B INTERFACE ENERGY σ_{AB}

The principle of the calculation is clear. Far away from the surface the form of the order parameter is known: the ABM state on one side and the BW state on the other (see Ref. 4 for a description of these states). The surface energy σ_{AB} is calculated as the additional free energy from the region where the order parameter changes continuously from one state to the other. There are two contributions to this energy. The first is the extra bulk free energy of the non-equilibrium states passed through. This is of order $f|F_s|t$, where f is the maximum energy passed through expressed as a fraction of the energy F_s stabilizing the A and B phases over the normal liquid, and t is the thickness of the transition region. The second contribution is the energy of spatial variations of the order parameter (the "bending energy") and is of order $|F_s|(\xi/t)^2 t$ with ξ the temperature dependent coherence length. Minimizing the sum of the two contributions gives $t \sim f^{-1/2}\xi$ and $\sigma_{AB} \sim 2f^{1/2}|F_s|\xi$. Clearly for the "best" path (i.e. the path minimizing the additional energy) f will be small, and t is many coherence lengths. A macroscopic approach is then justified.

The practical calculation is, of course, more complicated. Let us first consider the situation near to the polycritical point (PCP) where T_{AB} is in the Ginzburg-Landau region. The bulk free energy of an arbitrary state is then given by the usual expansion up to second order in the mean square energy gap $|\Delta|^2$:

$$F_s = N(0)[\ln(T/T_C)|\Delta|^2 + \frac{1}{2}\kappa\ \bar{\beta}|\Delta|^4] \tag{1}$$

where

$$\bar{\beta} = \frac{7\zeta(3)}{8\pi^2(k_B T_C)^2}$$

and

$$\kappa = (3/5)\sum_i a_i\ I_i$$

with I_i the five fourth-order invariants[5,6]

$$I_1 = |\Sigma d_{\alpha i}^2|^2 \quad I_2 = d_{\beta i}^* d_{\beta j}^* d_{\alpha i} d_{\alpha j} \quad I_3 = d_{\alpha i}^* d_{\beta i}^* d_{\alpha j} d_{\beta j}$$

$$I_4 = (\sum_{\alpha i} |d_{\alpha i}|^2) = 1 \qquad I_5 = d_{\alpha i}^* d_{\beta j}^* d_{\alpha j} d_{\beta i}$$

(summation over repeated indices is assumed) and a_i the state independent coefficients which must be found from experiment or from microscopic theory. (The notation we use is as Ref. 4, except we have transferred the factor 3/5 explicitly to κ. The I_i for various phases remain as given in Table I.) For the a_i we use the predictions of spin fluctuation theory,[7] Table I, which is known[8] to give reasonably good answers for quantities depending only on ratios of the energies of phases. The bending free energies are also simple here (we use the Anderson-Brinkman[8] form).

	a_1	a_2	a_3	a_4	a_5
Weak coupling BCS theory [4]	-1	2	-2	2	2
Spin fluctuation correction [7]	-.10δ	-.05δ	-.70δ	.20δ	-.55δ
Value at PCP δ = .465	-1.05	1.98	-2.33	2.09	1.74

Table I: Values of the constants a_i in weak coupling and spin fluctuation theory.[7] δ is a pressure dependent parameter equal to .465 at the PCP where $a_3 + a_4 + a_5 = a_1 + a_4 + \frac{1}{3}(a_2 + a_3 + a_5)$.

$$F_b = \frac{3}{5} N(0)\, \xi_0^2 [\,|\Delta_\alpha \bar{d}_{\beta i}|^2 + (\nabla_\alpha \bar{d}_{\beta i})(\nabla_\beta \bar{d}^*_{\alpha i}) + |\nabla_\alpha \bar{d}_{\alpha i}|^2] \qquad (2)$$

where $\bar{d} = \sqrt{|\Delta|^2}\, d$, and $\xi_0 = \xi(1 - T/T_C)^{1/2}$ is the pair radius:

$$\xi_0 = \frac{7\zeta(3)}{48\,\pi^2} \left(\frac{\hbar v_F}{k_B T_C}\right)^2$$

Clearly to minimize the energy $F_B + F_s$ integrated over the transition region with respect to the eighteen parameter space of the order parameter is exceedingly difficult. Instead we use physical arguments to suggest the best path, at least in a variational sense. We first state the path, and then motivate the choice.

We choose the path A - P - B, where P is the planar or "2-D" state (again see Ref. 4 for this nomenclature).

In weak coupling approximation the ABM state may be transformed continuously through degenerate states to the planar state (this may be thought of as rotating the $\hat{\ell}$ of the spin up pairs through π with respect to the $\hat{\ell}$ of the spin down pairs, which are uncoupled in the weak coupling energy expression). The planar state may then be transformed continuously through states of decreasing energy to the BW state. Strong coupling corrections make the ABM and BW energies equal to T_{AB}, and the peak height $f|F_S|$ is the planar - ABM energy difference, zero in weak coupling approximation (Fig. 1). The peak height is therefore purely a strong coupling correction, and small if these are not too large. We assume this is sufficient to define the path A - P - B to be a good trial function for the order parameter.

An apparent flaw in this argument at or above the PCP is that strong coupling corrections even in principle cannot be assumed arbitrarily small - at T_{AB} they must just cancel the weak coupling ABM - BW energy difference. Nevertheless Table II shows the peak height consequent on our assumption to be certainly much smaller than if we had chosen the other p-wave unitary state (the polar state[4]) to be the peak. Non-unitary stationary points also

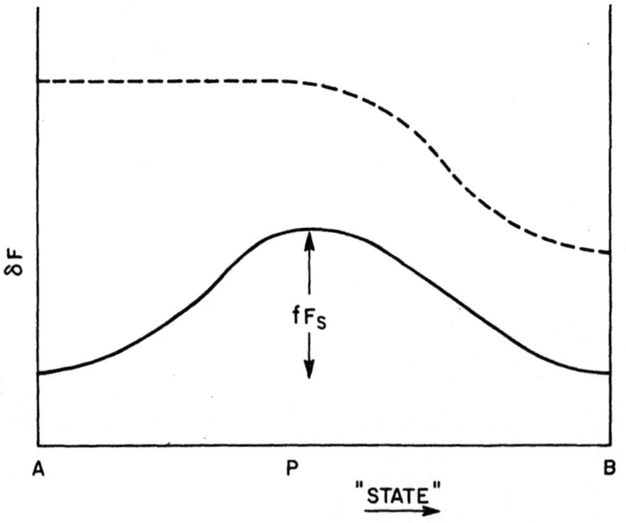

Figure 1. Bulk free energy as a function of the state passed through between A and B phases: broken line - weak coupling theory; continuous line - actual behavior with strong coupling corrections. The energy shifts are roughly to scale according to the spin fluctuation theory. Letters on the abscissa label the stationary states.

Table II. Values of κ (see Eq. 1) for the unitary phases using predictions of spin fluctuation theory.

	ABM	Planar	BW	Polar
Weak Coupling BCS Theory[4]	2	2	5/3	3
Spin Fluctuation Correction[7]	-1.05δ	$-.55\delta$	$-.33\delta$	-1.2δ
Value at PCP $\delta = .465$	1.51	1.74	1.51	2.44

probably have a higher energy,[8] but this has not been proved in detail. It can be seen from Table II that the justification of the assumption is that only the rather small ABM - BW weak coupling energy difference need be made up by strong coupling effects, which are therefore quite small. The argument is in fact most satisfactory far below the PCP, with A phase stabilized by a magnetic field. Here strong coupling effects are believed to become very small. The energy of the path A - P and the peak height are unaffected by the field; the energy P - B is largely magnetic and must be recalculated. This recalculation should not much affect the final result however if expressed as the ratio $\sigma_{AB}/|F_s|\xi$.

Thus we take as the variational form

$$\overleftrightarrow{d} = \frac{1}{\sqrt{2}} \begin{bmatrix} \cos\chi & 0 & 0 \\ 0 & \cos\theta\cos\chi & 0 \\ -i\sin\theta\cos\chi & 0 & \sqrt{2}\sin\chi \end{bmatrix} \qquad (3)$$

where $\chi = 0$, $\frac{\pi}{2} \geq \theta(x) \geq 0$ defines states from A to P and $\theta = 0$, $0 \leq \chi(x) \leq \sin^{-1}(1/\sqrt{3})$ defines states from P to B, and the surface is perpendicular to the x direction in orbit space. $|\Delta|^2$ is taken to be constant. This is not exactly true, but will have only a small effect on the final result, and considerably simplifies the working.

The form Eq. 3 in the energy expressions Eqs. 1 and 2 gives the additional free energy for each x:

$$A \rightarrow P: \quad \frac{\Delta F(x)}{|F_s|} = \frac{3}{5}\xi^2 \frac{\partial\theta}{\partial x}^{2} + (1+\cos 2\theta) \times \frac{[(2a_1+a_{235})\cos 2\theta + (2a_1+a_2-3a_{35})]}{8a_{345}} \tag{4}$$

$$P \rightarrow B: \quad \frac{\Delta F(x)}{|F_s|} = \frac{6}{5}\xi^2 (1+\sin^2\chi)\left(\frac{\partial\chi}{\partial x}\right)^2 + \frac{a_{235}(9\sin^4\chi - 6\sin^2\chi + 1)}{6(a_{14} + \frac{1}{3}a_{235})} \tag{5}$$

where $a_{235} = a_2 + a_3 + a_5$, etc. These energies must be integrated across the interface and then minimized with respect to variations of (x) and (x). The minimization can be done exactly to give an integral easily evaluated numerically.

With the coefficients a_i from Table I we find contributions to $\sigma_{AB}/|F_s|\xi$ of 0.74 from A → P and 0.36 from P → B, to give a total $\sigma_{AB} = 1.1|F_s|\xi$. For a given set of a_i the method gives an upper bound for σ_{AB}.

Unfortunately this estimate of σ_{AB} cannot be compared directly with experiments to date. Experiment[9] is at melting pressure where T_{AB} is not in the Ginzburg-Landau region and relevant free energies including strong coupling corrections are not known. The calculation applies near the PCP (in fact formally at the PCP, since only here is the A-B phase transition predicted by a free energy expansion up to only fourth order in $|\Delta|$). We have suggested[9] that the ratio $\sigma_{AB}/|F_s|\xi$ may be roughly constant along T_{AB}. The measured value T_{AB} at melting pressure is $\sigma_{AB}/|F_s|\xi = 0.7$, in good agreement considering the uncertainty in the extrapolation procedure.

We can also derive boundary conditions for the orientation of the A and B phase near and AB interface. It can be shown that any spin or orbit space rotation of Eq. 3 results in a greater bending energy contribution. There are therefore large energies, some fraction of σ_{AB}, acting on the orientation of A and B phases at the interface. In particular $\hat{\ell}$ in the A phase lies in the plane of the interface - in contrast to the result at solid surfaces. In addition there is a boundary condition involving the surface normal $\hat{s}$, the spin axis $\hat{d}$ in the ABM state, and the rotation matrix $\overleftrightarrow{R}(\hat{n},\theta)$ defining the BW phase (and specifying the relative orientation of spin and orbit space axes in Eq. 3):

$$(\hat{s} \cdot \overleftrightarrow{R}(\hat{n},\theta) \cdot \hat{d})^2 = 1$$

Usually θ should retain the bulk equilibrium value $\cos^{-1}(-\frac{1}{4})$ and $\hat{n}$ should lie in a direction such that a rotation through θ about $\hat{n}$ brings $\hat{s}$ onto $\pm\hat{d}$ (cf. the discussion at solid surfaces.[10,13]) It should be noted however that these boundary conditions, depending on the variational "wavefunction", are known with less confidence than σ_{AB}, the "eigenvalue".

CALCULATION OF $\sigma_B - \sigma_A$*

σ_B and σ_A are the extra energies from the distortion of the order parameter by the proximity of solid surfaces. In fact only the difference $\sigma_B - \sigma_A$ is of significance, since there are other much larger, but phase independent, interaction energies between liquid and solid. The energies are calculated in an analogous way to σ_{AB}: a form for the order parameter is taken that satisfies the boundary conditions at the surface, and tends to the equilibrium form at large distances. The total additional energy $F_B + F_S$ is then minimized.

Boundary conditions depending on the nature of the surface have been calculated microscopically.[12] For diffusely reflecting surfaces the condition is essentially $d_{\alpha i} = 0$ at the surface. For

*This has also been studied by Privorotskii[1] and Kuroda and Nagi.[11]

specularly reflecting surfaces the longitudinal component of the order parameter must still go to zero: $s_\alpha d_{\alpha i} = 0$; the condition on the transverse components is

$$(\hat{s} \cdot \nabla) d_{i} = b^{-1} d_{i} , \quad b \to \infty.$$

It is not clear to this author however if b is to be large compared with ξ (i.e. a boundary condition of zero normal derivative at the surface) or only compared with smaller pair radius ξ_0 (i.e. essentially no boundary condition on the transverse components on a scale ξ). This has very little consequence in practice on the calculation of $\sigma_B - \sigma_A$.

For the B phase near a surface it is assumed

$$\bar{d}(z) = \frac{|\Delta|}{\sqrt{3}} \begin{bmatrix} u(z) & 0 & 0 \\ 0 & u(z) & 0 \\ 0 & 0 & v(z) \end{bmatrix} \tag{6}$$

(again with spin and orbit space axes related by the equilibrium $\overleftrightarrow{R}(\hat{n},\theta)$) with v = 0 at z = 0 and u = v = 1 at large z. At specularly reflecting surfaces u need not be zero at the surface, and the order parameter here is the planar state[13] - the additional variation to reach an ABM phase would require a larger energy as in the discussion of σ_{AB} above. At diffusely reflecting surfaces u = 0. $|\Delta|$ in Eq. 6 is the energy gap in the bulk BW phase. Then the surface energy is

$$\frac{\sigma B}{|F_B|} = \int_0^\infty \left\{ \left(1 - \frac{2u + v}{3}\right)^2 + \frac{4}{9} \frac{\kappa_P - \kappa_{BW}}{\kappa_{BW}} \left(u^2 - v^2\right)^2 + \frac{6}{5}\xi^2 \left[\frac{2}{3}\left(\frac{\partial u}{\partial z}\right)^2 + \left(\frac{\partial v}{\partial z}\right)^2 \right] \right\} dz \tag{7}$$

where the integral is to be minimized with respect to variations of u(z), v(z) subject to the boundary conditions. Kuroda and Nagi[11] have done this numerically. We may evaluate σ_B from their Eq. 11 to give for $\sigma_B/|F_S|\xi$ (at the PCP) 0.54 for diffusely reflecting surfaces and 1.66 for specularly reflecting surfaces. The first of these results we can estimate fairly well by the simple trial function Eq. 6, but with u = 1 everywhere: this gives 0.58.

The ABM phase is not distorted at a specularly reflecting surface, providing $\hat{\ell}$ is along the surface normal, and $\sigma_A = 0$. At diffusely reflecting surfaces $\hat{\ell}$ remains perpendicular to the surface, but the magnitude of the order parameter must go to zero at the surface. We must then minimize

$$\frac{\sigma_A}{|F_A|} = \int_0^\infty \left[\frac{6}{5}\xi^2 \left(\frac{d\phi}{dz}\right)^2 + (1 - \phi^2)^2 \right] dz,$$

$$\phi(z) = \frac{|\Delta(z)|}{|\Delta(z\to\infty)|} \tag{8}$$

This can be done exactly[11] to give $(\sigma_A/|F_A|\xi) = \sqrt{32/15} \simeq 1.46$. Thus the final important result (at the PCP) is

$$(\sigma_B - \sigma_A)/|F_s|\xi = 0.54 \quad \text{for specularly reflecting surfaces.}$$

$$= 0.20 \quad \text{for diffusely reflecting surfaces.}$$

It is interesting to note the value of the ratio $(\sigma_B - \sigma_A)/\sigma_{AB}$ measured by experiments at melting pressure[14] is 0.38, compared with .50 calculated here for specular surfaces and .18 for diffuse surfaces. This ratio is simply cos γ, with γ the "contact angle" between AB interface and the solid surface when A and B coexist near a solid surface. The estimates of γ range from 60° to 80°, and the experimental value is 68°, with the A phase in the acute angle.

CONSEQUENCES ON NUCLEATION OF THE B PHASES

It has long been thought[15] that nucleation of B phase could not occur in the homogeneous bulk liquid. The calculations and experimental measurements of the surface energies confirm this. On the simplest picture of homogeneous nucleation, B phase is nucleated through the production by thermal fluctuations of a "bubble" of the phase sufficiently large to grow spontaneously. The critical radius of such a bubble is $r_C = 2\sigma_{AB}/\Delta F$, requiring an energy (surface plus volume)

$$\frac{2}{3}\pi\,(2\sigma_{AB})^3/(\Delta F)^2 \quad ,$$

where ΔF is the volume free energy stabilizing the B phase with respect to the A phase and driving the transition. ΔF obviously increases with the degree of supercooling, and can be measured experimentally.[9] Even for the very large supercooling ($\sim$400 μK) observed on first cooling at melting pressure[16] we find $r_C \sim 1.2$ and an energy $3 \times 10^6\ k_BT$. Clearly the Boltzmann factor effectively prohibits such a nucleation mechanism. Furthermore the result $\sigma_B - \sigma_A > 0$ suggests that solid surfaces do not aid nucleation. (Compare this with A phase nucleation on warming, where it is easy to imagine surface irregularities such that the energy gained from $\sigma_A - \sigma_B$ outweighs the energy lost at the AB interface, so that nucleation may occur with small or zero superwarming, as observed.[16,17]

There is insufficient space here to further speculate on solutions (e.g. Ref. 18) to this dilemma. Let us simply remark that the calculations and measurements of surface energies provide quantitative results (for example the rather large critical size r_C of a region of B phase, however produced, that will spontaneously grow) that must be taken into account in any theory of B phase nucleation.

A CAUTION ON USING BULK BENDING ENERGIES TO CALCULATE SURFACE EFFECTS

It is dangerous to assume in general that in calculating surface effects it is sufficient to retain bending energies simply from the expression valid in the bulk, Eq. 2.

Firstly, there is the difficulty that F_B is known with complete confidence only to within a pure divergence,[8]

$$\nabla_\alpha[\bar{d}_{\alpha i}(\nabla_\beta \bar{d}^*_{\beta i}) - \bar{d}_{\beta i}(\nabla_\beta \bar{d}^*_{\alpha i})] \tag{9}$$

equivalent to a surface term. This accounts for the different conclusions for cylindrical geometries reached by Refs. 2 and 3.

Secondly, there may be <u>explicit</u> surface energies in the gradients:

$$\sigma = K_1 \; s_\alpha \; \nabla_\alpha |\Delta|^2 + K_2[s_\alpha \bar{d}_{\beta i}(\nabla_\beta \bar{d}_{\alpha i})^* + \text{c.c.}] \tag{10}$$

where we have used the general boundary condition $s_\alpha d_{\alpha i} = 0$ to eliminate a third term. K_1, K_2 must be calculated from microscopic theory of the region within ξ_0 of the wall. Note that Eq. 9 merely gives an additional contribution to the K_2 term in Eq. 10.

Neither of these difficulties appear for diffusely reflecting surface, where $d_{\alpha i} = 0$ at the surface. The first term in Eq. 10 also does not appear for specularly reflecting surfaces if the strong boundary condition of zero normal derivative is used for the transverse components. In fact, this term may be included to account phenomenologically for the microscopic boundary conditions[12] at a general surface. The second term does not contribute at a <u>plane</u> surface. There seems, however, no reason to exclude it in general, and it may be important in the prediction of the stable phase in restricted, curved geometries. None of these energies contribute significantly to the calculation of $\sigma_B - \sigma_A$. Neither are they important in calculating the stability of textures varying over a scale much greater than ξ. Nevertheless the possible existence of such terms should be recognized in general.

ACKNOWLEDGEMENTS

The author wishes to acknowledge important contributions to the theory from W. F. Brinkman, and many indispensible conversations with D. D. Osheroff on all aspects of the problem.

References

(1) I. A. Privorotskii, Phys. Rev. B 12, 4825 (1975).

(2) G. A. Barton and M. A. Moore, J. Low Temp. Phys. 21, 489 (1975).

(3) L. J. Bucholtz and A. L. Fetter, (preprint).

(4) A. J. Leggett, Rev. Mod. Phys. 47, 331 (1975).

(5) W. F. Brinkman and P. W. Anderson, Phys. Rev. A 8, 2732 (1973).

(6) N. D. Mermin and G. Stare, Phys. Rev. Lett. 30, 1135 (1973).

(7) W. F. Brinkman, J. Serene and P. W. Anderson, Phys. Rev. A 10, 2386 (1974).

(8) P. W. Anderson and W. F. Brinkman, in "The Helium Liquids", edited by J. G. M. Armitage and I. E. Farqhar, (Academic Press, London, 1975).

(9) D. D. Osheroff and M. C. Cross (unpublished).

(10) H. Smith, W. F. Brinkman and S. Engelsberg, Phys. Rev. (to be published).

(11) Y. Kuroda and A. D. S. Nagi (preprint).

(12) V. Ambegaokar, P. G. de Gennes and D. Rainer, Phys. Rev. A 9, 2676 (1974).

(13) W. F. Brinkman, H. Smith, D. D. Osheroff and E. I. Blount, Phys. Rev. Lett. 33, 624 (1974).

(14) J. Landau, A. E. White and D. D. Osheroff (unpublished).

(15) A. J. Leggett (private communication).

(16) D. D. Osheroff: Ph.D. Thesis, Cornell University (unpublished).

(17) R. T. Johnson, R. L. Kleinberg, R. A. Webb and J. C. Wheatley: J. Low Temp. Phys. 18, 501 (1974).

(18) N. D. Mermin: Proceedings of the Sussex Symposium on ^{3}He (to be published).

*Postdoctoral Research Fellow.

NEUTRON INELASTIC SCATTERING FROM LIQUID ^3He*

K. Sköld† and C. A. Pelizzari

Solid State Science Division, Argonne National Laboratory
Argonne, Illinois 60439

Abstract: In two recent publications[1,2] we have reported on the results of neutron inelastic scattering studies of liquid helium-3 at 15 mK. These results are summarized in the present paper which also contains some additional data obtained at smaller wave-vectors.[3] It is shown that for $q \lesssim 1.4$ Å^{-1} the zero-sound mode and the spin fluctuation spectrum are observed as separate peaks in the scattering function. For $q \gtrsim 2$ Å^{-1} we obtain strong evidence for sharp two-phonon scattering at approximately twice the energy of the flat portion of the zero-sound dispersion curve.

I. INTRODUCTION

The macroscopic properties of liquid ^3He have been extensively studied in the last few years.[4] The efforts have resulted in a rather thorough understanding of all the major features which are found to be in essential agreement with the predictions of the Landau Fermi-liquid theory.[5] The underlying microscopic properties are, however, less well understood. In the corresponding case of liquid ^4He, a wealth of information in this regard has been supplied by results obtained from neutron inelastic scattering studies.[6] The

---

*Work performed under the auspices of the U.S. Energy Research and Development Administration.

†On leave from AB Atomenergi, Studsvik, Sweden and from the Royal Institute of Technology, Stockholm, Sweden.

large absorption cross-section of the helium-3 nucleus (11000 barns at 4 Å) has discouraged similar efforts in this case in the past. However, recent development of high flux reactors and of experimental techniques have now made such investigations feasible and several studies of this kind have already been reported.[1-3,7-9] In the present paper the experimental results obtained by the neutron scattering group at Argonne National Laboratory are summarized.[1-3] Recent theoretical efforts will be described in an accompanying paper[10] and we will therefore be concerned mainly with the presentation of the experimental results; we will also comment on the future propsects in this regard when appropriate.

Apart from some trivial factors the function observed in the neutron scattering experiment is

$$S(q,\omega) = S_c(q,\omega) + \frac{\sigma_I}{\sigma_c} S_I\ (q,\omega) \tag{1}$$

where σ_c and σ_I are the coherent and the incoherent scattering cross-sections respectively. Following the notation of Sears[11] we may express the scattering functions in terms of the number density $\rho(\vec{r},t)$ and the spin density $I(\vec{r},t)$ as

$$S_c(\vec{q},\omega) = \frac{I}{2\pi\rho} \iint d\vec{r}\ dt\ \exp\ [i(\vec{q}\cdot\vec{r}-\omega t)\ x$$

$$< \rho(0,0)\ \rho(\vec{r},t) > \tag{2}$$

and

$$S_I(\vec{q},\omega) = \frac{I}{2\pi\rho I\ (I + 1)} \iint d\vec{r}\ dt\ \exp\ [i(\vec{q}\ \vec{r} - \omega t)]\ x$$

$$< I(0,0) \cdot I(r,t) >$$

where $\vec{q}$ is a scalar quantity in the case of an isotropic system. The structure factor is obtained by integrating the scattering function over ω and is in the present case

$$S\ (q) = S_c\ (q) + \frac{\sigma_I}{\sigma_c} S_I\ (q)\ , \tag{3}$$

where

$$S_c(q) = 1 + \rho \int \exp\ (i\ \vec{q}\cdot\vec{r})\ [g(r) - 1]\ d\ \vec{r} \tag{4}$$

is the spin-averaged structure factor also measured by X-ray scattering. The corresponding spin-structure factor is measured by

$$S_I(q) = 1 + \frac{\rho}{2} \int \exp(i\, \vec{q} \cdot \vec{r})\, (g_{\uparrow\uparrow}(r) - g_{\uparrow\downarrow}(r))\, d\vec{r} \qquad (5)$$

with the relation $g(r) = \frac{1}{2}(g_{\uparrow\uparrow}(r) + g_{\uparrow\downarrow}(r))$. The neutron scattering method thus allows the study of all the correlation functions needed for a microscopic description of the system. It is paricularly useful in the present case that the spin-dependent scattering is comparable in strength to the coherent (spin-averaged) scattering ($\sigma_I = 1.2$ b, $\sigma_c = 4.9$ b as will be shown below). As $S_c(q)$ is known from X-ray scattering[12] the neutron scattering results can be used for a direct determination of static spin correlations from eq. (3) and eq. (5). The separation of the dynamic scattering functions will in general require a theoretical model for at least one of the two components.

II. EXPERIMENT

As mentioned above, a neutron scattering study of liquid ^{3}He is rendered very difficult by the large absorption cross-section of the ^{3}He nucleus; the observed intensity of scattered neutrons is in fact of the order of 300 times lower than in similar experiments on most liquids. This leads to difficulties with the counting statistics and also with the separation of the weak signal from the noise.

It is also very important in this case that scattering from the container material as well as other extraneous scattering processes be controlled carefully. In the present studies the intensity problem was overcome by using the statistical chopper time-of-flight technique;[13] this method is particularly useful in the case of an unfavorable signal to noise ratio. The problem with extraneous scattering was overcome by using a rather special geometry which allows this scattering to be determined with confidence. The sample was cooled in a ^{3}He - ^{4}He dilution refrigerator and the sample was monitored by two carbon resistors attached to the sample container. Full details of the experimental arrangements will be given in a forthcoming paper.[3]

Time-of-flight spectra were recorded simultaneously in 15 groups of detectors at scattering angles between 30.4° and 106.7° and the incident energy was 4.82 meV. The energy resolution was $\approx$ 0.3 meV.

III. RESULTS AND DISCUSSION

Examples of the results obtained after 10 days of data accumulation with sample in the container and 10 days with the container empty are shown in Fig. 1. The "empty run" is subtracted point by oint from the "full run" after normalization to equal number of counts in a beam monitor. The data shown in Fig. 1(a) and (b) correspond to $1.0\ \text{Å}^{-1} \lesssim q \lesssim 1.3\ \text{Å}^{-1}$ and $1.5\ \text{Å}^{-1} \lesssim q \lesssim 2.0\ \text{Å}^{-1}$ respectively (the value of q varies with the channel number in this representation) and it is worth noting that the small-q data show a two-peak structure while the large-q data show only one broad peak. The structure at small q, which later will be identified with the zero-sound mode and with the spin-fluctuation spectrum, is thus evident already in the raw data. The various correction procedures used to convert the observed spectra to the scattering function at selected values of q are described in Refs. 1 and 2 and will not be repeated here.

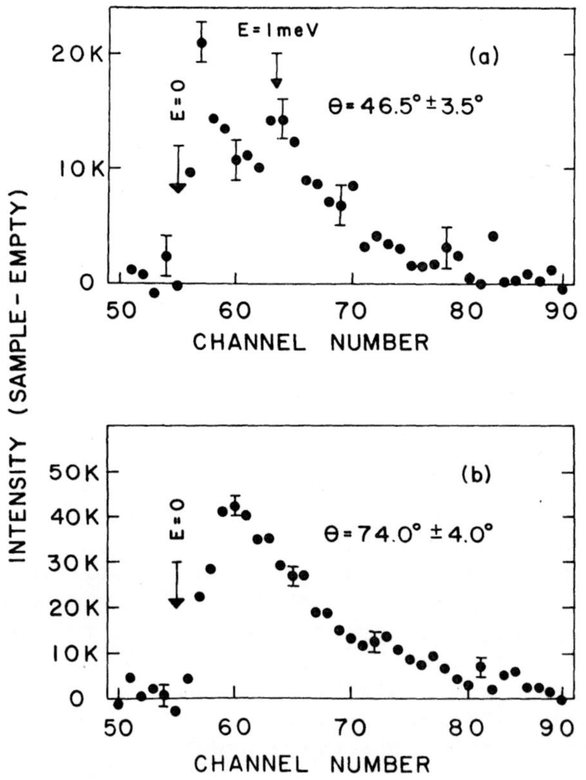

Figure 1. (a) and (b), examples of time-of-flight spectra at constant angle of scattering after subtraction of container scattering as explained in the text. Note that the statistical error is the same in all channels; this is a feature particular to the statistical chopper method.[13]

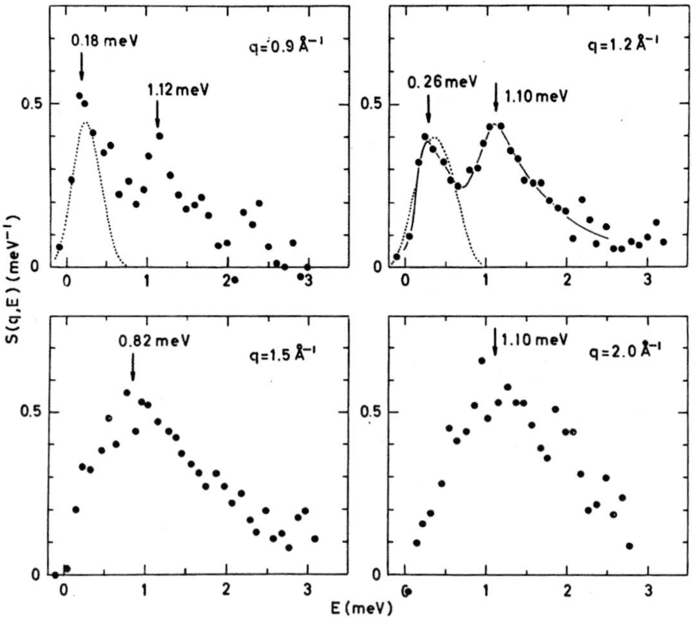

Figure 2. S(q,E) at selected values of q and normalized as in eq. (1). The arrows indicate the peak positions shown in Fig. 4. The dotted curves are the predictions for the non-interacting Fermi gas with $m^* = 3\ m_o$ and normalized with $\sigma_I/\sigma_c = 1.2/4.9$. The solid line for $q = 1.2\ Å^{-1}$ is a hand-drawn fit to the experimental points and is also shown in Fig. 5.

Examples of the resulting scattering function at 4 values of q are shown in Fig. 2. Briefly the total scattering function can be characterized like this: for $q \lesssim 1.4\ Å^{-1}$ the function shows two peaks of approximately comparable strength; for $q \gtrsim 1.4\ Å^{-1}$ the function consists of only one peak. Further analysis of the results, gives strong evidence for a sharp peak at ≈ 2 meV for $q \gtrsim 2.0\ Å^{-1}$. We interpret the low energy peak for $q \lesssim 1.4$ Å as spin-fluctuation scattering and the high energy peak ($E \approx 1.0$ meV) as the zero-sound mode. The peak at ≈ 2 meV is believed to be due to scattering by two zero-sound phonons. Before discussing these features of the scattering function further, it is useful to look briefly at the structure factor.

The total structure factor S(q) is obtained by numerical integration of the scattering function at constant q. The result is shown in Fig. 3 together with $S_c(q)$ obtained from X-ray scattering by Hallock[12] and the difference which is then $[\sigma_I/\sigma_c]S_I(q)$. Also shown is the result predicted for a non-interacting Fermi gas

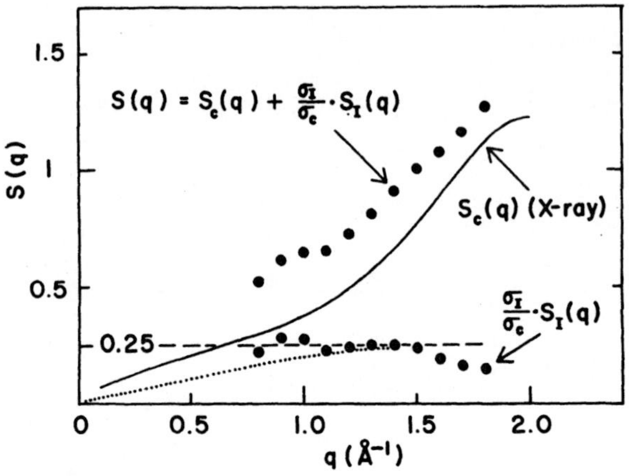

Figure 3. Experimental neutron structure factor, the X-ray structure factor[12] and the difference between the neutron and the X-ray results. The dashed line shows the average value of $\sigma_I/\sigma_c \cdot S_I(q)$ in the range $0.8\ Å^{-1} \leq q \leq 1.5\ Å^{-1}$ and the dotted curve shows the result for the non-interacting Fermi gas normalized to 0.25 for $q > 2\ q_F$.

(dotted curve) normalized to 0.25 for $q \gtrsim 2\ q_F$; 0.25 is the average value of $[\sigma_I/\sigma_c]\, S_I(q)$ for $0.8\ Å^{-1} \lesssim q \lesssim 1.5\ Å^{-1}$ and, assuming that this is the limiting value, we obtain $\sigma_I = (1.2 \pm 0.3)$b if $\sigma_I = (4.9 \pm 0.9)$b is assumed.[14] The value obtained for the spin-dependent contribution is approximately equal to the area of the low energy peak in S(q,E) for $q \lesssim 1.4\ Å^{-1}$; this, and also the fact that the contribution from $S_c(q,E)$ is known to be small[15] in this region of q and E, is the basis for our assignment of this peak as spin-fluctuation scattering. This has also been suggested recently by Glyde and Khanna.[16] The decrease of $[\sigma_I/\sigma_c]\, S_I(q)$ for $q > 1.6\ Å^{-1}$ could imply that we are not including the tail of the spectrum at these values of q and is therefore not necessarily significant.

The dotted curves in Fig. 2 show the spin-dependent scattering for a non-interacting Fermi gas with $m^* = 3\ m_o$ and normalized with $\sigma_I/\sigma_c = 0.25$. It is interesting to note that this result is in good agreement with the data at $q = 1.2\ Å^{-1}$, but predicts too low an intensity at $q = 0.9\ Å^{-1}$. The result obtained by Glyde and Khanna[10,16] from a RPA calculation and including in the interaction the first two asymmetric Landau parameters seem to explain the data at $q = 0.9\ Å^{-1}$ better than the result obtained from the non-interacting Fermi gas model. It would be of interest to obtain experimental results for smaller values of q to explore this further and we are presently planning an experiment which will measure S(q,E) down to $\approx 0.5\ Å^{-1}$.

We now turn our attention to the peak at $\gtrsim$ 1 meV for q $\lesssim$ 1.4 Å^{-1} which, we believe, represents the zero-sound mode. The energy at the peak is shown versus q in Fig. 4 together with the results obtained from a generalized polarization potential approach by Aldrich, Pethick, and Pines.[17] Also shown in Fig. 4 are the energy at the spin-fluctuation peak for small q and at the single broad peak for larger values of q as indicated by the arrows in Fig. 2. In addition to the results reported in Ref. 1, we also show some results obtained in a recent less accurate measurement[3] at smaller values of q.

The theoretical results within the particle-hole band actually represent the mean energy while the experimental points are the peak positions. The agreement between theory and experiment must be considered very promising, in particular in view of the fact that the calculations were made prior to our experiment. It is obvious from this and from the work of Aldrich[15] and of Aldrich, Pethick, and Pines[17] that already at this level of the theoretical development we can learn a great deal about the microscopic nature

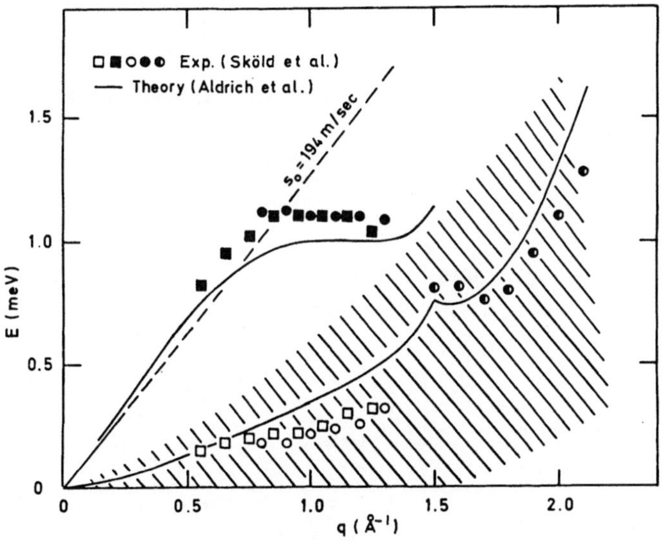

Figure 4. Peak position versus q for the zero-sound mode, for the spin-fluctuation scattering and for the broad single peak observed for q > 1.4 Å^{-1}. Circles are for the data reported in Ref. 1 and squares are for data in Ref. 3. The dashed line is the linear dispersion curve with c_0 = 194 m/sec (Ref. 18) and the solid line is the theoretical results by Aldrich, Pethick, and Pines.[17] Note that the theoretical curve within the particle-hole band shows the mean energy whereas the experimental points show estimated peak positions.

of normal ^3He. For a more complete discussion of the status of the theory we refer to the paper by Glyde and Khanna in this volume[10] and also Refs. 15 and 17.

It is interesting to compare the present results with those obtained by Stirling et al.[9] in particular with regard to the observation of a well-defined zero-sound mode in the present experiments. From the results obtained at 0.63 K these authors conclude, that a sharp mode of this kind is not observed for $q \geq 0.8$ Å^{-1}. In a recent paper by Glyde and Khanna[16] it is suggested that the results at 15 mK and the results at 0.63 K may be reconciled if finite temperature effects are included in the screened susceptibility function. Although their calculations show broadening of the collective mode at 0.63 K it is not entirely clear whether this is sufficient to explain the differences observed in the two experiments. In order to compare the observations directly, we show in Fig. 5 the experimental results by Stirling et al.[9] at the smallest value of q at which data are shown in Ref. 9 together with the corresponding results by Sköld et al.;[1] the results by Stirling et al.[9] are in arbitrary units and are normalized by eye to the solid curves in Fig. 2. The results shown in Fig. 5 indicate

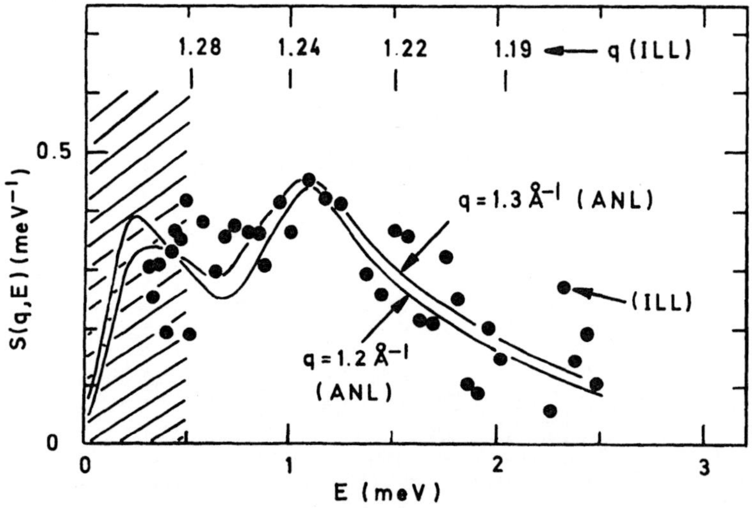

Figure 5. The data obtained by Stirling et al.[9], for $\theta = 53.5^{\circ}$ and $\lambda_{o} = 4.3$ Å (dots) together with the results obtained by Sköld et al.[1] for $q = 1.2$ Å^{-1} and $q = 1.3$ Å^{-1} respectively (solid curves). The solid curves are fitted by hand to the experimental points of Ref. 1 (see also Fig. 2). The q-values corresponding to the data of Ref. 9 are shown at the top of the figure. The hatched area shows the region where the data of Ref. 9 are uncertain due to difficulties with the subtraction of the container scattering as stated by the authors.

that, at least in this region of q, the two sets of data are reasonably consistent. We suggest, however, that the question of the damping of the zero-sound mode with temperature deserves further study both from an experimental and from a theoretical point of view.

We turn now to the question of the conjectured observation of shapr two-phonon scattering at large q. As explained in detail in Ref. 2, the statistics deteriorate rapidly with increasing energy transfer and are rather poor at $E \gtrsim 2$ meV. We observe, however, an indication of a peak at $E \gtrsim 2$ meV and $q = 2$ Å^{-1} in Fig. 2. In order to explore this further we must average the data over q; this, in effect, amounts to trading off resolution in q for improved statistics - again we refer to Ref. 2 for details and show only the results here. Fig. 6(a) and (b) show S(q,E) for $q = 2.0$ Å^{-1} and $q = 2.1$ Å^{-1} respectively and with no additional smoothing, i.e. the value of S(q,E) is obtained by cubic spline interpolation of the constant angle data for each E. Fig. 6(c) shows the result obtained for $q = 2.0$ Å^{-1} from a least squares fit of a cubic spline to all 15 angles for each value of E; this involves a smoothing of the data and the result should be viewed as the mean value of S(q,E) in some region around $q = 2$ Å^{-1}. However, the results seem to support the conclusion that a sharp peak develops at approximately twice the energy of the flat portion of the zero-sound dispersion curve for $q \geq 2$ Å^{-1}. We suggest that this peak is due to two-phonon scattering and note that such a sharp peak is a reasonable consequence of the extended flat region observed in the zero-sound dispersion curve. It is clearly of interest to study this effect further and to determine the energy as well as the spectral weight of the peak as function of q more precisely.

IV. CONCLUSIONS

Neutron inelastic scattering data are now being obtained with high enough accuracy to be directly relevant for the furthering of our understanding of the microscopic features of normal liquid ^{3}He It is particularly important that the neutron results not only yield information about the density fluctuations, but also about the dynamics and the structure of the spin arrangement. It is reasonable to expect such advances in the near future - both on the theoretical side and on the experimental side - that we will soon have as thorough an understanding of normal ^{3}He as is now the case for liquid ^{4}He. This can be expected to have important consequences for the fundamental description of the superfluid phases of ^{3}He.

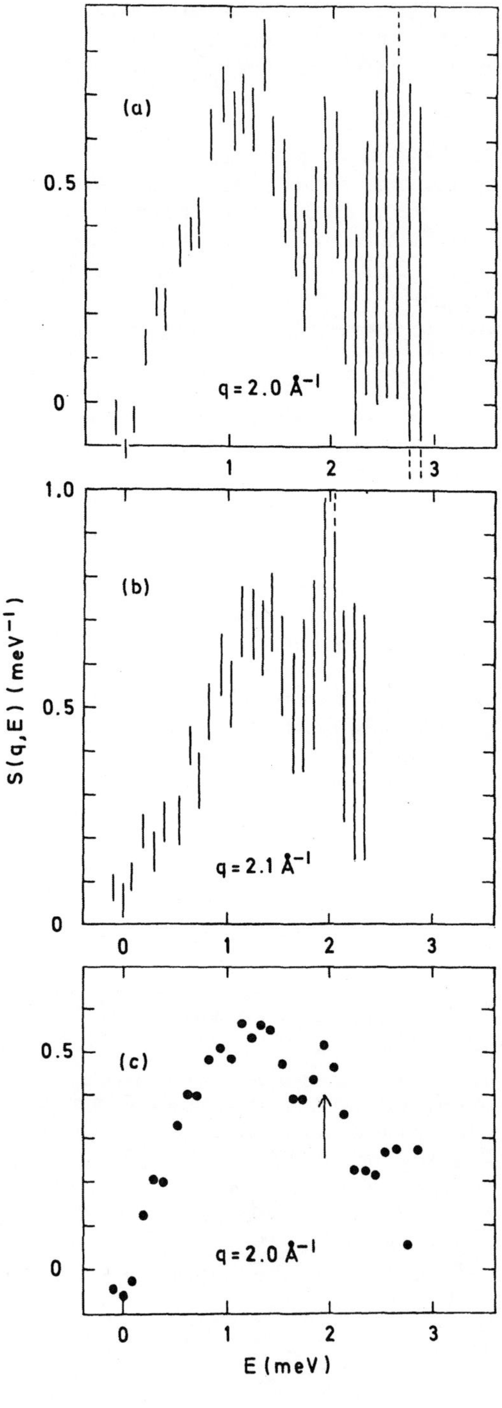

Figure 6. (a) S(q,E) for q = 2.0 A^{-1} obtained from the scattering function at constant angle by cubic spline interpolation at each energy. Vertical lines show the resulting statistical error. (b) Same as (a), but for q = 2.1 A^{-1}. (c) S(q,E) for q = 2.0 A^{-1} obtained from the scattering function at constant angle by least squares fitting of a cubic spline to all 15 angles at each energy. The arrow shows the extimated position of the two-phonon scattering peak.

References

(1) K. Sköld, C. A. Pelizzari, R. Kleb and G. E. Ostrowski, Phys. Rev. Lett. 37, 842 (1976).

(2) K. Sköld and C. A. Pelizzari, submitted for publication.

(3) K. Sköld, C. A. Pelizzari, R, Kleb and G. E. Ostrowski, to be published.

(4) J. Wheatley, Rev. Mod. Phys. 47, 415 (1975).

(5) D. Pines and P. Nozieres, "The Theory of Quantum Liquids", (Benjamin, New York, 1966).

(6) A. D. B. Woods and R. A. Cowley, Rep. Prog. Phys. 36, 1135 (1973).

(7) R. Scherm, W. G. Stirling, A. D. B. Woods, R. A. Cowley and G. J. Coombs, J. Phys. C: Solid State Phys. 7, L341 (1974).

(8) W. G. Stirling, R. Scherm, F. Volino, and R. A. Cowley, in Proceedings of the Fourteenth Int. Conf. on Low Temp. Phys., Ontaniemi, Finland, 1975, edited by M. Krusius and M. Vuorio

(8) (cont.) (North-Holland), Amsterdam, 1975), Vol. 1, p. 76.

(9) W. G. Stirling, R. Scherm, P. A. Hilton and R. A. Cowley, J. Phys. C: Solid State Phys. 9, 1643 (1976).

(10) H. R. Glyde and F. C. Khanna, these proceedings.

(11) V. F. Sears, J. Phys. C: Solid State Phys. 9, 409 (1976).

(12) R. B. Hallock, J. Low Temp. Phys. 9, 109 (1972).

(13) K. Sköld, Nucl. Instrum. Methods 63, 114 (1968); D. L. Price and K. Sköld, Nucl. Instrum. Methods 82, 208 (1970).

(14) T. A. Kitchens, T. Oversluizen, L. Passel, and R. A. Schermer, Phys. Rev. Lett. 32, 791 (1974).

(15) C. H. Aldrich III, Ph.D. thesis, University of Illinois, Urbana, 1974 (unpublished).

(16) H. R. Glyde and F. C. Khanna, Phys. Rev. Lett. 37, 1692 (1976).

(17) C. H. Aldrich III, C. J. Pethick and D. Pines, Phys. Rev. Lett. 37, 845 (1976).

(18) W. R. Abel, A. C. Anderson and J. C. Wheatley, Phys. Rev. Lett. 17, 74 (1966).

EXCITATIONS IN NORMAL LIQUID ^{3}He

H. R. Glyde

Department of Physics, University of Ottawa
Ottawa, Canada K1N 6N5

F. C. Khanna

Atomic Energy of Canada Limited, Chalk River Nuclear Laboratories
Chalk River, Ontario, Canada K0J 1J0

Abstract: A simple RPA model employing the Landau quasiparticle-quasihole interaction is used to describe the density and spin-density excitations observed in recent neutron scattering measurements on normal liquid ^{3}He. The model provides a good description of the excitations observed at T = 0.015 K and suggests that the scattering intensity may be significantly temperature dependent.

INTRODUCTION

While much interest has been focused on the superfluid phases of liquid ^{3}He, new results on neutron scattering from normal liquid ^{3}He have appeared.[1-3] These experiments provide the first information on the dynamics of the density and spin-density excitations at finite wave vector in normal liquid ^{3}He. Specifically, the initial experiments of Scherm *et al.*[1] and Stirling *et al.*[2] on ^{3}He at T = 1.4 K and 0.63 K, respectively, suggest that the excitations have a very broad frequency response with no well-defined excitations for $Q \geq 12$ nm^{-1}. On the other hand, the measurements of Sköld *et al.*[3] (9 nm^{-1} $\leq Q \leq$ 20 nm^{-1}) on liquid ^{3}He at T = 15 mK suggest (for $Q \leq 15$ nm^{-1}) a well-defined spin-density excitation at low frequency, which might be identified with a paramagnon model,[4-6] and a well-defined density excitation at higher frequency which might be identified with zero-sound[7-10] like propagation. Some calculations of the observed scattering intensity have also been presented.[2,11,12]

In this note we propose a simple random-phase approximation (RPA) model of the density and spin-density excitations, based on the Landau theory of Fermi Liquids,[13] aimed at describing the neutron scattering results. In particular, the quasiparticle-quasihole interaction in the model is identified with the phenomenological Landau parameter description of this interaction. The spin-symmetric part of the interaction used in the model contains, in addition, one parameter, K, which is fixed by assuming that the single particle-hole contribution to density fluctuations of wave vector $Q = 15\ \mathrm{nm}^{-1}$ exhausts one-half of the f-sum rule, an assumption based on comparison with neutron scattering observations. Since the Landau parameters are previously established by fits to specific heat, sound velocity and magnetic susceptibility data,[14] the model contains only one free parameter K.

The Landau theory was proposed (and can be justified[15] on a microscopic basis) to describe the excitations of the normal liquid ^{3}He in the long wave-length approximation ($Q \to 0$). The neutron measurements probe these excitations for wave-vectors up to $Q \sim 3\ k_F$ ($k_F \simeq 7.8\ \mathrm{nm}^{-1}$ for liquid ^{3}He at 1 atmos.). Since the present model may be regarded as a simple extension of the Landau theory, the first purpose here is to test, by comparison with experiment, whether it is in any way reasonable to extend this theory to describe excitations at finite Q. The second purpose is to investigate the temperature dependence of the dynamic structure factor $S(Q,\omega)$ in an attempt to reconcile the differing results at $T = 0.015$ K and at $T \gtrsim 0.6$ K.

THE MODEL

The observed neutron scattering intensity is proportional to the dynamic form factor[16]

$$S(Q,\omega) = \frac{\sigma_c}{\sigma} S_c(Q,\omega) + \frac{\sigma_i}{\sigma} S_i(Q,\omega) \ . \tag{1}$$

Here $\hbar Q$ ($\hbar\omega$) is the momentum (energy) transferred from the neutron to the liquid. This form factor has a coherent part

$$S_c(Q,\omega) = \frac{1}{N} \int_{-\infty}^{\infty} \langle \rho(-Q,0)\rho(Q,t)\rangle e^{-i\omega t}\, dt \ , \tag{2}$$

proportional to the correlation in the Qth Fourier component of the total particle density at time t, $\rho(Q,t) = \rho_\uparrow(Q,t) + \rho_\downarrow(Q,t)$, and an incoherent part,

$$S_I(Q,\omega) = \frac{4}{N} \int_{-\infty}^{\infty} \langle G_Z(Q,0) G_Z(Q,t) \rangle e^{-i\omega t} dt \tag{3}$$

which for an isotropic system is proportional to the correlations in the Qth Fourier component of the spin density, $G_Z(Q,t) = 1/2(\rho_\uparrow(Q,t) - \rho_\downarrow(Q,t))$. Here $\rho_\uparrow(\rho_\downarrow)$ is the density of particles with spin up (down) $\sigma = \sigma_c + \sigma_i$ is the sum of the coherent (σ_c) and the incoherent (σ_i) scattering cross section, and N is the total number of particles.

The simple model consists firstly of calculating the dynamic susceptibility, χ, corresponding to both S_c and S_I within the RPA. These are

$$\chi_c^{RPA}(Q,\omega) = \frac{\chi^o(Q,\omega)}{1 + (\hbar\Omega_o)^{-1} v_s(Q) \chi^o(Q,\omega)}$$

$$\chi_I^{RPA}(Q,\omega) = \frac{\chi^o(Q,\omega)}{1 + (\hbar\Omega_0)^{-1} v_a(Q) \chi^o(Q,\omega)} \tag{4}$$

which are related to S by

$$S(Q,\omega) = 2[n(\omega) + 1] \operatorname{Im} \chi(Q,\omega). \tag{5}$$

Here $\chi^o(Q,\omega)$ is the non-interacting Fermi liquid susceptibility summed over both spin states, $v_s = v_{\uparrow\uparrow} + v_{\uparrow\downarrow}$ and $v_a = v_{\uparrow\uparrow} - v_{\uparrow\downarrow}$ are the spin-symmetric and spin-antisymmetric components of the particle-hole interaction, respectively, $n(\omega)$ is the Bose function and Ω_0 is the volume per ^{3}He atom.

Secondly, we may rearrange the RPA expressions, for example for the coherent susceptibility, as

$$\chi_c^{RPA}(Q,\omega) = \chi^o(Q,\omega)[1 - (\hbar\Omega_0)^{-1} v_s(Q) \chi_c^{RPA}(Q,\omega)]. \tag{6}$$

If we take the limit of $\chi^{o}(Q,\omega)$ as $Q \to 0$, this reduces to the Landau transport equation for the response $\chi_c(Q,\omega) = \hbar\,\delta\rho(Q,\omega)/\delta u(Q,\omega)$ of the total density $\rho(Q,\omega)$ to an external potential u provided that v is related to the Landau parameters via[17]

$$(\hbar\Omega_0)^{-1} v_s = \left(\frac{dn}{d\omega}\right)^{-1} \left[F_0^s + \frac{F_1^s}{1 + \frac{F_1^s}{3}} s^2 \right] \tag{7a}$$

$$(\hbar\Omega_o)^{-1} v_a = \left(\frac{dn}{d\omega}\right)^{-1} \left[F_0^a + \frac{F_0^a}{1 + \frac{F_1^a}{3}} s^2 \right] \tag{7b}$$

Eq. (7b) follows from an equivalent limit for $\chi_I(Q,\omega)$. Here $s = (\omega/Q)/v_F$ is the ratio of the phase to the Fermi velocity and $(dn/d\omega) = \hbar(dn/d\varepsilon) = 3\hbar/2\varepsilon_F)$ is the density of frequency states per ^{3}He atom at the Fermi surface. The $F_0^s(F_0^a)$ and $F_1^s(F_1^a)$ are the dimensionless spin-symmetric (spin-antisymmetric) Landau parameters for angular momentum components $\ell=0$ and $\ell=1$, respectively. Higher components are assumed zero. The reduction also requires that m is replaced by the effective mass $m^* = m(1 + F_1^s/3)$ in $\chi^o(Q,\omega)$. In this way v and m* are determined in the model by their $Q \to 0$ limits

The RPA $S_c(Q,\omega)$ calculated via (5) from χ_c^{RPA} with v_s in (7a) satisfies the f-sum rule

$$\int_{-\infty}^{+\infty} \frac{d\omega}{2\pi}\, \omega\, S_c(\vec{Q},\omega) = -\lim_{\omega\to\infty} \frac{\omega^2}{2}\, \chi_c^{RPA}(\vec{Q},\omega) = \frac{\hbar Q^2}{2m}\,. \tag{8}$$

The χ^{RPA} accounts for only the single quasiparticle - quasihole excitations. Since at $Q \to 0$ only single particle-hole excitations are excited, it is proper that these excitations exhaust the whole sum rule in that limit. However, at finite Q, the observed scattering[1-3] clearly contains a large contribution from multi particle-hole excitations as well as from single particle-hole excitations. In order that $_c^{RPA}$ exhaust only part of the f-sum rule at finite Q, it is necessary to modify v_s in (7a) as Q increases.

For a translationally invariant system, the multi-particle-hole contribution[18] to $S_c(Q,\omega)$, $S_c^{MPH}(Q,\omega)$, is believed to be proportional to $Q^4/\bar{\omega}$ at low Q, where $\bar{\omega}$ is some average frequency. Thus S_c^{MPH} makes a contribution to the f-sum rule $\sim Q^4$. To account for this, and to preserve the f-sum rule, we subtract an equivalent amount $\sim Q^4$ from the single particle-hole contribution to the f-sum rule. This requires that the $S_c^{RPA}(\vec{Q},\omega)$ satisfy the modified sum rule

$$\int_{-\infty}^{\infty} \frac{d\omega}{2\pi} S_c^{RPA}(Q,\omega)\omega = \frac{\hbar Q^2}{2m}(1 - KQ^2) \tag{9}$$

where K is a constant. This follows from (8) provided we multiply v^s in (7a) by a function $\beta(Q)$, i.e.,

$$v^s(Q) = v^s\beta(Q) \tag{10}$$

where

$$\beta(Q) = [1 - \frac{3}{F_1}(\frac{KQ^2}{1-KQ^2})] . \tag{11}$$

This has the effect of reducing the interaction as Q increases.

Comparison of preliminary calculations[18] with experiment suggested that $S_c^{RPA}(Q,\omega)$ accounted for approximately one-half of the observed scattered intensity sum rule in the region of $Q \gtrsim 15\ nm^{-1}$. The remainder was assumed to be the multiparticle-hole excitation contribution. Thus we, rather arbitrarily, select the value $Q = 15\ nm^{-1}$ as the point where S_c^{RPA} contributes one-half to the f-sum rule which from (9) requires $K = 1/2(15)^{-2}nm^2$.

The incoherent part, however, satisfies the sum rule

$$\int_{-\infty}^{\infty} \frac{d\omega}{2\pi} S_I^{RPA}(Q,\omega)\omega = \frac{\hbar Q^2}{2m^*}(1 + \frac{F_1^a}{3})$$

which accounts for only $\sim 1/4$ of the total, $\hbar Q^2/2m$. We believe that this violation of the f-sum rule is due, in part, to the neglect of the multiparticle-hole excitations and we do not modify the interaction v^a.

The $Q \to 0$ limit and the above single assumption fixing K entirely determine $v(Q)$. To calculate χ^o in the RPA expressions we have used the finite temperature result derived by Khanna and Glyde.[19] The parameters F_1^s, F_0^s and F_D^a are determined from macroscopic measurements of the specific heat, the sound velocity and static spin susceptibility, respectively,[14] and F_1^a from the sum rule.

$$\sum_{\ell} \left[\frac{F^{s}}{1+\frac{F}{(2\ +1)}} + \frac{F^{a}}{1+\frac{F^{a}}{(2\ +1)}} \right] = 0$$

which follows from the antisymmetry of the two particle interaction. We have used σ_c = 4.9 b and σ_i = 1.2 b.[20]

Most of the calculations for the dynamic structure factor are done for liquid ^{3}He at a pressure P = 28.4 kPa.

THE RESULTS

In Figure 1 the S(Q,ω) calculated for T = 15 mK employing the T = 0 K Landau parameters in Table 1 is shown. The S(Q,ω) has been folded with a Gaussian of FWHM = 0.33 meV to simulate the effect

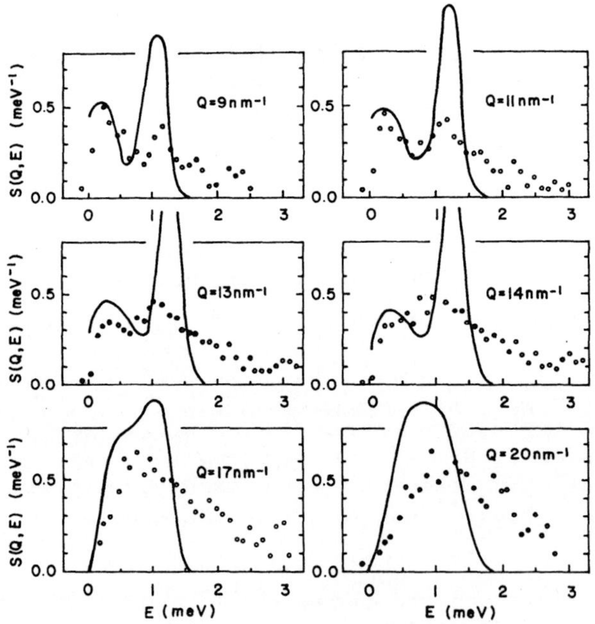

Figure 1. The dynamic form factors, S(Q,E), at 15 mk: ——— calculated S(Q,E) folded with a Gaussian to simulate instrument resolution width of 0.33 meV, °°°°°° experimental points of Sköld <u>et al</u>. (1976).

Table 1. Landau parameters[1]: 1) ref. 14; a) 0.28 atm.; b) 27 atm.

Pressure	F_0^s	F_1^s	F_0^a	F_1^a
28.4 kPa[a]	10.77	6.3	-0.70	-0.76
2.74 Mpa[b]	74.4	13.9	-0.74	-0.51

of the instrumental energy resolution width. The points are the observed values of Sköld *et al*. The incoherent $S_I(Q,\omega)$ part of $S(Q,\omega)$ is concentrated in a resonance at low energy transfer. This resonance is usually denoted as the paramagnon mode. The width of this resonance is much less than the instrumental width so that for practical purposes it may be regarded as a phonon-like mode, an interpretation which is useful in discussion the paramagnon contribution to the superfluid properties of ^{3}He.[6] The position and intensity of this mode agrees well with the observed values.

The calculated coherent $S_c(Q,\omega)$ is concentrated almost entirely in the large, zero sound peak for $Q \lesssim 14\ \text{nm}^{-1}$. The position of this peak (the zero sound mode spectrum) agrees quite well with the observed value.[21] However, while the present model suggests that most of the coherent intensity should lie within this mode, the observed intensity is widely spread and extends to much higher energy transfers. This suggests that the zero sound mode is significantly broadened by decay to multi-particle-hole excitations which are not considered in the present model. The present model includes decay of the zero-sound mode to the single particle-hole states (Landau damping) only, which is negligible for $Q \lesssim 15\ \text{nm}^{-1}$, but thereafter the decay width becomes so large that the zero sound mode is reduced to an over-damped mode.

Figure 2 shows the integrated intensities or static form factors

$$S_c(Q) = \int_0^\infty \frac{d\omega}{2\pi}\, S_c(Q,\omega)$$

and

$$S_I(Q) = \int_0^\infty \frac{d\omega}{2\pi}\, S_I(Q,\omega) .$$

The calculated $S_c(Q)$ is somewhat larger than the experimental results for $Q \leq 14\ \text{nm}^{-1}$. At $Q \gtrsim 18\ \text{nm}^{-1}$ the calculated $S_c(Q)$ reduces

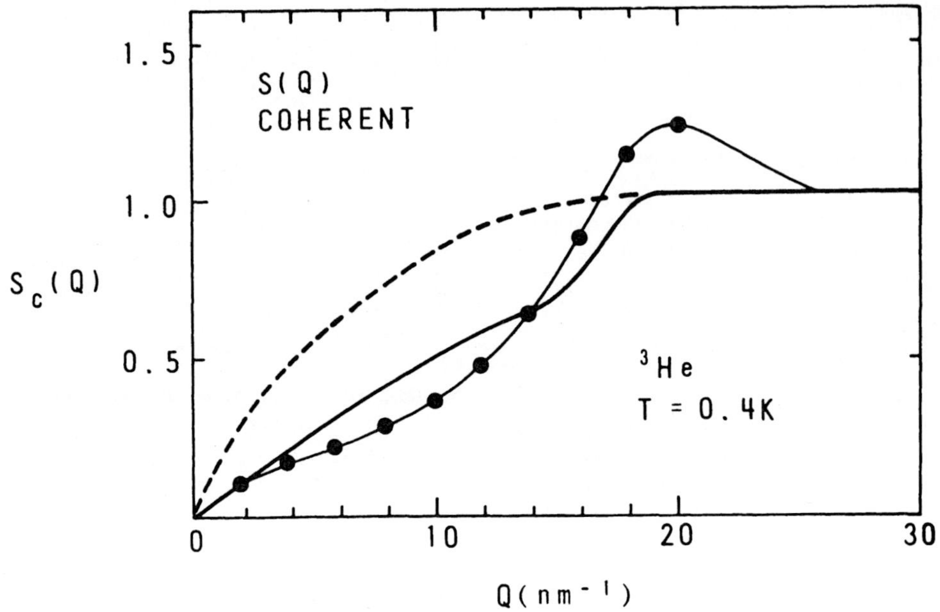

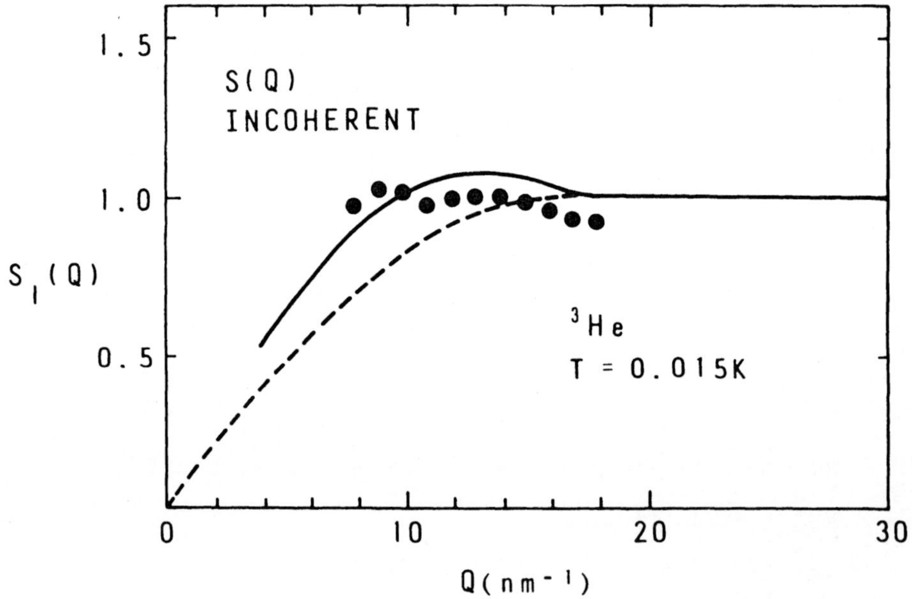

Figure 2. The static structure factor S(Q). The solid lines are the present RPA calculations, the dashed lines are $S_o(Q)$. The dots and line are the smoothed, X-ray $S_c(Q)$ values of Hallock (1972) at T = 0.41 K and the heavy dots are the extracted $S_I(Q)$ values of Sköld et al. (1976).

to the noninteracting value (dashed line) since $\beta(Q) = 0$ for $Q \gtrsim 18\ nm^{-1}$. For entirely independent spins, the incoherent static structure factor should become $S_I(Q) \to 1$. From the lower part of figure 2 we see there is little spin correlation for $Q \gtrsim 7\ nm^{-1}$ in liquid ^{3}He.

In Fig. 3 we show the zero sound mode spectrum calculated for ^{3}He at P = 2.74 MPa (27 atm) using the appropriate phenomenological Landau parameters (table 1). The zero sound mode is sufficiently separated from the single particle-hole spectrum that should it have a roton-like minimum, this minimum may be observable.

In order to investigate the possible temperature dependence of $S(Q,\omega)$ we have allowed both $\chi^o(Q,\omega)$ and the quasi-particle interaction (m^* and $v(Q)$) to vary with temperature. The dependence of m^* on temperature is difficult to determine with any confidence. We have therefore taken three plausible values of $m^*/m = 2.8$, $m^*/m = 2.3$ and $m^*/m = 1.8$ as the effective mass at T = 0.3 K. This temperature was chosen as the plausible final upper limit to the validity[22] of the Landau theory. The Landau parameters corresponding to these effective masses are listed in Table 2.

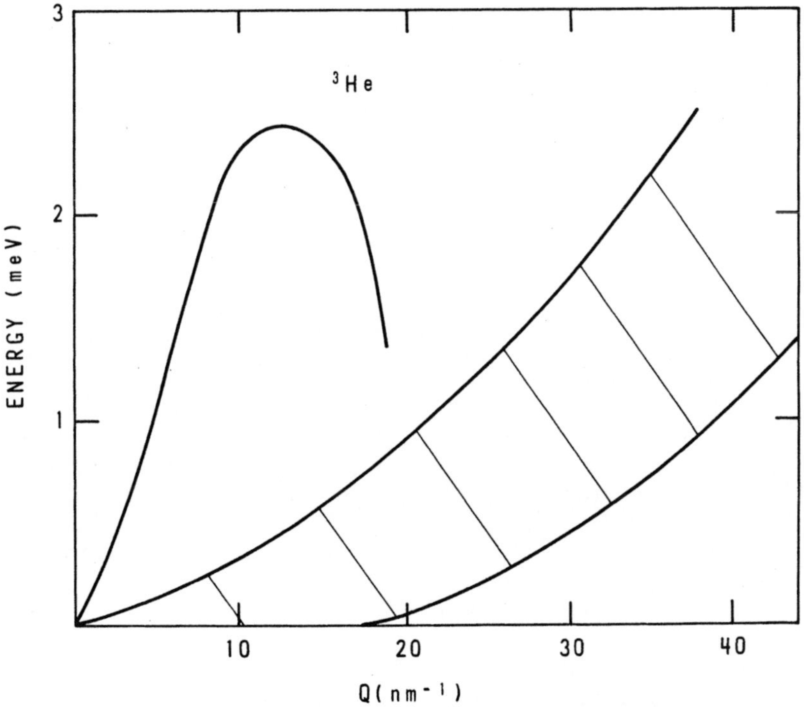

Figure 3. The calculated zero sound and particle-hole spectrum at T = 0 K and p = 2.74 M Pa.

Table 2. Landau parameters at finite temperature at 28.4 kPa pressure.

Temperature	F_0^s	F_1^s	F_0^a	F_1^a
m*=1.8	6.43	2.4	-0.75	1.10
0.3°K m*=2.3	7.21	3.9	-0.68	-0.38
m*=2.8	10.0	5.4	-0.61	-1.28
0.6°K m*=1.8	6.43	2.4	-0.6	-0.57

The change in $S(Q,\omega)$ at $Q = 6\ nm^{-1}$ between T = 0.015 K and T = 0.3 K (for the three selected m*/m) is shown in Fig. 4. There we see that there is essentially no loss of structure with temperature for this Q value. Figure 5 shows the same variation for $Q = 14\ nm^{-1}$. There we see that the structure is completely removed for m*/m = 1.8 at T = 0.3 K. This variation suggests that at larger Q, $Q \gtrsim 10\ nm^{-1}$ say, the effect of temperature could be very important and might explain why no well-defined structure was observed by Stirling _et al_. for $Q \gtrsim 12\ nm^{-1}$ at T = 0.63 K.

Finally, to compare directly with the data of Stirling _et al_. we choose m*/m = 1.8 at T = 0.63 K and calculate the scattered intensity expected in the constant angle scans in Figs. 6 and 7. In view of the uncertainty about the magnitude of the particle-hole interaction and of the effective mass, the present calculations at T = 0.63 K may be viewed as qualitative. But the reasonable agreement with the general trends of the experiment suggests the excitations over a wide range of Q can be described by a single m*/m. At $Q \gtrsim 14\ nm^{-1}$ the calculated $S(Q,\omega)$ in this model reduces simply to that for a non-interacting gas of quasiparticles of m*/m = 1.8 so that the agreement also suggests the real interaction is small. At $Q \geq 25\ nm^{-1}$ an even smaller m*/m is suggested.

A preliminary discussion of this work has already appeared[21] and a full exposition, including comparison with previous theoretical work will appear in the Canadian Journal of Physics.

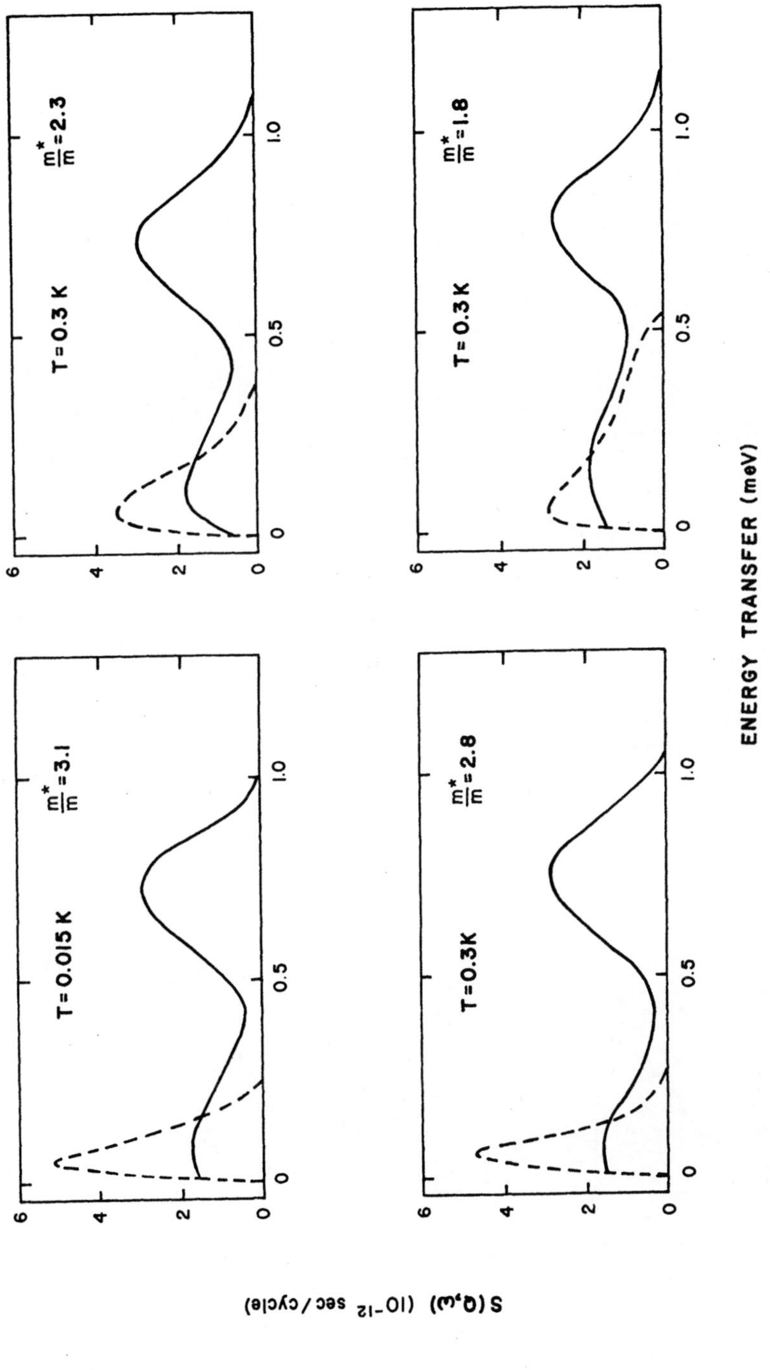

Figure 4. The variation of $S(Q,\omega)$ (in units of 10^{-12} sec.) between T = 0.015 K and T = 0.3 K for the range of values of m*/m at T = 0.3 K suggested by Fig. 8 for Q = 6 nm^{-1}. The dashed line shows the unfolded incoherent $S_I(Q,\omega)$.

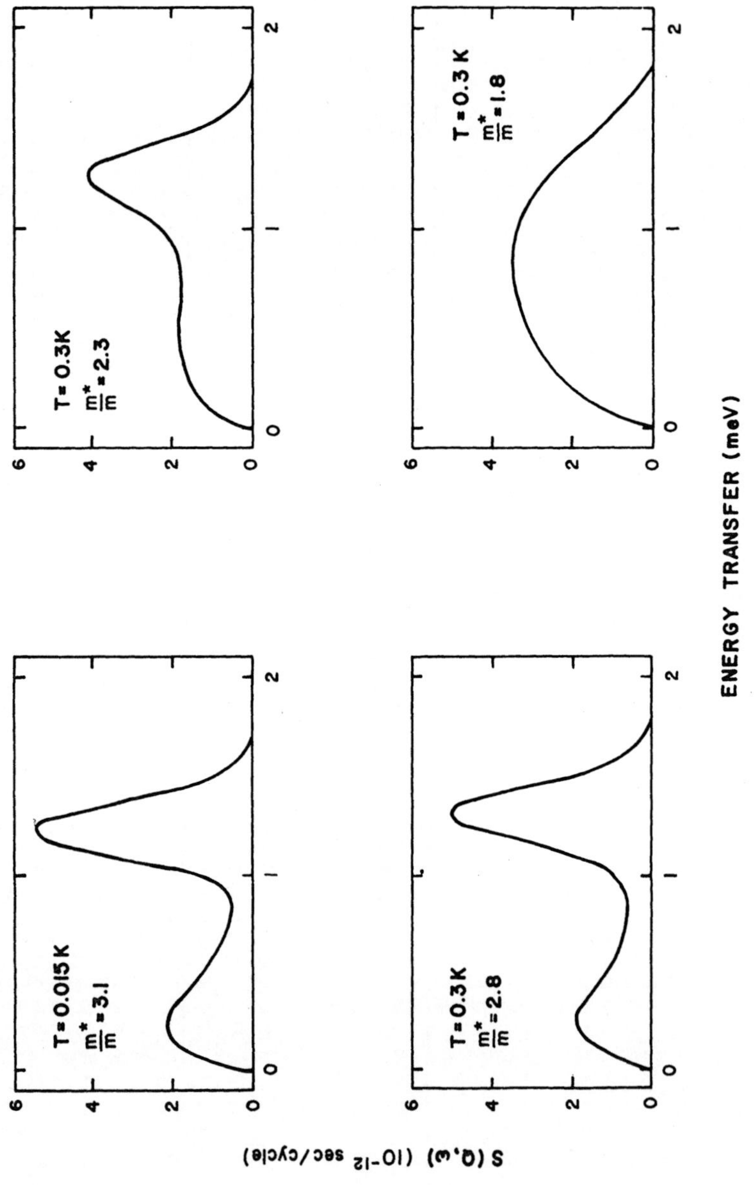

Figure 5. Same as Fig. 4, for Q = 14 nm^{-1}.

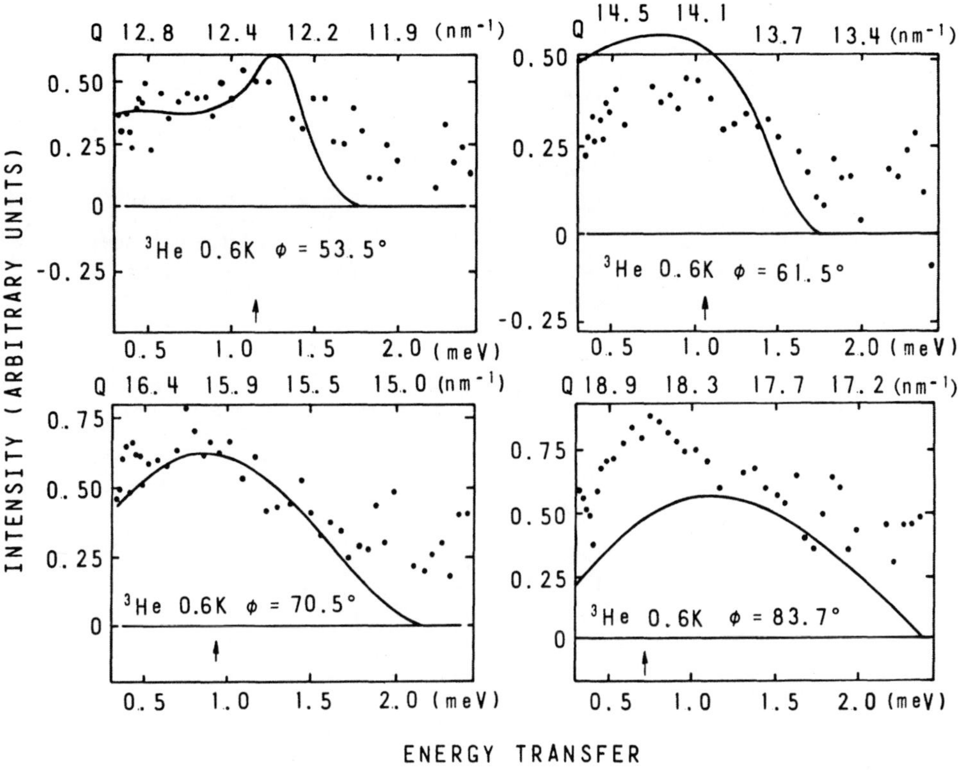

Figure 6. The scattered intensity observed by Stirling et al. (1976) (dots) in constant scattering angle scans for incident neutron wavelength λ = 0.43 nm at T = 0.63 K. The solid lines are the present calculated intensities for T = 0.63 K and m^*/m = 1.8 folded with a Gaussian to simulate the observed instrument energy resolution having FWHM = 0.33 MeV. The scale at the top of the graph shows the corresponding wave vector transfer Q.

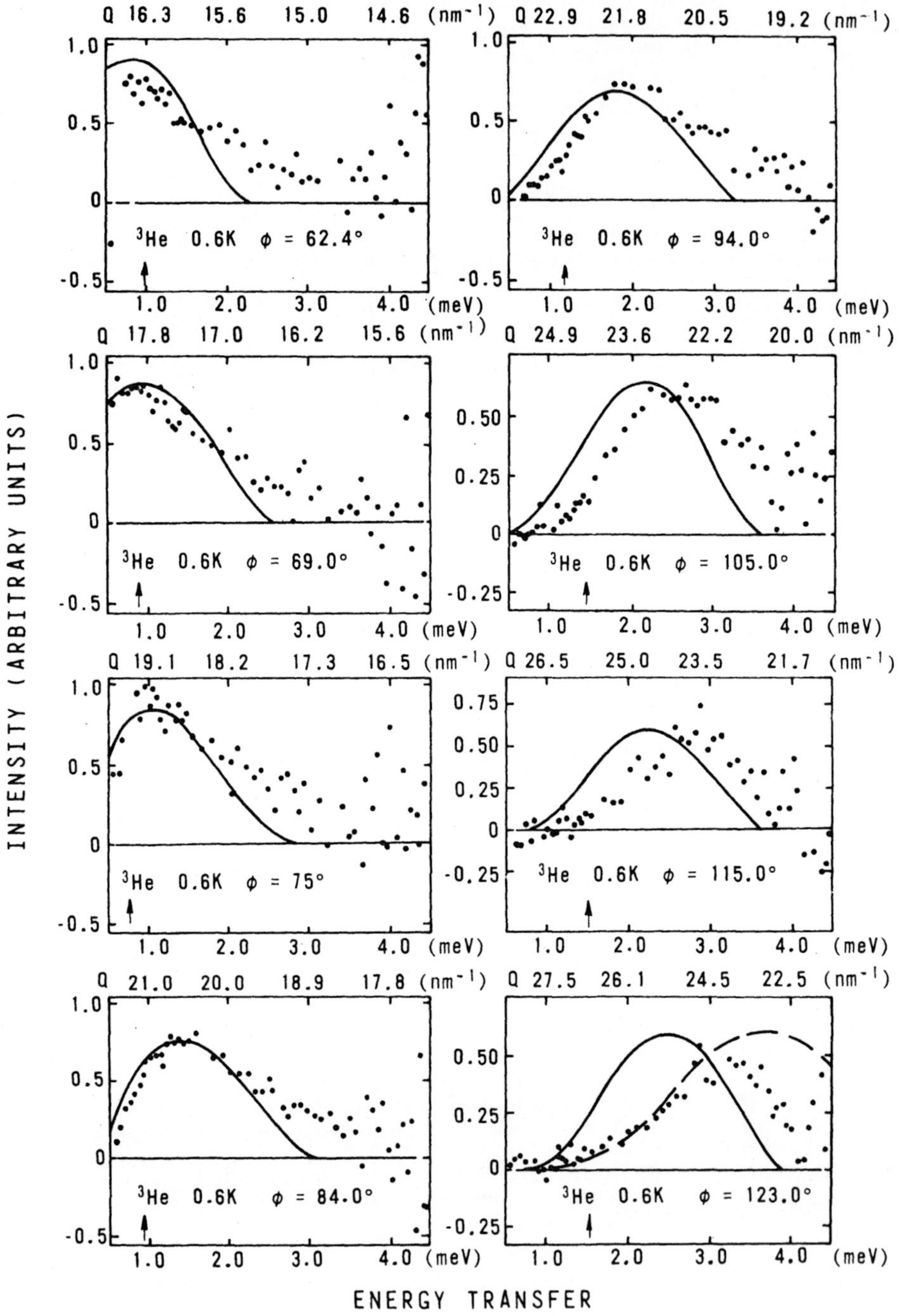

Figure 7. Same as Fig. 6 for $\lambda = 0.38$ nm for which FWHM = 0.48 meV. The single dashed line shows the non-interacting gas scattering intensity ($m^*/m = 1.0$).

CONCLUSION

We find that a simple description based on Landau quasi-particle interaction within the simple RPA provides a reasonable description of the excitations in liquid ^{3}He at low temperature for $Q \leq 2k_F$ (15 nm^{-1}). Since this lies well outside the expected region of validity of the Landau theory, a microscopic explanation of the model is needed. A detailed calculation of the width of the zero sound mode would spread the dynamic response over a much wider frequency range and hence bring it into better agreement with experiment.

The model suggests that $S(Q,\omega)$ shows significant temperature dependence for $Q \gtrsim 10$ nm^{-1}. It would be most interesting to test these predicitions explicitly by experiment. Finally, the extent of the agreement with the T = 0.63 K neutron data for $Q \gtrsim 12$ nm^{-1} employing a non-interacting gas of quasiparticles of $m^*/m = 1.8$ suggests that the interaction is sufficiently weak that a microscopic calculation of the interaction of finite Q and finite T might be successful.

ACKNOWLEDGMENTS

It is a pleasure to acknowledge valuable discussions of this subject with R. A. Cowley, S. W. Lovesey, R. Scherm, V. F. Sears and A. D. B. Woods.

References

(1) R. Scherm, W. G. Stirling, A. D. B. Woods, R. A. Cowley and G. J. Coombs, J. Phys. C7, L341 (1974).

(2) W. G. Stirling, R. Scherm, P. A. Hilton and R. A. Cowley, J. Phys. C9, 1643 (1976).

(3) K. Sköld, C. Pelizzari, R. Kleb and G. E. Ostrowski, Phys. Rev. Lett. 37, 842 (1976).

(4) N. Berk and J. R. Schrieffer, Phys. Rev. Lett. 17, 750 (1966).

(5) S. Doniach, Proc. Phys. Soc. 91, 86 (1967).

(6) A. J. Leggett, Rev. Mod. Phys. 47, 331 (1975).

(7) L. D. Landau, Sov. Phys. JETP 5, 101 (1957).

(8) B. E. Keen, P. W. Mathews and J. Wilkes, Phys. Lett. 5, 5 (1963).

(9) W. R. Abel, A. C. Anderson and J. C. Wheatley, Phys. Rev. Lett. 17, 74 (1966).

(10) D. Pines, in "Quantum Fluids", edited by D. Brewer, (North-Holland, Amsterdam, 1966).

(11) C. H. Aldrich, C. J. Pethick and D. Pines, Phys. Rev. Lett. 37, 845 (1974); C. H. Aldrich, Ph.D. Thesis (Univ. of Illinois) 1974.

(12) S. W. Lovesey, J. Phys. C8, 164a (1975).

(13) L. D. Landau, Sov. Phys. JETP 3, 920 (1956).

(14) J. C. Wheatley, Rev. Mod. Phys. 47, 415 (1975).

(15) P. Nozières, "Theory of Interacting Fermi Systems" (W. A. Benjamin, New York, 1964).

(16) V. F. Sears, J. Phys. C9, 409 (1976).

(17) S. Babu and G. E. Brown, Ann. Phys. 78, 1 (1973).

(18) H. R. Glyde and F. C. Khanna, Proc. Gatlinburg Conf. on neutron scattering (1976); Ed. R. Moon, Conf. 760601-P1.

(19) F. C. Khanna and H. R. Glyde, Can. J. Phys. 54, 648 (1976).

(20) V. F. Sears and F. C. Khanna, Phys. Lett. B 56, 1 (1975).

(21) H. R. Glyde and F. C. Khanna, Phys. Rev. Lett. 37, 1692 (1976).

(22) V. Emery, Ann. Phys. 28, 1 (1964).

TRANSVERSE ZERO SOUND IN NORMAL ^3He*

Pat R. Roach

Argonne National Laboratory, Argonne, Illinois 60439

J. B. Ketterson

Northwestern University, Evanston, Illinois 60201 and
Argonne National Laboratory, Argonne, Illinois 60439

Abstract: We have measured the complex acoustic shear impedance of liquid ^3He by observing the decay of a transiently excited AC cut quartz transducer. Comparison with recent theories suggests that our results at low temperatures can only be due to the excitation of transverse zero sound. We have also observed the direct transmission of transverse excitations between two closely spaced transducers in the vicinity of 3 mK. These results are analyzed in terms of the propagation of transverse zero sound although theory suggests that a single-particle contribution must also be present.

I. INTRODUCTION

Following the development of his theory of a Fermi liquid in 1957,[1] Landau predicted the existence of two collective excitations in normal ^3He at very low temperatures;[2] these excitations are referred to as longitudinal (symmetric) and transverse

---

*Work supported by the U. S. Energy Research and Development Administration and the National Science Foundation under Grant No. DMR-74-12186.

(asymmetric) zero sound. The onset of the longitudinal zero sound mode was first observed by Wilks and coworkers[3] while the complete temperature dependence of the velocity and attenuation was first studied by Abel, Anderson, and Wheatley;[4] the transverse mode has not been observed previously.

A requirement for the propagation of zero sound is that its velocity, V, be greater than the Fermi velocity, V_F. The strength of the Landau-quasiparticle interaction parameters, F_i, determines V and for the transverse mode the condition $V > V_F$ requires $F_1 > 6$ for the case $F_i = 0$, $i \geq 2$; experimental data indicate $F_1 > 6$ for all pressures; reliable data on higher F_i are not available.

The zero (or collisionless) sound regime is characterized by the condition $\omega\tau \gg 1$ where ω is the angular sound frequency and τ is a quasiparticle collision time; in the degenerate regime $\tau \propto T^{-2}$. As the temperature is increased, a hydrodynamic regime is entered where $\omega\tau \ll 1$. The modification of the transverse mode at higher temperatures is radically different from the longitudinal case: the transverse wave goes over into the classical damped viscous shear mode with a complex propagation constant, $K \equiv k + i\alpha$, given by

$$K = (1 + i) \left(\frac{\omega\rho}{2\eta}\right)^{1/2} ; \qquad (1)$$

here ρ is the density and η is the viscosity. This mode is diffusive and does not propagate in the usual sense.

For the case $F_2 = 0$, the attenuation of transverse zero sound was first calculated by Corruccini, Clarke, Mermin and Wilkins[5] and the full dispersion relation for all values of $\omega\tau$ was later calculated by Lea, Birks, Lee and Dobbs.[6] The case $F_2 \neq 0$ has been considered by Fomin[7] and most recently, using a two-time relaxation time approximation for the collision integral, by Flowers and Richardson.[8] These calculations all indicate that the attenuation of transverse zero sound will be several orders of magnitude higher than that of longitudinal zero sound and that its velocity will be only slightly higher than the Fermi velocity.

In order to get around the problem of the very high attenuation in the liquid, we have adopted a slightly indirect method of studying sound propagation; we determine the complex acoustic shear impedance of the liquid by studying the decay time and the frequency shifts of the ringing of a transiently excited AC cut quartz transducer immersed in ^{3}He. In a classical liquid the acoustic impedance is defined by $Z = \rho V$ where ρ is the liquid density and V is the complex acoustic velocity, $V = \omega/K$. If K is written $K = k + i\alpha$ then the acoustic impedance becomes

$$\frac{Z}{\rho} \quad \frac{R}{\rho} + i\frac{X}{\rho} = \frac{\omega k}{k^2 + \alpha^2} - i\,\frac{\omega\alpha}{k^2 + \alpha^2} \qquad (2)$$

In ^{3}He the situation is more complicated and one can expect the acoustic impedance at very low temperatures to result from both the excitation of zero sound and the excitation of single quasiparticles. There have recently appeared calculations of the acoustic shear impedance in ^{3}He by Fomin[9] who considers the low temperature limit ($\omega\tau >> 1$), and by Flowers, Richardson and Williamson[10] who give results for the entire range of $\omega\tau$. These calculations both give the result that the limiting low temperature value of the acoustic impedance is considerably less than the value ρv_F, the value expected classically for a sound mode traveling at a velocity $V \sim v_F$.

II. EXPERIMENTAL TECHNIQUE

We now briefly review the experimental technique used to measure the acoustic impedance. A coaxially plated AC cut quartz transducer, loaded only by the ^{3}He, is excited by an rf pulse whose width is long compared with the rise time, τ_R, of the ^{3}He loaded transducer; following the removal of the excitation pulse the phase and amplitude at any point in the decay of the ringing of the transducer are measured by nulling (at the receiver input) that point in the decaying signal with a second rf signal of variable phase and amplitude. A block diagram of the electronics is shown in Fig. 1.

The behavior of the ringing signal is given by

$$\psi(t) = \psi(0)\ \exp[-(t/\tau_R) - i(\omega + \Delta\omega)t]$$

where

$$\frac{1}{\tau_R} = \frac{2R_H}{\pi R_Q}\,\omega_0 \quad \text{and} \quad \Delta\omega = \frac{2X_H}{\pi R_Q}\,\omega_0\ ; \qquad (3)$$

here $f_0 = \omega_0/2\pi$ is the fundamental frequency of the transducer, $\omega = (2n+1)\omega_0$ where n is the harmonic of the transducer being excited, R_H and X_H are the real and imaginary parts of the ^{3}He acoustic impedance, Z_H, and R_Q is the (real) acoustic impedance of the quartz (our expression for τ_R and $\Delta\omega$ assume $Z_H << R_Q$). The

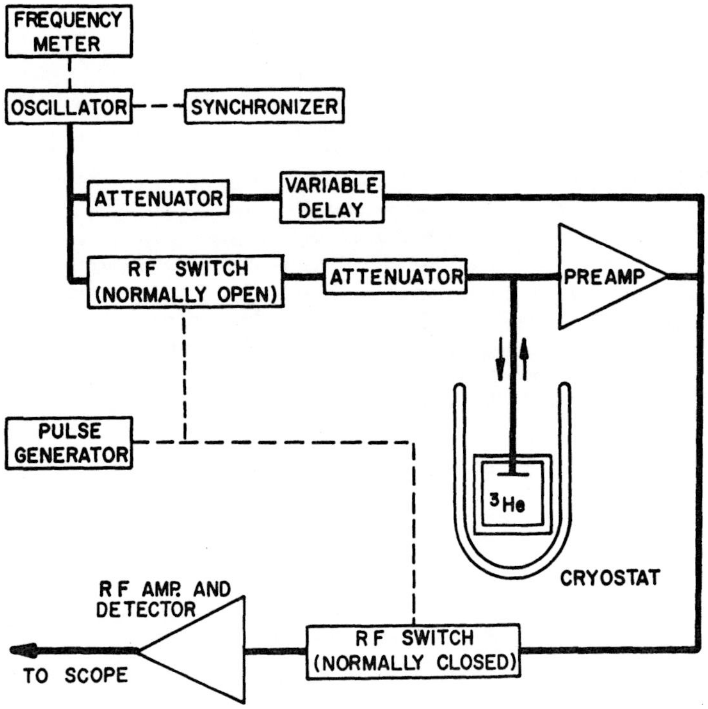

Figure 1. Block diagram of the electronics used to measure the acoustic shear impedance of liquid ^{3}He.

existence of spurious effects on the amplitude and phase of the ringing signal made it necessary to normalize the data. The decay was measured at high temperatures where the viscous damping is negligible and the spurious decay time, τ_S, determined. At lower temperatures, where ^{3}He loading effects appear, the total decay time, τ_T, was measured and τ_R was deduced from $1/\tau_T = 1/\tau_R + 1/\tau_S$.

In the experiment studying direct transmission of transverse excitations, a pair of AC cut quartz transducers was used to generate and detect the transverse waves. Because of the very high attenuation predicted for transverse zero sound in ^{3}He, the path length employed was 0.0025 cm, provided by a fine platinum wire separating the two transducers. Some bowing of the transducer is possible as well as a compression of the wire spacer; both of these effects can lead to some uncertainty in the path length. Due to the extremely short wavelength of the transverse mode (5×10^{-4} cm at 12 MHz) nonparallelism between the transducers can reduce the amplitude of the signal. By viewing with coherent light two optical flats separated by a wire spacer, it was determined that parallelness of the order of a fringe could be achieved. With the path length employed there is no advantage in using a pulse technique since a time separation between the feedthrough and the signal is not possible; thus a continuous, amplitude-modulated, rf signal was applied to the first transducer. The output of the second transducer was applied to an rf amplifier-detector and then to a lock-in amplifier which was, in turn, driven by the modulation frequency, f_m. The modulation frequency of 20 Hz was such that $f_m \ll 1/\tau_R$ (where τ_R is the ringing time of the transducers) and thus the transducer can follow the rf envelope. A certain amount of electromagnetic coupling or feedthrough exists between the receiving and transmitting transducers; for 12 MHz the feedthrough was the same order of magnitude as the zero sound signal at 2.5 mK. The feedthrough was nulled by passing am-modulated rf from the oscillator through a phase shifter and attenuator and adding it to the receiver input. Some temperature dependence of the feedthrough is expected due to the temperature dependence of the electrical impedance of the transducers (which in turn arises from the ^3He acoustic impedance); this affects the magnitude of the rf signal in the cell and thus the feedthrough. However, this small effect on the feedthrough should be completely temperature independent below $\sim$ 5 mK according to our acoustic impedance measurements. Some spurious signal from a small amount of longitudinal zero sound could also exist. This effect is also expected to be negligible below 5 mK. Therefore, we nulled the detected signal near this temperature and the signal which grows rapidly out of the noise below about 4 mK can only arise from transverse excitations being transmitted through the liquid. In order to measure the attenuation, a second attenuator-phase shifter combination (in parallel with the previously mentioned one) was used to null the transverse signal. In order to increase the signal to noise, an unmodulated rf bias of appropriate phase and amplitude was also added to the receiver input.

III. EXPERIMENTAL RESULTS AND DISCUSSION

Figures 2(a) and 2(b) show, for a pressure of 23 bar, the values of R_H/ρ and X_H/ρ as a function of temperature for frequencies of 36, 60 and 108 MHz, determined from measurements of $1/\tau_R$ and $\Delta\omega$, respectively. The values of X_H/ρ have been normalized such that the results for the three frequencies all extrapolate to zero at high temperature since this is the theoretically predicted behavior. In the time since the initial presentation of these results,[11] several calculations have appeared[9,10] which give predictions for the behavior of the acoustic shear impedance. The curves in Fig. 2 are the results of the calculations by Flowers _et al._[10] using $F_2 = 0$. The values of τ used in their calculations were converted to T by using the viscous relaxation times given by Wheatley[12] for low temperatures and by interpolating between the

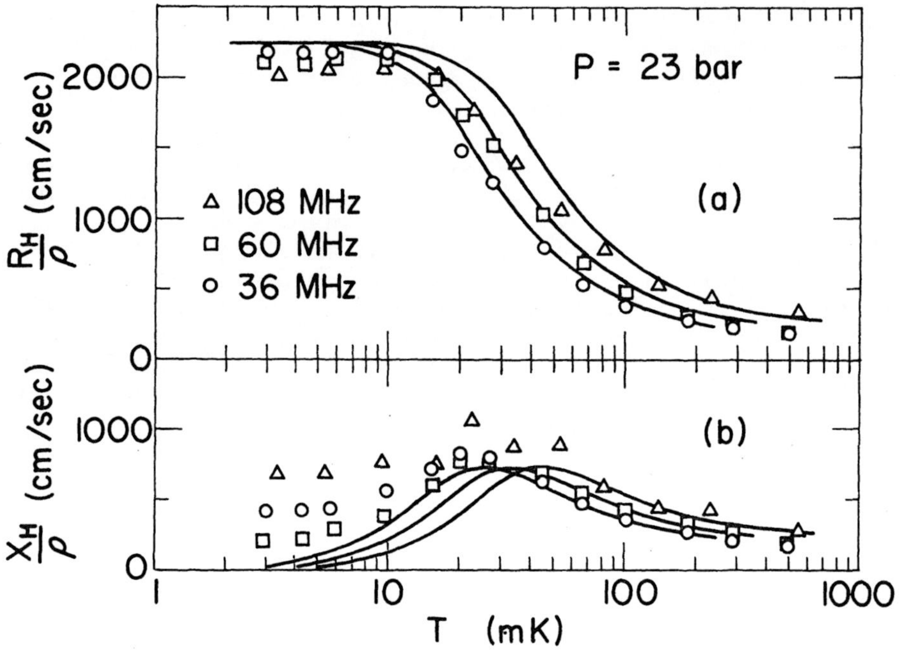

Figure 2. The temperature dependence of the measured values of (a) R_H/ρ and (b) X_H/ρ for a pressure of 23 bar. The solid lines show the behavior predicted in Ref. 10.

viscosity data of Bertinat et al.[13] at low pressure and of Alvesalo et al.[14] at the melting pressure to deduce $\tau(T)$ at higher temperatures. It is seen that the data for R_H/ρ are in reasonably good agreement with theory. Both suggest that the limiting value of R_H/ρ is 2000-2200 cm/sec, considerably less than the value of v_F = 3440[12] cm/sec which would be expected classically. The values of X_H/ρ are in reasonable agreement with theory only at high temperatures. At low temperatures the measured values of X_H/ρ are considerably larger than the theoretically predicted values. The origin of this discrepancy is not understood at this time, but we suspect that it is due to spurious, temperature-dependent effects that are not properly handled by our simple normalization procedure. Nevertheless, it will be noted (as emphasized by Fomin[9]) that an example of the characteristic crossings seen in the theoretical curves can also be seen in the 36 and 60 MHz data at around 30 mK.

Figure 3 shows the results of the measurements of the direct transmission of transverse excitations through the liquid ^3He. As in all acoustic experiments where only the relative attenuation

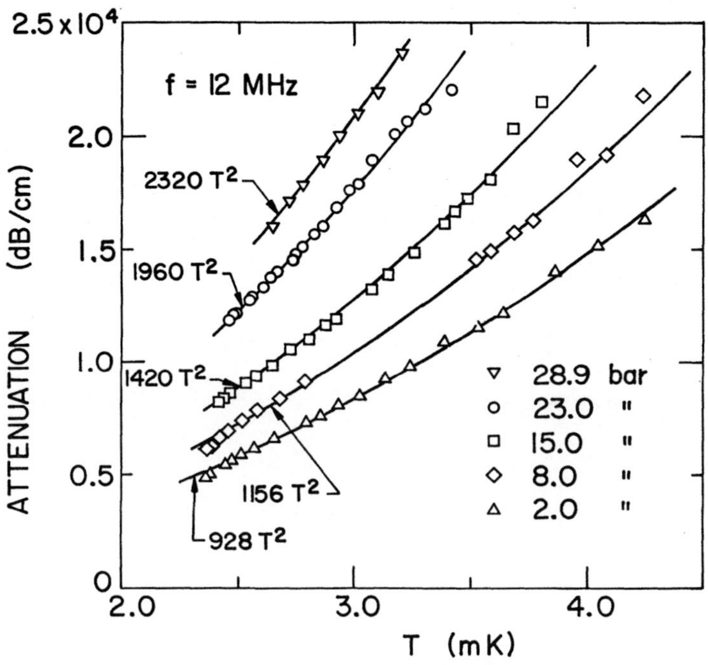

Figure 3. The temperature dependence of the attenuation of transverse excitations for pressures of 2.0, 8.0, 15.0, 23.0 and 28.9 bar. The fitted coefficient of the T^2 temperature dependence is shown next to the data of each pressure.

can be determined, a normalization was required. The normalization was accomplished by requiring T^2 dependence for the attenuation. A smoothly varying behavior of the extrapolated zero temperature normalization constant with pressure provided an overall consistency check on the data. The lines through the data are the T^2 curves fitted to the data to determine the coefficients listed with the data for each pressure. Figure 4 shows these T^2 -coefficients plotted vs. pressure. The solid curves in this figure are the theoretical attenuation coefficients for transverse zero sound for different values of the parameter F_2. These calculated curves are based on a single-relaxation-time-approximation expression for the dispersion relation[7,8,11] and use values tabulated by Wheatley[12] for v_F, τ, and F_1. The dashed curve gives the attenuation coefficient at the limiting values of F_1 and F_2 beyond which the dispersion relation has no real solutions. With these assumptions and approximations (and neglecting any contribution from single quasiparticle excitations), one would conclude that the value of F_2 varies from 1.5 at 2 bar to -1.2 at 29 bar. Flowers and Richardson[8] calculate the behavior of the shear impedance and of transverse zero sound using two relaxation times in order to

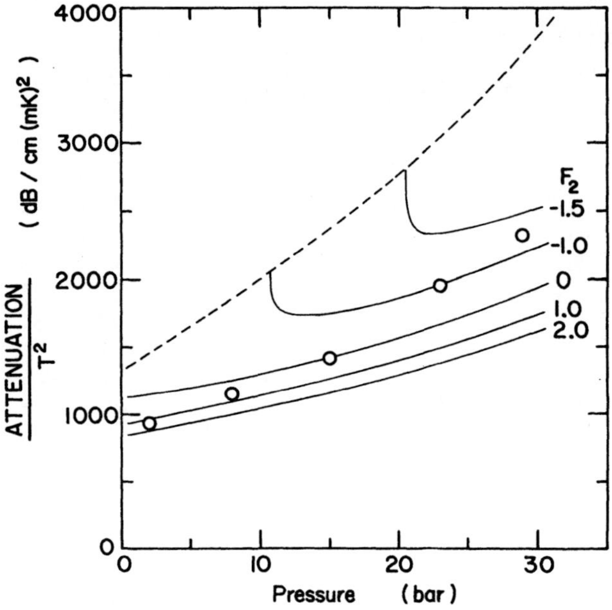

Figure 4. The pressure dependence of the attenuation coefficients of the transverse excitations. The solid lines are theoretical values of the attenuation coefficient of transverse zero sound for various values of F_2.

study the behavior for all values of $\omega\tau$. Using this approach, with a relaxation time describing the transverse zero sound taken to be about 1/3 that describing the high temperature viscosity data (used in Fig. 4), one finds that the theoretical curves in Fig. 4 are all raised much higher than the data. In addition, the detailed shape of the curves of constant F_2 are somewhat different. The higher attenuation predicted for transverse zero sound in this approach might suggest that the signal detected contains a large single quasiparticle contribution. The single quasiparticle contribution would be expected to have an attenuation described by the dashed line in Fig. 4. Due to some uncertainty in the experimental path length and due to some uncertainty in the parameters determining the attenuation of quasiparticles in the theory, it is possible that our data are not inconsistent with the dashed line of Fig. 4; the pressure dependence certainly seems similar. Altogether, these questions raise some doubt about the identification of our transmission results as due entirely to transverse zero sound. However,our acoustic impedance results are not open to the same criticism since attenuation of the sound mode has little effect on R_H at low temperatures. In fact, the large value of R_H/ρ at low temperatures can only be explained as arising from the excitation of transverse zero sound since the value would be 84% lower if only quasiparticles were being excited.[10]

References

(1) L. D. Landau, Zh. Eksp. Teor. Phys. 30, 1058 (1956); Soviet Phys. JETP 3, 591 (1956).

(2) L. D. Landau, Zh. Eksp. Teor. Phys. 32, 59 (1957); Soviet Phys. JETP 5, 101 (1957).

(3) B. E. Keen, P. W. Matthews, and J. Wilks, Phys. Letters 5, 5 (1963); D. S. Betts, B. E. Keen, and J. Wilks, Proc. Roy. Soc. A 289, 34 (1965); I. J. Kirby and J. Wilks, Phys. Letters A 24, 60 (1966).

(4) W. R. Abel, A. C. Anderson, and J. C. Wheatley, Phys. Rev. Letters 17, 74 (1966).

(5) L. R. Corruccini, J. S. Clarke, N. D. Mermin, and J. W. Wilkins, Phys. Rev. 180, 225 (1969).

(6) M. J. Lea, A. R. Birks, P. M. Lee, and E. R. Dobbs, J. Phys. C: Solid State Phys. 6, L226 (1973).

(7) I. A. Fomin, Ah. Eksp. Teor. Phys. 54, 1881 (1968); Soviet Physics JETP 27, 1010 (1968).

(8) E. G. Flowers and R. W. Richardson, (to be published).

(9) I. A. Fomin, Pis'ma Zh. Eksp. Teor. Phys. 24, 90 (1976).

(10) E. G. Flowers, R. W. Richardson, and S. J. Williamson, Phys. Rev. Letters 37, 309 (1976).

(11) Pat R. Roach and J. B. Ketterson, Phys. Rev. Letters 36, 736 (1976).

(12) J. C. Wheatley, Rev. Mod. Phys. 47, 415 (1975).

(13) M. P. Bertinat, D. S. Betts, D. F. Brewer, and G. J. Butterworth, J. Low Temp. Phys. 16, 479 (1974).

(14) T. A. Alvesalo, H. K. Collan, M. T. Loponen, O. V. Lounasmaa and M. C. Veuro, J. Low Temp. Phys. 19, 1 (1975).

MOTION OF IONS IN NORMAL AND SUPERFLUID ^{3}He

Juhani Kurkijärvi†

Research Institute for Theoretical Physics
University of Helsinki, 00170 Helsinki, Finland

Alexander L. Fetter

Institute of Theoretical Physics
Department of Physics, Stanford University
Stanford, Ca. 94305

Abstract: We discuss the drag force on an ion in low temperature ^{3}He in the normal state and in the superfluid state. We give numerical results for the nonlinear high velocity regime of the mobility in the superfluid B phase in the elastic scattering approximation.

Ions have a rich history in the study of the structure and dynamics of the liquid helium.[1,2,3] The discovery of superfluid ^{3}He[4] renewed interest in the motion of ions in pure ^{3}He[5,6] where experimental data are available[6,7] down to 1 mK on negative ions

†Permanent address: Department of Technical Physics and Low Temperature Laboratory, Helsinki University of Technology, 02150 Espoo, Finland.

in normal ^{3}He and in the superfluid A and B phases.

The theory of ion motion in superfluid ^{3}He is a challenging and multifaceted problem. At low temperatures the drag force on an ion is mainly due to collisions with Landau quasiparticles.[8] Unlike in impurity scattering in metals, however, the ion here has a finite mass which leads to inelastic scattering, and energy loss to the quasiparticles at low enough temperatures. This should eventually block the scattering and lead to infinite ion mobility since the quasiparticle states below the fermi energy are already occupied. Josephson and Lekner[9] have developed a self-consistent formalism that includes the coupled dynamics of the ion and the quasiparticles, a formalism that contains subtle approximations and leaves us with the problem of coming to grips with the dynamics of the ion in the background of the quantum fluid. Finally, the collisions process between a quasiparticle and the ion is delicate in the superfluid phases and a nontrivial calculation[10] is needed to produce a finite linear mobility at all, in the absence of recoil effects.

It has become interesting, in connection with the superfluid phases of ^{3}He, to formulate a theory of ion mobility including nonlinear effects in the ion velocity $\mathbf{v}$.[10] We therefore generalize the Josephson-Lekner formalism to take into account the full $\mathbf{v}$-dependence and incorporate the superfluid features in a preliminary weak scattering approximation.[11] The additional coherence factors produce a complicated structure which together with the inelastic effects associated with ionic recoil present a difficult problem. In the case of negative ions in ^{3}He, however, a self-consistent study of the nonlinearities in the normal phase indicate that the purely elastic scattering model is probably adequate also in the superfluid phases not too far from T_c. We compute the full velocity dependent mobility in the B-phase in the elastic scattering approximation. For a given relationship between the gap and the reduced temperature we give the mobility as a function of the normal state mobility at T_c and the temperature implying a simple scaling law for different pressures. Our results are in good qualitative agreement with experiment.

The system of interest is an ion at $\mathbf{r}_0$, moving with a drift velocity $\mathbf{v}$ through stationary ^{3}He, treated as a weakly interacting gas of Landau quasiparticles. Since the quasiparticles have long mean free paths at low temperatures, kinetic theory provides a reasonable description of the interaction with the ion, which is assumed to occur through a potential $V(\mathbf{r} - \mathbf{r}_0)$.

$$\hat{H}_1 = V^{-1} \sum_{\mathbf{k}\mathbf{k}'\alpha} a^{\dagger}_{\mathbf{k}'\alpha} a_{\mathbf{k}\alpha} V_{-\mathbf{K}} e^{i\mathbf{K}\cdot\mathbf{r}_0} , \tag{1}$$

where V is the quantization volume, $a^{\dagger}_{k\alpha}$ is the creation operator for a Landau quasiparticle with wave vector $\underset{\sim}{k}$ and spin projection α, $V_{\underset{\sim}{K}}$ is the Fourier transform of V(r),

$$\hbar \underset{\sim}{K} \equiv \hbar(\underset{\sim}{k} - \underset{\sim}{k}') \tag{2}$$

is the momentum transfer to the ion. The "Golden rule" then gives the transition probability per unit time and the force

$$\frac{d\underset{\sim}{P}_{qp}}{dt} = \frac{2\pi}{\hbar} \Sigma\, \hbar(\underset{\sim}{k}_i - \underset{\sim}{k}_f) |\langle f|H_1|i\rangle|^2\, \delta(\varepsilon_f - \varepsilon_i) \tag{3}$$

where ε_f and ε_i denote the total final and initial energies including both the ion and the quasiparticles.

In treating the superfluid, the first step is the introduction of new quasiparticle operators $\{\gamma_{\underset{\sim}{k}\alpha}\}$ that are linear combinations of the original operators $a_{\underset{\sim}{k}\alpha}$ and $a^{\dagger}_{-\underset{\sim}{k}\alpha}$.

$$\begin{aligned} a_{\underset{\sim}{k}\alpha} &= u_{\underset{\sim}{k}\alpha\beta}\, \gamma_{\underset{\sim}{k}\beta} + v_{\underset{\sim}{k}\alpha\beta}\, \gamma^{\dagger}_{-\underset{\sim}{k}\beta} \\ a^{\dagger}_{-\underset{\sim}{k}\alpha} &= u^{*}_{-\underset{\sim}{k}\alpha\beta}\, \gamma^{\dagger}_{-\underset{\sim}{k}\beta} + v^{*}_{-\underset{\sim}{k}\alpha\beta}\, \gamma_{\underset{\sim}{k}\beta} \end{aligned} \tag{4}$$

where repeated indices are summed over the two spin components. The associated 2 x 2 matrices $u_{\underset{\sim}{k}}$ and $v_{\underset{\sim}{k}}$ have the explicit representation[12]

$$\begin{aligned} u_{\underset{\sim}{k}\alpha\beta} &= \delta_{\alpha\beta}(E_{\underset{\sim}{k}} + \xi_k)[(E_{\underset{\sim}{k}} + \xi_k)^2 + |\Delta_{\underset{\sim}{k}}|^2]^{-1/2} \quad , \\ v_{\underset{\sim}{k}\alpha\beta} &= \Delta_{\underset{\sim}{k}\alpha\beta}[(E_{\underset{\sim}{k}} + \xi_k)^2 + |\Delta_{\underset{\sim}{k}}|^2]^{-1/2} \quad , \end{aligned} \tag{5}$$

where $\xi_k = \hbar^2(k^2 - k_F^2)/2m,$

$$E_{\underset{\sim}{k}} = (\xi_k^2 + |\Delta_{\underset{\sim}{k}}|^2)^{1/2} \tag{6}$$

and the gap matrix $\Delta_{\underset{\sim}{k}\alpha\beta}$ is assumed unitary. Note that $u_{\underset{\sim}{k}}$ and $v_{\underset{\sim}{k}}$ are symmetric matrices with $u_{\underset{\sim}{k}} = u_{-\underset{\sim}{k}}$ and $v_{\underset{\sim}{k}} = -v_{-\underset{\sim}{k}}$ the last because of the p-wave character. A straightforward calculation averaging over initial states yields the net force on the quasiparticles

$$-\frac{d\underset{\sim}{P}_{qp}}{dt} = \frac{\pi}{\hbar V^2} \sum_{\underset{\sim}{k}\underset{\sim}{k}'} \underset{\sim}{K}|V_K|^2 \Big\{[S_{\underset{\sim}{v}}(\underset{\sim}{K},E/\hbar - E'/\hbar)f(1-f')$$

$$+ S_{\underset{\sim}{v}}(\underset{\sim}{K},E'/\hbar - E/\hbar)f'(1 - f)]$$

$$\times \operatorname{tr}[(u^*_{\underset{\sim}{k}}u_{\underset{\sim}{k}'} - v^*_{\underset{\sim}{k}'}v_{\underset{\sim}{k}})(u^*_{\underset{\sim}{k}'}u_{\underset{\sim}{k}} - v^*_{\underset{\sim}{k}}v_{\underset{\sim}{k}'})]$$

$$+ [S_{\underset{\sim}{v}}(\underset{\sim}{K},E/\hbar + E'/\hbar)ff' + S_{\underset{\sim}{v}}(\underset{\sim}{K},-E/h - E'/h)$$

$$\times (1 - f)(1 - f')]$$

$$\times \operatorname{tr}[(u^*_{\underset{\sim}{k}}v_{\underset{\sim}{k}'} + u^*_{\underset{\sim}{k}'}v_{\underset{\sim}{k}})(u_{\underset{\sim}{k}}v^*_{\underset{\sim}{k}'} + u_{\underset{\sim}{k}'}v^*_{\underset{\sim}{k}})]\Big\} \tag{7}$$

where the first set of terms arises from scattering of quasi-particles and the second from creation and annihilation of quasi-particle pairs. Here $E(E')$ denotes $E_{\underset{\sim}{k}}(E_{\underset{\sim}{k}'})$, f is the Fermi function, and $S_{\underset{\sim}{v}}(\underset{\sim}{K},\omega)$ is the dynamic structure factor for the moving ion

$$2\pi\, S_{\underset{\sim}{v}}(\underset{\sim}{K},\omega) = \int_{-\infty}^{\infty} dt\, e^{i\omega t}\, F_{\underset{\sim}{v}}(\underset{\sim}{K},t), \tag{8a}$$

$$F_{\underset{\sim}{v}}(\underset{\sim}{K},t) = \langle e^{-i\underset{\sim}{K}\cdot\underset{\sim}{r}_0(t)}\, e^{i\underset{\sim}{K}\cdot\underset{\sim}{r}_0(0)}\rangle\ . \tag{8b}$$

We now follow Josephson and Lekner in assuming that, as a consequence of the properties of Galilean transformations,

$$S_{\underset{\sim}{v}}(\underset{\sim}{K},\omega) = S_0(\underset{\sim}{K},\omega - \underset{\sim}{K}\cdot\underset{\sim}{v})\ . \tag{9}$$

$S_0(\underset{\sim}{K},\omega)$ also obeys the detailed-balanced condition

$$S_0(-\underset{\sim}{K},-\omega) = e^{-\beta h\omega}\, S_0(\underset{\sim}{K},\omega) \tag{10}$$

where S_0 is actually independent of the direction $\hat{K}$.

A little manipulation with Eqs. (9) and (10) simplifies Eq. (7) to the fundamental form

$$\frac{d\underset{\sim}{P}_{qp}}{dt} = \frac{2k_F^2}{\pi} \int \frac{d\Omega}{4\pi} \frac{d\Omega'}{4\pi} \underset{\sim}{K} \frac{d\sigma}{d\Omega} I \tag{11}$$

where we have transformed the sums over $\underset{\sim}{k}$ and $\underset{\sim}{k}'$ into integrals and used the Golden Rule to identify the differential cross section

$$\frac{d\sigma}{d\Omega} = \frac{N(0)^2\pi^2}{k_F^2} \, |V_K|^2 \tag{12}$$

with $N(0) = m^* k_F/2\pi^2\hbar^2$ the density of states for one spin projection. The remaining quantity I in Eq. (11) contains all the specific superfluid properties

$$\begin{aligned} I = (e^{\beta\hbar \underset{\sim}{K}\cdot \underset{\sim}{v}}-1) \int_0^\infty d\xi \int_0^\infty d\xi' \{ [& S_0(\underset{\sim}{K}, E/\hbar - E'/\hbar - \underset{\sim}{K}\cdot\underset{\sim}{v}) \\ & \times f(1-f') \\ & + S_0(\underset{\sim}{K}, E'/\hbar - E/\hbar - \underset{\sim}{K}\cdot\underset{\sim}{v}) \, f'(1-f)] \\ & \times [1 - (EE')^{-1} \, \mathrm{Re}(\underset{\sim}{\Delta}_{\underset{\sim}{k}}\cdot\underset{\sim}{\Delta}^*_{\underset{\sim}{k}'})] \\ & + [S_0(\underset{\sim}{K}, E/\hbar + E'/\hbar - \underset{\sim}{K}\cdot\underset{\sim}{v}) ff' \\ & + S_0(\underset{\sim}{K}, -E/\hbar - E'/\hbar - \underset{\sim}{K}\cdot\underset{\sim}{v})(1-f)(1-f')] \\ & \times [1 + (EE')^{-1} \, \mathrm{Re}(\underset{\sim}{\Delta}_{\underset{\sim}{k}}\cdot\underset{\sim}{\Delta}^*_{\underset{\sim}{k}'})]\} \ , \end{aligned} \tag{13}$$

where the coherence factors have now been evaluated explicitly with Eq. (5) and the customary p-wave "vector" representation[13]

$$\Delta_{\underset{\sim}{k}\alpha\beta} = i(\underset{\sim}{\Delta}_{\underset{\sim}{k}}\cdot\underset{\sim}{\sigma}\sigma_2)_{\alpha\beta} \ . \tag{14}$$

Terms proportional to $\xi\xi'$ have been dropped because they vanish on integration over ξ and ξ'. Equations (11) and (13) together constitute the desired application of the Josephson-Lekner formalism. They include terms of all order in $\mathbf{v}$, and the familiar linear behavior is obtained by the approximation $\exp(\beta\hbar\mathbf{K}\cdot\mathbf{v}) - 1 \simeq \beta\hbar\mathbf{K}\cdot\mathbf{v}$, with $\mathbf{v} = 0$ in the remaining factors. Furthermore, the p-wave character of the gap appears in each of the coherence factors.

The Golden Rule calculation implies, strictly speaking, that the correlation function of Eq. (8) is that of the free ion. In Eq. (12), however, we implicitly identify $|V_K|^2$ with the absolute square of the T-matrix. In the same way we will consider $S_0(K,\omega)$, just as Josephson and Lekner, as the scattering function of the ion interacting with the quasiparticle gas, which is exact only in the purely elastic scattering model with the ion moving unperturbed at a constant velocity. This approximation is very different from the conventional one made in slow neutron scattering, for instance, in that the wave length associated with the quasiparticle is not large compared with the size of the ion; in fact, the opposite is true.

The large momentum transfer, on the order of k_F, is the reason for which the coupling with the order parameter is weak in the superfluid phase and the contribution of order parameter modes is negligible as distinct from the long wave length sound propagation problem.[14] With the correlation function approach of Ambegaokar, de Gennes and Rainer[15] one can also show[16] that the effect of any texture distortion around the ion on the mobility is negligible.

Let us consider the drag force in the normal state where the gap function vanishes identically.

$$I_n = (e^{\beta\hbar\mathbf{K}\cdot\mathbf{v}}-1) \iint_{-\infty}^{\infty} d\xi d\xi' \; S_0(K,\xi/\hbar-\xi'/\hbar-\mathbf{K}\cdot\mathbf{v}) f(\xi) \times [1-f(\xi')]. \tag{15}$$

Use of Eq. (8a) and the detailed balance now allows a direct evaluation of the integrals over ξ and ξ'.

$$I_n = \frac{\pi}{2\beta^2} \tanh(\tfrac{1}{2}\beta\hbar K\cdot v) \int_{-\infty}^{\infty} \frac{dtF_0(K,t)}{\cosh^2(\pi t/2\beta\hbar)} \times \left[1 + \frac{\sin^2(\frac{1}{2}\mathbf{K}\cdot\mathbf{v}t)}{\sinh^2(\pi t/2\beta\hbar)}\right] . \tag{16}$$

An expansion to first order in $\underset{\sim}{v}$ reproduces the Josephson-Lekner relation for the low field mobility. The idea of the Josephson-Lekner approach is to note that an ansatz for the frequency dependence of the linear mobility is sufficient to determine $F_0(K,t)$ in the (probably excellent) Gaussian approximation which then gives the static mobility via Eq. (16) and the ansatz can be used again until self-consistency. It is interesting to notice that Eq. (16) leads to a linear mobility for all v if elastic scattering is assumed, i.e. $F_0(K,t) = 1$. As one expects the Josephson-Lekner formalism to work at best for almost elastic scattering, we consider the case of a diffusing ion and substitute $F_0(K,t) = \exp(-K^2D|t|)$ into Eq. (10) assuming D, or the normal state mobility divided by the electron charge μ_n/e via the Einstein relation $D = \mu_n k_B T/e$, to be small,

$$I_n \approx \hbar^2 \underset{\sim}{K}\cdot\underset{\sim}{v}[1 - \frac{2\ln 2}{\pi}\beta h K^2 D + (\beta\hbar\underset{\sim}{K}\cdot\underset{\sim}{v})^2 \frac{\beta\hbar K^2 D}{6\pi}(\ln 2 - \frac{9\zeta(3)}{2\pi^2})] . \qquad (17)$$

The term $\hbar^2\underset{\sim}{K}\cdot\underset{\sim}{v}$ arises from $F_0(K,t) = 1$ and the rest from the integral carried out for $F_0(K,t)-1$. We have expanded to first order in D and second order in $\underset{\sim}{K}\cdot\underset{\sim}{v}$ to get the first nonlinear correction in v. After the angular integrations of Eq. (11) one has for the drag force

$$e\varepsilon = \frac{dP_{qp}}{dt} = np_F\sigma_1 v\ [1 - \frac{4\ln 2}{\pi}\beta\hbar k_F{}^2 D\frac{\sigma_2}{\sigma_1} + \frac{2}{5\pi}(\ln 2 - \frac{9\zeta(3)}{2\pi^2})(\beta\hbar k_F v)^2\ \beta\hbar k_F{}^2 D\frac{\sigma_3}{\sigma_1}]. \qquad (18)$$

where $n = k_F{}^3/3\pi^2$ is the number density of ^{3}He and σ_i denotes a particular weighted mean of the differential cross section

$$\sigma_i = \int d\Omega(d\sigma/d\Omega)(1 - \cos\Theta)^i. \qquad (19)$$

One notices the important role of the dimensionless parameter $\beta\hbar k_F{}^2 D$ or $\hbar k_F{}^2\mu_n/e$. If it gets too large, Eq. (18) leads to a negative mobility signaling the failure of the diffusion model. Via the second term it also determines the velocity at which nonlinear effects should be seen in the normal state mobility. What is satisfactory is that the diffusion model reproduces the observed temperature independent mobility[6,7] of negative ions, just as if inelasticity were not there.

For negative ions at 18 bar $\hbar k_F{}^2 \mu_n/e = 0.1$ justifying, in some sense, the diffusion approximation. The onset of nonlinearities in the normal state mobility is also not in disagreement with the prediction of Eq. (18) which requires a 10% quadratic correction for $v \gtrsim 22$ cm sec^{-1} at 3mK. We note, however, that the positive ion mobility at $\gtrsim 100$ mK exceeds that of the negative ion by a factor ≈ 3 at the vapor pressure,[17] so that the situation may be considerably more complicated for positive ions. Deep in the superfluid phases, where the mobility is large, the inelastic effects seem to be inescapable.

The diffusion model is a particularly simple example of the self-consistent loop implied by the Josephson-Lekner formalism. Its predictions would not change if the correct small time behavior of $F_0(K,t)$ were included as required by the theory of Brownian motion. It is an empty description, however, as it offers no mechanism of transition to a recoil dominated regime other than the warning through a growing $\hbar k_F{}^2 \mu_n/e$. Where inelastic effects do matter, a more sophisticated analysis of the dynamics of the ion in the background of a degenerate quantum fluid is required. A single relaxation time model for the frequency dependent mobility, in particular, is not sufficient[18] as it fails to account for the strictly temperature independent mobility observed at low temperatures.

The small value of the parameter $\hbar k_F{}^2 \mu_n/e$ for negative ions in the normal phase makes one believe that the elastic model might still be valid in the superfluid phases not too far from T_c, and we now consider this case, setting $S_0(K,\omega) = \delta(\omega)$ and ignoring the finite width of order $k_F{}^2 D$.

In the superfluid phase, the presence of the energy gap introduces considerable complications. The velocity $\underset{\sim}{v}$ always occurs in the combination $\hbar \underset{\sim}{K} \cdot \underset{\sim}{v}$, $\gtrsim p_F v$. In the normal state, $k_B T/p_F$ sets the scale of velocity, but the superfluid has an additional characteristic velocity $\Delta(T)/p_F$, which can be larger or smaller than $k_B T/p_F$. The linear mobility regime is restricted to velocities smaller than both these values, and the behavior in other situations requires the more general formalism developed in Sec. 2.

We narrow our scope still more and treat only ^{3}He-B, where the vector Δ_k has the form $\Delta R\hat{k}$,[13] with R an orthogonal 3 x 3 matrix and $\Delta\sim$a temperature dependent constant and the quasiparticle energies isotropic $E_k = (\xi_k{}^2 + \Delta^2)^{1/2}$. The coherence factors remain anisotropic, with Eq. (16b) given by

$$(EE' \mp \Delta^2 \hat{k} \cdot \hat{k}')/EE'. \tag{20}$$

In the isotropic B phase $d\mathbf{P}_{qp}/dt$ lies along the direction $\hat{v}$, with only the magnitude to be evaluated. Thus we may consider the scalar product $\mathbf{v}\cdot d\mathbf{P}_{qp}/dt$ which must be independent of $\hat{v}$. We perform an explicit average over this direction, evaluating the quantity

$$\langle \mathbf{K}\cdot\mathbf{v}I\rangle \equiv (4\pi)^{-1} \int d\Omega_{\hat{v}}\, \mathbf{K}\cdot\mathbf{v}I \tag{21}$$

that appears in the scalar product of Eq. (13) with $\mathbf{v}$. In the present case of elastic scattering, Eq. (29) permits an immediate evaluation

$$\langle \mathbf{K}\cdot\mathbf{v}I\rangle = (\hbar K v)^{-1}\, J \tag{22}$$

with J defined by

$$J = 4\Delta^3 \int_1^\infty x\, dx\, \Theta(\hbar Kv-2\Delta x)$$

$$\times \int_0^{x-1} \frac{dy(x^2-y^2+\cos\theta)[\tanh\frac{1}{2}\beta\Delta(y+x)-\tanh\frac{1}{2}\beta\Delta(y-x)]}{[(x+1)^2-y^2]^{1/2}[(x-1)^2-y^2]^{1/2}}$$

$$+ 4\Delta^3 \int_0^{\hbar Kv/2\Delta} y\, dy \int_{y+1}^\infty \tag{23}$$

$$\times \frac{dx(x^2-y^2-\cos\theta)[\tanh\frac{1}{2}\beta\Delta(x+y)-\tanh\frac{1}{2}\beta\Delta(x-y)]}{[x^2-(y+1)^2]^{1/2}[x^2-(y-1)^2]^{1/2}}$$

where $x = (2\Delta)^{-1}(E+E')$, $y = (2\Delta)^{-1}(E-E')$ and the integration in the xy plane is confined to the right of the lines $y = \pm(x-1)$. Further analytical progress can be made with Eq. (21).[11] In Fig. 1, however, we display the numerically evaluated results from Eq. (23) and compare them with experiment. The theoretical curves nicely reproduce the transition to the pair breaking regime where the velocity again begins to increase parallel to the normal state mobility curve. At high v there seem to be no striking inelastic effects present in the experimental results.

We thank Dierk Rainer for enlightening discussions and Henrik Vidberg for help in computing.

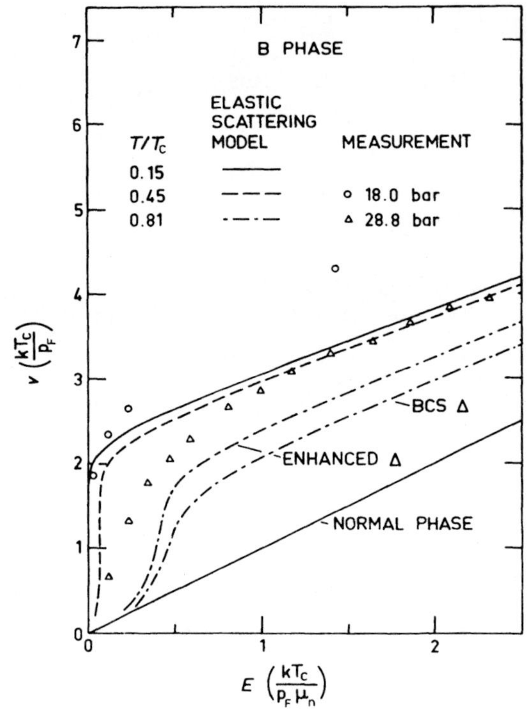

Figure 1. Velocity of ion plotted against electric field as evaluated numerically from Eq. (23) and compared with experiment at 18.0 bar and 28.8 bar. Enhanced Δ means $1.18\times\Delta_{BCS}$, gap enhanced to reproduce the specific heat discontinuity at 28.8 bar.

References

(1) A. L. Fetter, in "The Physics of Liquid and Solid Helium," edited by H. K. Bennemann and J. B. Ketterson (Wiley, New York, 1976), pp. 242-305.

(2) K. W. Schwarz, in "Advances in Chemical Physics," edited by I. Prigogine and S. A. Rice (Wiley, New York, 1975), Vol. XXXIII, pp. 1-49.

(3) L. Meyer and F. Reif, Phys. Rev. 110, 279 (1958).

(4) D. D. Osheroff, W. J. Gully, R. C. Richardson, and D. M. Lee, Phys. Rev. Lett. 29, 920 (1972).

(5) R. M. Bowley, J. Phys. C 9, L151 (1976).

(6) A. I. Ahonen, J. Kokko, O. V. Lounasmaa, M. A. Paalanen, R. C. Richardson, W. Schoepe, and Y. Takano, Phys. Rev. Lett. 37, 511 (1976).

(7) A. I. Ahonen, J. Kokko, M. A. Paalanen, R. C. Richardson, W. Schoepe, and Y. Takano, to be published in J.L.T.P.

(8) G. Baym, R. G. Barrera, and C. J. Pethick, Phys. Rev. Lett. 22, 20 (1969).

(9) B. D. Josephson and J. Lekner, Phys. Rev. Lett. 23, 111 (1969).

(10) G. Baym, C. J. Pethick, and M. Salomaa, to be published. The weak scattering approximation leads to a vanishing linear mobility because of the diverging density of states at the gap edge. A full treatment of the T-matrix removes this singularity as shown by the above mentioned authors.

(11) A. L. Fetter and J. Kurkijärvi, (to be published).

(12) A. J. Leggett, Rev. Mod. Phys. 47, 331 (1975).

(13) R. Balian and N. R. Werthamer, Phys. Rev. 131, 1553 (1963).

(14) J. Serene, 1973, thesis, Cornell University (unpublished).

(15) V. Ambegaokar, P. G. de Gennes and D. Rainer, Phys. Rev. A 9, 2676 (1974).

(16) D. Rainer, private communication.

(17) M. Kuchnir, J. B. Ketterson, and P. R. Roach, J. Low Temp. Phys. 19, 531 (1975).

(18) R. M. Bowley, J. Phys. C 4, L207 (1971).

THE MOBILITY OF NEGATIVE IONS IN SUPERFLUID ^{3}He

A. I. Ahonen, J. Kokko, O. V. Lounasmaa, M. A. Paalanen,
R. C. Richardson, W. Schoepe, and Y. Takano

Low Temperature Laboratory, Helsinki University of Technology
SF-02150 Espoo 15, Finland

Abstract: We have found that the mobility of negative ions increases rapidly below T_c in both superfluid ^{3}He phases. The ratio μ/μ_N of superfluid to normal mobility is larger in the B phase than in the A phase. A critical velocity consistent in magnitude with the Landau limit for pair breaking has also been observed. In the normal fluid we find a temperature independent mobility between 40 mK and T_c for all pressures between 0 and 28 bar. The increase of μ_N with increasing pressure is in agreement with the bubble model for the negative ion.

The motion of an ion in a quantum liquid is governed by its interaction with the elementary excitations of the liquid.[1] In superfluid ^{4}He ion experiments revealed a rich variety of information about the excitation spectrum and about the hydrodynamic structures such as quantized vortices. The low-temperature mobility of an ion in normal ^{3}He and, in particular, in the superfluid phases should also yield relevant information about the interaction of the impurity with the quasiparticles. In this work we have extended previous studies[2,3] of the normal ^{3}He fluid down to T_c and have surveyed the ion behavior in the superfluid phases. A brief communication of our work was published recently,[4] and a detailed account will be presented elsewhere.[5]

Liquid ^{3}He was cooled in a nuclear demagnetization cryostat. The temperature was measured with a platinum NMR thermometer.[6] In the A phase we also used the transverse frequency shift of the ^{3}He NMR signal to determine the temperature.[7]

We have used a time of flight method to measure the mobility of the negative ions. A cloud of electrons from a pulsed field emission tip was shaped into a narrow packet by means of a gate and then driven across a 5 mm drift space with various electric fields. The traveling time of the electrons was measured with a fast rise electrometer and a signal averager. Care was taken to keep the heating of the ^{3}He by the ion current negligibly small. For further experimental details see Refs. 4 and 5.

RESULTS

In the normal phase we found a temperature independent mobility between 40 mK and T_c for all pressures between 0 and 28 bars. The mobility increases with pressure as shown in Fig. 1.

In the superfluid phases we have found that the drift velocity v of the ions varies linearly with the driving field E only for low velocities. At higher velocities v is nonlinear. In the linear regime, where a field independent mobility $\mu = v/E$ can be defined, we found a rapidly increasing mobility as the temperature is lowered (see Fig. 2). In the B phase the increase is faster than

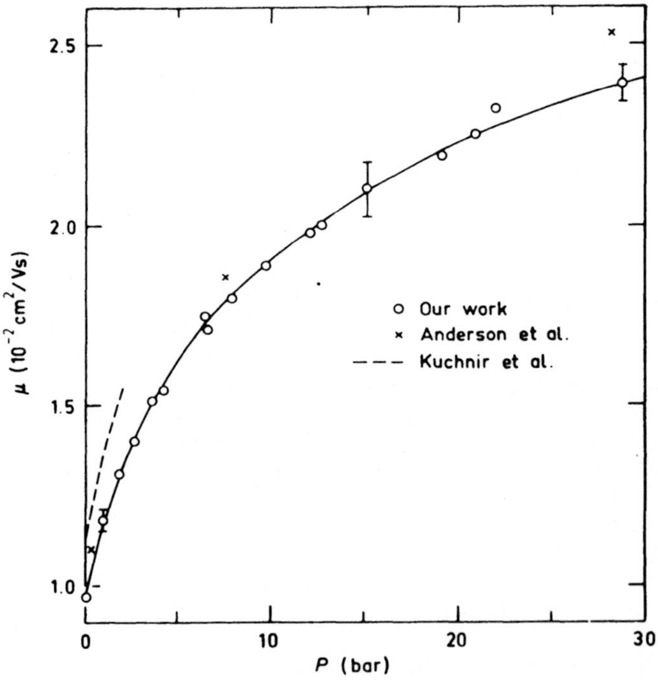

Figure 1. Low temperature mobility of negative ions in normal ^{3}He (in the plateau regime) versus pressure.

in the A phase. Within our resolution the ratio of the superfluid to normal mobility μ/μ_N is pressure independent in the B phase (Fig. 3). In the A phase the high resolution of the "shift thermometer" provided us with more accurate data than those obtained in the B phase. As shown in Fig. 4 we obtained a pressure dependence of μ/μ_N in the A phase which is outside of our error bars.

In Fig. 5 the nonlinear behavior of the drift velocity as a function of the electric field is illustrated. In normal ^{3}He at 3 mK a slight deviation from linearity is visible for fields only above ca. 300 V/cm. As the temperature was decreased below the superfluid transition, the onset of nonlinearity occurred at smaller fields. At our lowest temperatures, $T/T_c < 0.5$, the drift velocity became nonlinear at even the lowest field for which we could detect the arriving charges, 2 V/cm.

DISCUSSION

Our data in the normal ^{3}He fluid agree with previous measurements[2,3] which were performed above 17 mK. Our result that the plateau of constant mobility extends all the way down to T_c can be

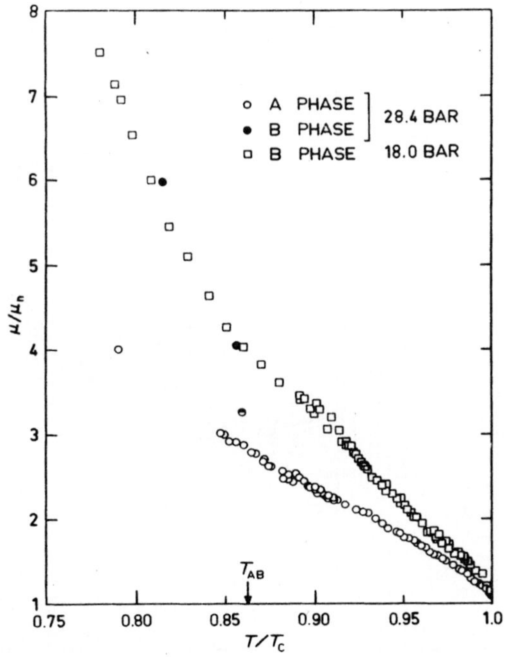

Figure 2. The ratio of superfluid to normal mobility versus reduced temperature.

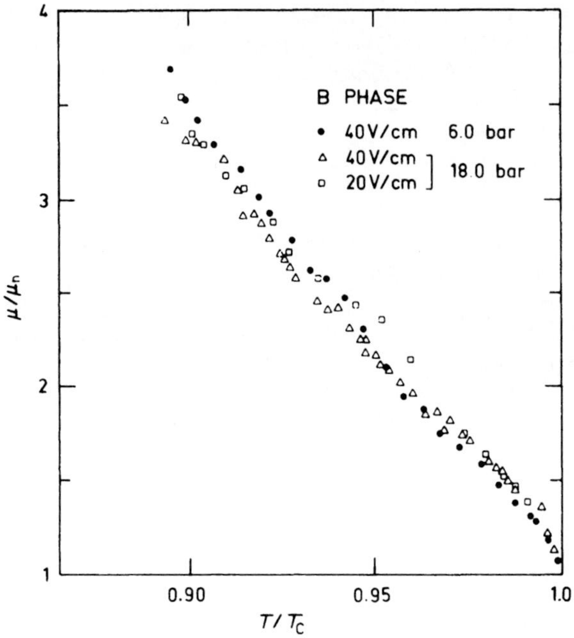

Figure 3. Ratio of B phase to normal mobility at two different pressures versus reduced temperature.

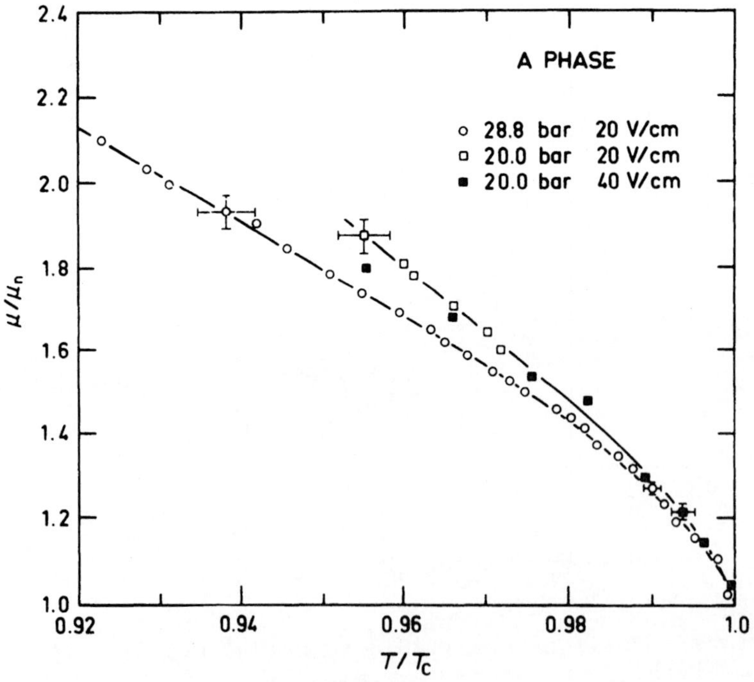

Figure 4. Ratio of A phase to normal mobility at two different pressures versus reduced temperature.

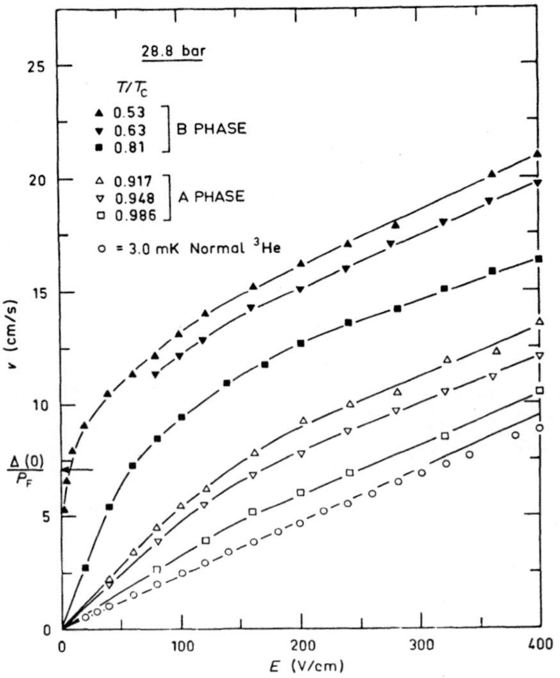

Figure 5. Drift velocity in normal and superfluid ^{3}He versus electric field.

discussed within the framework of the Josephson-Lekner theory[8] which had predicted an essentially constant mobility down to very low temperatures. Ultimately, however, the mobility was thought to diverge as T^{-2} because of the decreasing number of empty states into which quasiparticles can scatter. Obviously, the temperatures at which the normal fluid mobility becomes temperature dependent should be given by the theory to be lower than T_c. So far, the model to which the Josephson-Lekner formalism has been applied does not seem to solve this problem completely, in particular at higher pressures. As recognized by Josephson and Lekner, the decisive point for the temperature variation of the mobility is the question whether the ion recoils as a free particle in a scattering event (leading to the T^{-2} dependence) or whether it undergoes a diffusive motion. In the extreme case of a very small diffusion coefficient, the mobility reduces to the simple and temperature independent elastic scattering result:

$$\mu_N = e(n\sigma p_F)^{-1}$$

where e is the electronic charge, n the ^{3}He number density, σ the scattering cross section, and p_F the ^{3}He Fermi momentum. Recently, Fetter and Kurkijärvi[9] have extended the theory to finite diffusion coefficients and drift velocities. Within their model the zero field mobility is slightly higher (e.g. 14% at a pressure of 18 bar) than the elastic scattering result.

From the measured values of μ_N we can determine the pressure dependence of σ and furthermore that of the radius R of the negative ion by setting $\sigma = \pi R^2$, which is valid for $k_F R >> 1$. The resulting pressure dependence of R is in excellent agreement with the bubble model for the negative ion.[5]

Finally, the onset of a nonlinear behavior of the drift velocity at high electric fields in normal ^{3}He may be compared with the results of the Fetter-Kurkijärvi calculations.[9] At 3 mK the observed nonlinearity (Fig. 5) agrees roughly with their result. However, in our range of electric fields the effect is too small to make a quantitative test of the theory. In particular, we cannot confirm the predicted temperature dependence of the nonlinear behavior in the normal liquid.[9]

The most striking feature of our data is the steep increase of the ion mobility as the temperature falls below T_c. In addition, nonlinear drift velocities become very pronounced. At lower temperatures the onset of nonlinearity occurred already at small electric field intensities which prevented us from obtaining data in the linear regime below ca. $T/T_c \sim 0.8$. Besides the limited temperature interval, a detailed analysis of the mobility data is complicated by following two reasons. (1) In the A phase our data are more accurate because of the higher temperature resolution. However, in our geometry the drift velocity $\vec{v}$ was perpendicular to the static magnetic field thus leaving the angle between the $\vec{l}$ vector and $\vec{v}$ undefined. (2) In the isotropic B phase where the analysis is thought to be simpler, the accuracy of the data is hampered by the poor temperature resolution.

In spite of these limitations, our data in the linear regime reveal a number of interesting features. Early theories[10,11] on ion mobility in superfluid ^{3}He suggested that well below T_c the mobility should rise as

$$\mu/\mu_N \propto \exp(\Delta/k_B T).$$

Our data indicate that the increase of the mobility is actually much steeper. The very recent calculations by Baym, Pethick and Salomaa indeed yield a more rapidly increasing superfluid mobility.[12] Their results in the B phase are close to the experimental data; for further details, see Ref. 12. In addition, the lower mobility in the A phase is consistent with their work though numerical results are not yet available.

Our data in the A phase are sufficiently accurate to allow an analysis close to T_c where Δ is proportional to $(1-T/T_c)^{1/2}$. We find that for $T/T_c > 0.97$ our data can be fitted to the equation

$$\left(\frac{\mu_N}{\mu} - 1\right)^2 = a(1 - T/T_c)$$

where the coefficient $a = 4.1 \pm 0.5$. Quantative understanding of this result must await further calculations of the A phase mobility. Since μ/μ_N should be governed by the magnitude of the energy gap, we have looked for a pressure dependence of μ/μ_N. In the B phase we could not find any within our resolution (Fig. 3). In the A phase we found a slight increase of μ/μ_N with decreasing pressure (Fig. 4) which we do not understand as strong coupling effects should lead to an opposite behavior.

The onset of the nonlinear behavior of the drift velocities is certainly due to pair breaking effects. When the drift velocity exceeds the Landau limit $v_c = \Delta/p_F$ the collision energy is sufficient for breaking superfluid pairs. The observed deviation from linear behavior indeed occurs at velocities of the magnitude of v_c ($v_c \simeq 7$ cm/s at T = 0 and 28 bars). Furthermore, when $v > v_c$ the drag force approaches the normal state value. Recent calculations[9] show that a quantitative theoretical description of the complete v(E) curves is difficult. But our observations of a deviation from linearity at $v \simeq v_c$ and of a normal state slope for $v >> v_c$ can be understood qualitatively.

In summary, we have only made a brief survey of the negative ion motion in superfluid ^{3}He. We feel that further development of the theory and more experimental work will ultimately lead to new information about superfluid ^{3}He. The question of an anisotropic mobility in the A phase and the behavior of the much lighter positive ion species should be investigated in future experiments.

References

(1) For a review on ions in helium see the paper by A. L. Fetter in "The Physics of Liquid and Solid Helium", edited by K. H. Bennemann and J. B. Ketterson (Wiley, New York, 1976).

(2) A. C. Anderson, M. Kuchnir, and J. C. Wheatley, Phys. Rev. 168, 261 (1968).

(3) M. Kuchnir, P. R. Roach, and J. B. Ketterson, Phys. Rev. A 2, 262 (1970).

(4) A. I. Ahonen, J. Kokko, M. A. Paalanen, R. C. Richardson, W. Schoepe, and Y. Takano, Phys. Rev. Lett. 37, 511 (1976).

(5) A. I. Ahonen, J. Kokko. O. V. Lounasmaa, M. A. Paalanen, R. C. Richardson, W. Schoepe, and Y. Takano, to be published.

(6) The cryogenic system is described by A. I. Ahonen, P. M. Berglund, M. T. Haikala, M. Krusius, O. V. Lounasmaa, and M. A. Paalanen, Cryogenics 16, 521 (1976).

(7) A. I. Ahonen, M. Krusius, and M. A. Paalanen, Journal of Low Temp. Phys. 25, 421 (1976).

(8) B. D. Josephson and J. Lekner, Phys. Rev. Lett. 23, 111 (1969).

(9) A. L. Fetter and J. Kurkijärvi, to be published.

(10) T. Soda, Prog. Theor. Phys. 53, 903 (1975).

(11) R. M. Bowley, J. Phys. C 9, L 151 (1976).

(12) G. Baym, C. Pethick, and M. Salomaa, to be published.

THE MOTION AND MOBILITY OF IONS IN SUPERFLUID ^{3}He

Toshio Soda

Department of Physics, Tokyo University of Education
Tokyo, 112 Japan

Abstract: The mobility of ions in superfluid ^{3}He-B and -A is calculated near the transition temperature and at absolute zero, taking into account the ionic recoil and using the low temperature approximation for the Van Hove scattering function of an ion.

The mobility of ions has been calculated by the present author[1] and Bowley[2] by assuming the scattering of the quasi-particles to be elastic and the ion to be recoiless. They showed an increase of the mobility below the transition temperature. The experiment by Ahonen, Kokko, Lounasmaa, Paalanen, Richardson, Schoepe and Takano[3] discovered much larger increases; more so for the B-phase than the A-phase below the transition temperature. In this work we take into account the recoil effect of the ion in the calculation of the mobility, similar to the treatment of Bowley.[4]

The rate of the momentum transfer at an ion with velocity $\vec{V}$ by the scattering of the quasi-particles in superfluid ^{3}He is given by

$$\frac{d\vec{P}}{dt} = 2 \sum_{k_i} \sum_{k_f} \Gamma_{-v}(k_i,k_f)\hbar(\vec{k}_i-\vec{k}_f)n(k_i)[1-n(k_f)] , \qquad (1)$$

where $\Gamma_{-v}(k_i, k_f)$ corresponds to the transition probability for the quasi-particle to go to state $|\vec{k}_f, E_f>$ from $|\vec{k}_i, E_i>$ with the background fluid moving with velocity $-\vec{V}$ and n_i is the Fermi distribution function of the quasi-particle. If the distance moved by the ion during a scattering is small compared to the inverse of the momentum transfer, $q = |\vec{k}_i - \vec{k}_f|$, we can express $\Gamma_{-v}(k_i, k_f)$ in terms of the Van Hove scattering function $S_v(q, \omega = \hbar^{-1}(E_i - E_f))$ and the scattering amplitude $T(k_i, k_f)$, as follows,

$$\Gamma_{-v}(k_i,k_f) = \frac{2\pi}{\hbar^2}\left(\frac{2\pi\hbar^2}{m}\right)^2 \left|T(k_i,k_f)\right|^2 S_v(q,\omega)\rho(k_f). \tag{2}$$

Here, $\rho(k_f)$ is the density of states of quasi-particle states at k_f and m is the mass of ^{3}He. By assuming $|T(k_i,k_f)|^2 = |T(k_f,k_i)|^2$ and the detailed balance condition obeyed by $S_v(q,\omega)$, we simplify the expression of the momentum transfer (1) to

$$\frac{d\vec{P}}{dt} = \frac{1}{\hbar(2\pi)^3}\int d\Omega_i \int d\Omega_f \int dE_i \frac{E_i}{\varepsilon_i} \int dE_f \frac{E_t}{\varepsilon_f} k_i k_f S_v(q,\omega)$$

$$|T(k_i,k_f)|^2 \hbar\vec{q}[1-\exp(-\beta\hbar\vec{q}\vec{V})]n(k_i)(1-n(k_f)) \ , \tag{3}$$

where ε_i is the energy, measured from the Fermi energy, of the quasi-particle in the normal state and Ω_i and Ω_f are the solid angles of $\vec{k}_i$ and $\vec{k}_f$.

The rate of momentum transfer (3) balances with the electric field $e\vec{E}$ and we equate it with $e\vec{V}/\mu$, thus defining the mobility μ. Thus the mobility μ is given by

$$e\mu^{-1} = \frac{e\vec{E}\vec{V}}{V^2} = \frac{1}{(2\pi)^3 h}\int d\Omega_i \int d\Omega_f \int dE_i \frac{E_i}{\varepsilon_i} \int dE_f \frac{E_f}{\varepsilon_f} k_i k_f S_v(q,\omega)$$

$$|T(k_i,k_f)|^2 n(k_i)(1-n(k_f)) \frac{(\vec{q}\cdot\vec{V})}{V^2} (1-\exp(-\beta h\vec{q}\vec{V})). \tag{4}$$

With M as an ionic mass, $S_V(q,\omega)$ is given by

$$S_V(q,\omega) = \left(\frac{M}{2\pi q^2}\right)^{1/2} \exp\left[-\frac{\beta M}{2q^2}\left(\omega - \vec{q}\vec{V} - \frac{q^2}{2M}\right)^2\right] . \tag{5}$$

If we use the Born approximation for $|T(k_i, k_f)|$ in the superfluid state, we have a threshold singularity at the gap edge $E = |\Delta|$; this difficulty is rescued by using the t-matrix for $T(k_i, k_f)$ throughout. We write $|T(k_i, k_f)|^2$ as a product for the following coherent factor, applicable both for ^{3}He-A and -B.

$$I(k_i,k_f) = \frac{1}{2}\left[1 + \frac{\varepsilon_i}{E_i}\frac{\varepsilon_f}{E_f} - \mathrm{Re}\,\frac{\Delta^*_{k_i}}{E_i}\frac{\Delta_{k_f}}{E_f}\right] . \tag{6}$$

and the square of the normal fluid scattering amplitude $|T^n(k_i, k_f)|^2$, denoted as $\sigma(k_i, k_f)$ which is related to the normal differential cross section. By integration of Eq. (4) over E_i, expansion of $S_v(q, \omega)(1 - \exp(-\beta h\vec{q}\vec{V}))$ in powers of V, and the assumption that $\sigma(q, k_f)$ is independent of E_i, the mobility becomes [with $W = \beta(E_i-E_f)$]

$$e\mu^{-1} = \frac{1}{8\pi^3}\left(\frac{\beta M}{2\pi}\right)^{1/2}\int d\Omega_i \int d\Omega_f \int_0^\infty dW\, \frac{e^{W/2}}{1-e^W}\,\ln\left|\frac{1+e^{-\beta\Delta_i-W}}{1+e^{-\beta\Delta_i+W}}\right| \frac{(\vec{q}\vec{V})^2}{V^2}$$

$$\times\, k_i k_f\, \sigma(q,k_f)\, e^{\left[-\frac{MW^2}{2\beta q^2} - \frac{\beta q^2}{8M}\right]} \left\{1 + \frac{(\vec{q}\vec{V})}{q^2} W + \left(\frac{\beta\vec{q}\vec{V}}{2}\right)^2 \left[\frac{1}{3!} - \right.\right.$$

$$\left.\left.\frac{2M}{\beta q^2} + \frac{1}{2}\left(\frac{2MW}{\beta q^2}\right)^2\right] + \ldots \right\} . \tag{7}$$

We first treat the case of the B phase. We note that the momentum transfer satisfies the following relationships, $q^2 = k_i{}^2 + k_f{}^2 = 2k_i k_f \cos\theta$ and $q\,dq = k_i k_f \sin\theta d\theta$, where θ is the angle between k_i and k_f. Near T_c the mobility satisfies the equation,

$$e\mu^{-1} = \frac{h}{3\pi}\left(\frac{M\beta}{2\pi}\right)^{1/2}\int_0^{2k_F(T)} dq\, q^2\, e^{-\frac{\beta h^2 q^2}{8M}} \int_0^\infty dW\, e^{-\frac{MW^2}{2\beta q^2}} \left\{1+\left(\frac{\beta\vec{q}\vec{V}}{2}\right)^2\left[\frac{1}{3!}\right.\right.$$

$$\left.\left. - \frac{2M}{\beta q^2} + \frac{1}{2}\left(\frac{2MW}{\beta q^2}\right)\right]+\ldots\right\} \left[\frac{W}{e^{W/2} - e^{-W/2}} - \beta\Delta\,\frac{1}{e^{W/2} + e^{-W/2}}\right] \sigma(q,k_f). \tag{8}$$

with k_F the Fermi momentum. The dominant part of the W integration comes when the values of exp $- MW^2/2\beta q^2$ is large. Thus we find that the quantity in the bracket of Eq. (8) is $1 - 1/2\beta\Delta$. Therefore, if μ_N be the normal mobility,

$$\mu_B{}^{-1} = \mu_N{}^{-1}\left(1 - \frac{1}{2}\beta\Delta\right) , \tag{9}$$

which is the same result as Bowley's treatment of a recoiless ion near T_c. Near $T = 0$, we substitute $Y^2 = \beta\hbar^2 q^2/2M$ and extend the upper limit of $Y = \sqrt{\frac{\beta\hbar^2}{2M}}\, 2k_F$ to infinity, because of the extreme low temperature, in Eq. (7), keeping only the lowest power in V and find

$$e\mu^{-1} = \frac{\hbar}{3\pi}\left(\frac{M}{h^2\beta}\right)^2 \frac{4}{\pi^{1/2}} \int_0^\infty dW \frac{e^{w/2}}{e^w-1} \ln \frac{1+e^{-\beta\Delta+W}}{1+e^{-\beta\Delta-W}}$$

$$\times\int_0^\infty dYY^2\, e^{-\frac{Y^2}{4}-\frac{W^2}{4Y^2}}\, \sigma(Y,q). \tag{10}$$

The Y integration yields $2\pi^{1/2}(1 + 1/2W)\, e^{-1/2W}$ and we can set

$$\ln \frac{1+e^{-\beta\Delta+W}}{1+e^{-\beta\Delta-W}} \simeq e^{-\beta\Delta}(e^W - 1)/e^W + 1)$$

to find the following behavior of μ near $T = 0$ (assuming $\sigma(y,k_f)$ to be a constant, σ, independent of all angles):

$$e\mu_B^{-1} = \frac{\hbar}{3\pi}\left(\frac{M}{\hbar^2}\right)^2 \frac{4}{\pi}\left(\ln 2 + \frac{\pi^2}{24}\right)(k_B T)^2 \exp\left(-\frac{\Delta}{k_B T}\right)\sigma\,. \tag{11}$$

This also agrees with the recoiless treatment by Bowley.[2]

Now we treat the case of the A phase, where the energy gap is given by $|\Delta(\theta_i)| = \sqrt{3/2}\,\Delta\sin\theta_i$ and θ_i is the angle between the vector $\vec{k}_i$ and the orbital vector $\vec{I}$. Near T_c it turns out that we obtain the mobility expression by replacing Δ in Eq. (8) by $1/2[\Delta(\theta_i) + \Delta(\theta_f)]$ and recovering one more angle integral $(4\pi)^{-1}\int d\Omega_i$ to the integral of Eq. (8). The angle, θ, between k_i and k_f satisfies $\theta = \theta_i - \theta_f$. If $\vec{V}$ is parallel to the orbital vector $\vec{I}$, then $1/2[\Delta(\theta_i) + \Delta(\theta_f)] = 1/2\sqrt{3/2}\,\Delta[\sin\theta_i + \cos\theta\sin\theta_i - \sin\theta\cos\theta_i]$. Making the angle average, we find that the mobility behaves in the weak coupling limit as

$$\mu_{A\parallel}^{-1} = \mu_N^{-1}\left(1 - \frac{1}{2}\left(\frac{\pi}{4}\sqrt{\frac{3}{2}}\right)\beta\Delta\right) = \mu_N^{-1}(1 - 0.481\beta\Delta). \tag{12}$$

On the other hand when $\vec{V}$ is perpendicular to $\vec{I}$, then $\frac{1}{2}(\Delta(\theta_i)+\Delta(\theta_f))$ $= 1/2\sqrt{3/2}\,\Delta(\cos\theta_i+\cos\theta\cos\theta_i+\sin\theta_i\sin\theta)$ and, after the appropriate averaging over angle, we find

$$\mu_{A\perp}^{-1} = \mu_N^{-1}\ (1 - \frac{1}{2}(\frac{3\pi^2}{64}\sqrt{\frac{3}{2}})\ \beta\Delta) = \mu_N^{-1}(1 - 0.283\beta\Delta) \qquad (13)$$

In the strong coupling treatment, the coefficients in front of $\beta\Delta$ in Eqs. (12) and (13) becomes, for example, 0.517 and 0.342 respectively for the pressure of 28.4 bar. Near T = 0 the behavior of μ_A^{-1} obeys a power law in T instead of the exponential behavior of the B-phase. This case is not realistic.

The experimental behavior of the mobility does not exactly reproduce the calculated results of Eq. (9) and Eq. (12) or Eq. (13). The discrepancy may be due to the neglected angle and energy dependence of the differential cross section.

References

(1) T. Soda, Proc. 14th Int. Conf. on Low Temperature Physics, 1, 13 (1975), and Prog. Theor. Phys. 53, 903 (1975).

(2) R. M. Bowley, J. Phys. C 9, L151 (1976).

(3) A. I. Ahonen, J. Kokko, O. V. Lounasmaa, M. A. Paalanen, R. C. Richardson, W. Schope and Y. Takano, Phys. Rev. Lett. 37, 511 (1976).

(4) R. M. Bowley, J. Phys. C 4, 853 (1971).

PRELIMINARY MEASUREMENTS OF THE LOW TEMPERATURE ION MOBILITY IN NORMAL LIQUID ^3He*

Paul D. Roach and J. B. Ketterson

Northwestern University, Evanston, Illinois 60201 and
Argonne National Laboratory, Argonne, Illinois 60439

Pat R. Roach

Argonne National Laboratory, Argonne, Illinois 60439

Abstract: The temperature dependence of the mobility of positive and negative ions in normal liquid ^3He between $\sim$ 2.5 mK and 600 mK and at pressures from the vapor pressure to 23 bar are reported.

There is an increasing interest in the use of ions to probe the properties of liquid ^3He. Part of this interest stems from their potential in probing the superfluid state measurements of the negative ion mobility in the superfluid A and B phases have appeared recently.[1] The normal state ion mobility is also of interest. Part of this interest arises from the rather deep mathematical problems that are encountered in attempting to provide a rigorous discussion of this problem.[2] In addition, for temperature in the vicinity of 0.1 K, the positive ion mobility showed a peculiar dependence on the measurement conditions in one experiment[3] and an unusual peak in two others.[4,5]

---

*Work supported by the U. S. Energy Research and Development Administration and the National Science Foundation under Grant DMR-74-13186.

As a part of a long range program aimed at studying very low temperature ion transport in ^3He, we have measured the normal state positive and negative ion mobility in ^3He (a heat leak present during these experiments prohibited measurements in the superfluid phases). The ions were produced by a group of four tungsten field emission tips prepared using standard techniques.[6] The ion cell was contained in a threaded epoxy appendix which formed the bottom of an adiabatic demagnetization chamber containing cerium magnesium nitrate (CMN); the main cell has been discussed elsewhere.[7,8] The ion cell itself consisted of the following elements: (1) the tungsten field emission tips, (2) an extraction grid, (3) the gate grids, (4) the drift path and drift path electrodes, (5) the collector screen grid, and (6) the collector. A schematic drawing of the ion cell electrodes and the electronics is shown in Fig. 1. All grids were made from gold screen having a 82% transparency. The use of multiple tips was quite important, especially in the positive ion studies; while a relative potential of 400 volts provided a negative ion current of 10^{-11} amps, a relative potential of 1100 volts was required to achieve the same current for positive ions. At the higher potentials required for the positive ions, the tip life was greatly shortened and when the emission of one tip stopped, another would take over. All four tips were consumed during the positive ion studies; anticipating this, we collected the positive ion data after studying the negative ions. Mobilities were measured using the time-of-flight technique over a path length of 0.81 cm. The emission tip was biased at a voltage slightly smaller than that required to produce a significant emission. To produce an ion burst, an additional voltage pulse was applied to the tip. A short time after this voltage pulse was turned off, the resulting ion burst arrived at the gate grids. At this time the gate was closed briefly by the application of a voltage pulse causing reverse gate bias. This gate closure caused a "notch" in the ion current pulse; the arrival of this notch at the collector was used for the precise determination of the time of flight. The potentials were adjusted so that the electric field was constant in the region between the extraction grid and the collector. In particular, two annular electrodes in the drift space were biased so that the field was uniform in this region. Most of the measurements were taken at a field of 182 volts/cm. After traversing the drift space and passing through the screen grid, the ions were collected, amplified by a fast electrometer circuit, and fed into a signal averager. The signal to noise was such that a single burst was observable and only a relatively small number of pulses was required to make a mobility measurement. No detectable heating was caused by the ion current pulses, where the energy was estimated to be typically 3×10^{-3} ergs per pulse. The relative accuracy of our measurements is approximately $\pm$ 2% although the absolute accuracy is somewhat less ($\sim \pm$ 5%). Temperatures were determined by measuring the susceptibility of the CMN demagnetiza-

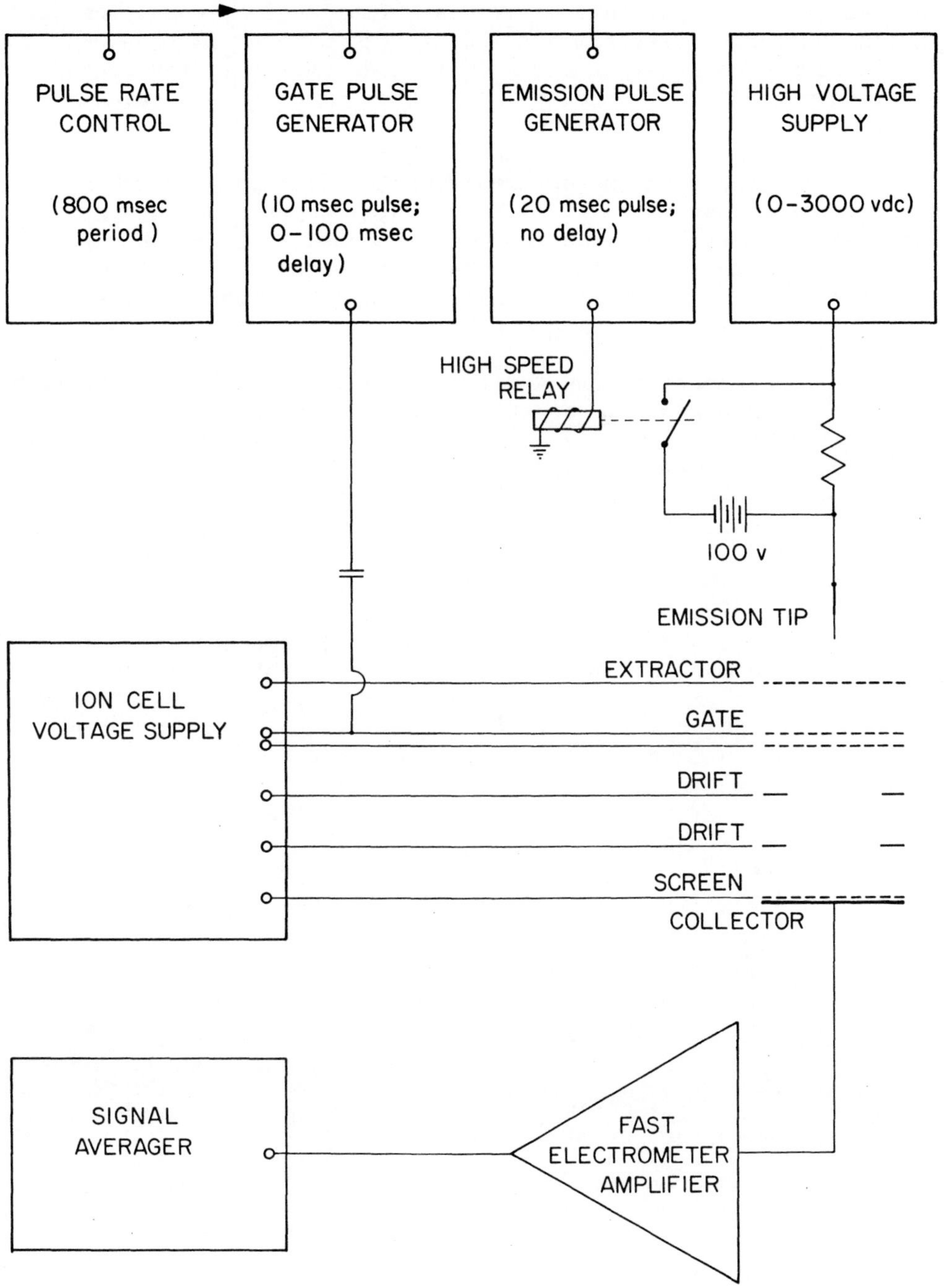

Figure 1. Schematic drawing of ion cell and associated electronics.

tion salt. The temperature scale was established by comparing with previous ultrasonic experiments[5] using this demagnetization cell and a separate calibration was not carried out. For this reason we have assigned an error of ± 10% to the temperature in these measurements.

Figure 2 shows our measurements of the negative ion mobility at the vapor pressure and at 23 bar. The low pressure data are consistent with earlier measurements performed at this laboratory.[9] In particular, the mobility is temperature independent at low temperatures. Our data are also consistent with the normal state measurements of the Finnish group.[1]

Our measurements of the temperature dependence of the positive ion mobility at a pressure of 60 Torr are shown in Fig. 3. For temperatures in the vicinity of 0.1 K the new data show a radically different temperature dependence from that observed in two previous sets of measurements taken at this laboratory.[4,5] Our new data are partially consistent with the lower extreme of a peculiar "double valued" set of measurements taken by Anderson, Kuchnir, and Wheatley.[3] The present data we observed to be independent of the electric fields below 200 volts/cm; in addition, the mobility was observed to be reproducible on warming and cooling the sample (the latter was made possible by our use of a mechanical heat switch[9]). The source of the discrepancy between the various data sets is not known at this time and because of this we see no compelling reason to favor one data set over another.

For temperatures somewhat below 0.1 K our data appear to merge with those of Anderson, _et al_. At still lower temperature the mobility becomes temperature independent; this feature has not been observed previously and it is for this reason we are reporting these preliminary measurements at this time.

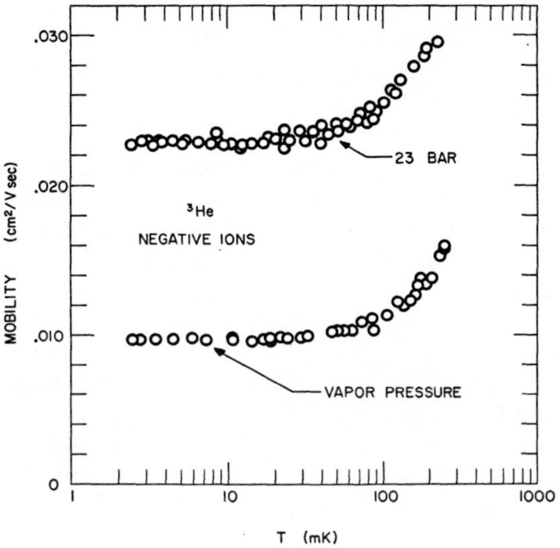

Figure 2. Negative ion mobility in liquid ^{3}He at the vapor pressure and at 23 bar.

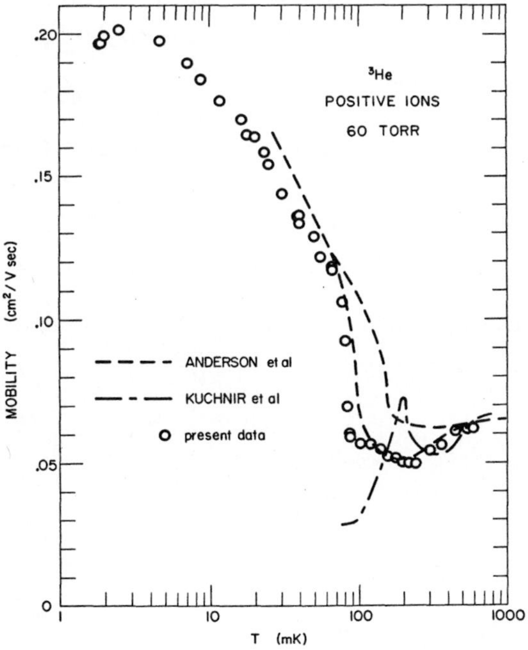

Figure 3. Positive ion mobility in liquid ^{3}He at pressure of 60 torr (0.08 bar).

References

(1) A. I. Ahonen, J. Kokko, O. V. Lounasmaa, M. A. Paalanen, R. C. Richardson, W. Schoepe, and Y. Takano, (to be published).

(2) B. D. Josephson and J. Lekner, Phys. Rev. Letters 23, 111 (1970); R. M. Bowley, J. Phys. C 4, L207 (1971); for a clear qualitative discussion of this problem, see A. L. Fetter, Vortices and Ions in Helium, in "The Physics of Liquid and Solid Helium, Part I", edited by K. H. Bennemann and J. B. Ketterson (John Wiley, New York, 1976).

(3) A. C. Anderson. M. Kuchnir, and J. C. Wheatley, Phys. Rev. 168, 261 (1968).

(4) M. Kuchnir, J. B. Ketterson, and P. R. Roach, J. Low Temp. Phys. 19, 531 (1975).

(5) M. Kuchnir, J. B. Ketterson, and P. R. Roach, in Proc. 13th Intern. Conf. on Low Temp. Phys., edited by K. D. Timmerhaus, W. J. O'Sullivan, and E. F. Hammel (Plenum, New York, 1974), Vol. 1, p. 474.

(6) K. M. Bowkett and D. A. Smith, "Field-Ion Microscopy" (North-Holland Publ. Co., Amsterdam, 1970), p. 57.

(7) J. B. Ketterson, P. R. Roach, B. M. Abraham, and Paul D. Roach, in "Quantum Statistics and the Many Body Problem", edited by S. B. Trickey, W. P. Kirk, and J. W. Dufty (Plenum Press, New York, 1975), p. 35.

(8) P. R. Roach, J. B. Ketterson, B. M. Abraham, J. Monson, and P. D. Roach, Rev. Sci. Instr. 46, 207 (1975).

(9) M. Kuchnir, P. R. Roach, and J. B. Ketterson, Phys. Rev. 2, 262 (1970).

PHASE TRANSITIONS IN LIQUID CRYSTALS*

Chia-Wei Woo

Department of Physics and Astronomy
Nortwestern University
Evanston, Illinois 60201

I. INTRODUCTION

Let me begin with a few words which will be discouraging to ^{3}He physicists. In contrast to superfluid ^{3}He, liquid crystals are plentiful, inexpensive, and copious in variety. By synthesis, one can readily create new species. There are a large number of phases, each with its own peculiar properties. Experiments are done at ordinary temperatures, using easily accessible magnetic fields and relatively unsophisticated apparatus. All the challenges reside in the physics of the substances themselves, rather than the techniques required for unveiling the physics. Note that I have said a mouthful without even mentioning technological or biomedical applications.

Now, it is really quite misleading to speak of liquid crystals as substances. A liquid crystal is a <u>phase</u> in which a substance can exist. Ordinarily we think of thermodynamic systems as made up of spherically symmetric molecules, or point particles. Thus we are concerned only with <u>spatial</u> order, which gives us gases, liquids and solids. When we go beyond translation to introduce new degrees of freedom, we find it necessary to define other kinds of order and phases. (The best known example is ferromagnetism.)

---

*Work supported in part by the National Science Foundation through grant No. DMR73-07659.

Let us, then, observe molecules which do not possess spherical symmetry. Take, for example, a rod. Even if we concede cylindrical symmetry, there will be two new orientational degrees of freedom: $\hat{\Omega} \equiv (\theta,\phi)$ with respect to some symmetry-breaking axis. If the interaction between the rods depends on $\hat{\Omega}$, spontaneous orientational ordering can set in under suitable conditions. Thus one can have - by lowering temperature or increasing pressure or both - phases which are solid-like, in that there exists macroscopic spatial order in the system, but orientationally disordered. These are "plastic crystals", or more appropriately, ODC for "orientationally disordered crystals". Alternatively, one can have phases which are orientationally ordered but spatially at least in part disordered. These are "liquid crystals." (Another unfortunate nomenclature.) Both types of phases occur in the intermediate region between isotropic-liquid and solid, and are often for this reason known as mesophases.

So, while the organic chemistry of liquid crystals may seem prohibitive to us, there is nothing mysterious in the physics.

Nematic liquid crystals possess only orientational order. Molecules in such a system point at, and fluctuate about, a definite direction (designated in the literature as the "director"). Smectic liquid crystals are also orientationally ordered, but in addition possess a layered structure, thus incomplete spatial order. The most commonly seen cholesteric liquid crystals are just twisted nematics - twisted about an axis normal to the director. Thus the director is planar, and rotates periodically along a pitch axis. The pitch often falls in the visible spectrum, and is easily controllable by thermodynamic and electmagnetic means, thus giving rise to the optical properties much exploited in display devices.

Given the limitations of time, in this talk I shall discuss only one problem, that of the isotropic-nematic transition, as an example of the kind of problems that one tackles in studying phase transitions in liquid crystals. The example chosen requires only elementary thermodynamical and statistical concepts, and is readily digestible for an audience outside the field.

II. PHENOMENOLOGY

It is informative to begin with a phenomenological theory.

Phase transitions can be described quite adequately in terms of order parameters. (An order parameter is a macroscopic quantity defined such that it, e.g. vanishes in the disordered phase and takes on finite values in the ordered phase.) In the Landau theory, the Helmholtz free energy F at and about the transition temperature T_c is expanded in powers of the order parameter, σ. For a second-order transition, one writes approximately:

$$F = F_o + A\sigma^2 + C\sigma^4 .$$

As is obvious from Fig. 1, T_c occurs at $(\partial^2 F/\partial\sigma^2)$ [illegible] $= 0$. Thus one can write A as $a(T-T_c)$ and leave F in the form containing empirical parameters a, T_c, and C. From F all thermodynamic and phase transition properties can be calculated. Note that the volume dependence has been left out completely, and temperature dependence has been allowed only through the parameter A. One can, of course, make the theory more sophisticated at the expense of usefulness.

For application to the isotropic(I)-nematic(N) transition, one first defines an orientational order parameter. In the simplest form, one takes $\langle\cos^2\theta\rangle$(instead of $\langle\cos\theta\rangle$ in order to observe inversion symmetry; see Fig. 2). Or more conveniently,

$$\sigma = \langle P_2(\cos\theta)\rangle \equiv \int f(\theta) P_2(\cos\theta) d\hat{\Omega} \tag{1}$$

$$= \begin{cases} 0, & \text{completely disordered} \\ \ldots & \\ 1, & \text{completely ordered} \end{cases}$$

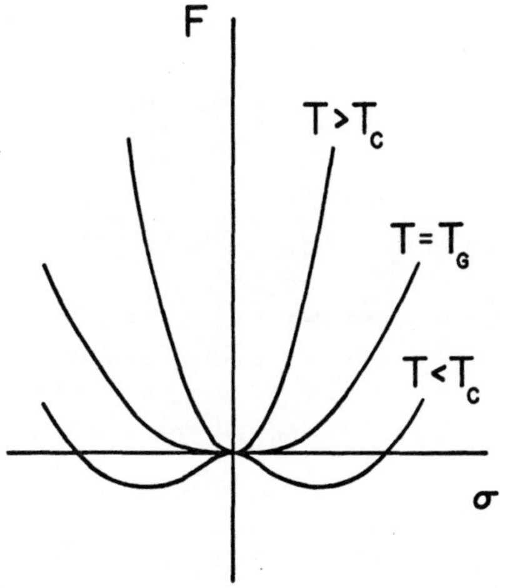

Figure 1

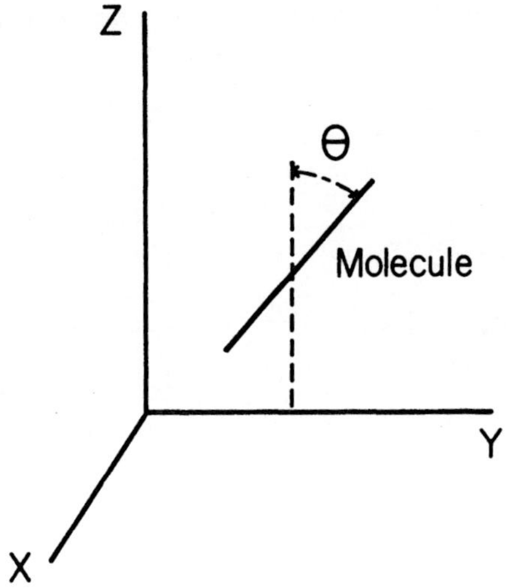

Figure 2

where f(θ) denotes the normalized distribution function.

Such an order parameter can be measured by means of NMR or other methods. In most cases it is observed to jump from 0 to 0.3-0.4 as T crosses the transition temperature T_{IN} from above. We learned, then, that the transition is first-order and that fluctuations are quite significant in the (ordered) nematic phase.

In the Landau-deGennes theory of I-N transitions, one write approximately:

$$F = F_o + A\sigma^2 + B\sigma^3 + C\sigma^4 \; .$$

One observes from Fig. 3 that at and slightly below T_{IN}, the isotropic phase remains metastable until T reaches a limiting temperature T^*. The latter plays the role of a maximum supercooling temperature. Geometry tells us that at $T = T^*$, $(\partial^2 F/\partial\sigma^2)_{\sigma=0} = 0$. Thus one can write:

$$F(\sigma) = F_o + a(T-T^*)\sigma^2 + B\sigma^3 + C\sigma^4 \; , \qquad (2)$$

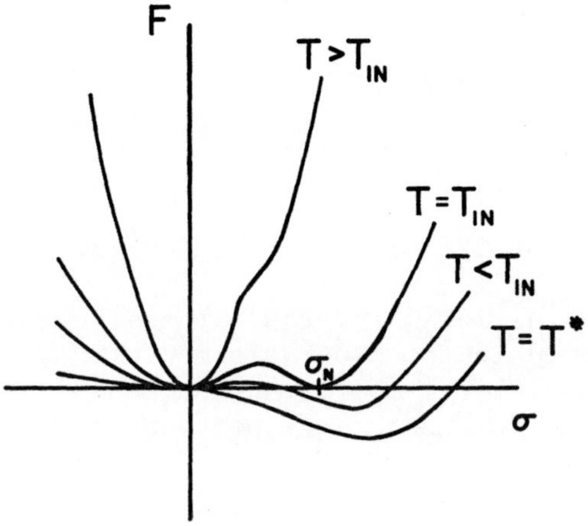

Figure 3

in the same spirit as Landau's theory for second-order transitions. The 4 parameters a, T^*, B, and C can now be determined empirically.

(i) At $T = T_{IN}$, $F(0) = F(\sigma_N)$, which implies that $T_{IN} = T^* + \frac{B^2}{4aC}$. (3)

(ii) At $T = T_{IN}$, $_N = \frac{-3B + \sqrt{9B^2 - 32a(T_{IN} - T^*)C}}{8C} = \frac{-B}{2C}$. $(B < 0)$ (4)

(iii) The entropy change

$$\Delta S_{IN} = S_I(T_{IN}) - S_N(T_{IN}) = -\left.\frac{\partial F}{\partial T}\right|_{\sigma=o} + \left.\frac{\partial F}{\partial T}\right|_{\sigma=\sigma_N}$$

$$= a\sigma_N^2(T_{IN}) + \left.\frac{\partial F}{\partial \sigma}\right|_{\sigma=\sigma_N} \bullet \frac{\partial \sigma}{\partial T} = \frac{aB^2}{4C^2} ,$$

and hence the latent heat is given by $\Delta H_{IN} = T_{IN}\Delta S_{IN} = \frac{aB^2}{4C^2} T_{IN}$. (5)

(iv) Define T_ξ as the temperature at which light scattering intensity in the pretransitional isotropic region diverges. deGennes showed (see below) that

$$T_\xi = T^*. \tag{6}$$

The quantities on the left of Eqs. (3)-(6) are all measurable. Consequently, the empirical parameters in Eq. (2) can be determined directly.

III. DE GENNES' DERIVATION OF EQ. (6)

In the pretransitional isotropic phase, consider nematic embryos being formed in the isotropic medium. Use a truncated form of Eq. (2) as a local approximation, and include an "elastic distortion" term. The free energy per particle becomes:

$$F(\vec{r}) \approx F_o + a(T-T^*)\sigma^2(\vec{r}) + K[\nabla\sigma(\vec{r})]^2. \tag{7}$$

Thus,

$$F = \int_V \rho F(\vec{r})d\vec{r} = NF_o + \rho V \sum_{\vec{q}} [a(T-T^*)+Kq^2]\sigma q^2 ,$$

and

$$\langle\sigma_q^2\rangle \equiv \int \sigma_q^2 \, e^{-F/kT} \prod_{\vec{q}} d\sigma_q \Big/ \int e^{-F/kT} \prod_{\vec{q}} d\sigma_q = \frac{kT}{2\rho V} \frac{1}{a(T-T^*)+Kq^2} ;$$

or

$$\langle\sigma(0)\sigma(r)\rangle = \frac{kT}{8\pi\rho K} \frac{e^{-r/\xi(T)}}{r} , \text{ with } \xi(T) = [\frac{K}{a(T-T^*)}]^{\frac{1}{2}} .$$

Finally, the light scattering intensity is given by

$$I \propto \int_\infty \langle\sigma(0)\sigma(r)\rangle d\vec{r} = \frac{kT}{2\rho a(T-T^*)} , \tag{8}$$

which diverges as $T \to T^*$. By definition, then, $T_\xi = T^*$.

IV. MOLECULAR MODEL THEORIES

There are some serious difficulties with the Landau-deGennes theory. First of all, F is not the proper thermodynamic potential to use for a first-order transition since volume changes discontinuously across the transition. (Even though the volume change $\Delta v_{IN}/v_I$ is small, it is still the leading term.) The volume dependence cannot simply be omitted from F. Next, since σ_N even at $T=T_{IN}$ is not small, the validity of truncating the series as in Eq. (7) - or even as in Eq. (2) - is in doubt. Most disturbing of all, even quantities which are fundamental to the theory can come out wrong. For example, according to the theory,

$$\sigma_N(T^*) = \frac{-3B}{4C} , \quad \text{and thus} \quad \frac{\sigma_N(T^*)}{\sigma_N(T_{IN})} = \frac{3}{2} . \qquad (9)$$

In reality, measurements show that since $|T_\xi - T_{IN}|$ is of the order of 1-3 degrees, $\sigma_N(T_\xi)$ can differ from $\sigma_N(T_{IN})$ by at most a few percent. To achieve the result of Eq. (9), $|T_\xi - T_{IN}|$ would have to be some 10-30 degrees!

What do we do then? Our answer is that even for a phenomenological theory a molecular model is needed.

In this talk I do not intend to discuss real statistical mechanical calculations. The molecular theory introduced is for providing a model, on the basis of which we shall conduct an empirical analysis. For some years now physicists in liquid crystals research have under-rated the importance of a good understanding of the microscopic picture. I wish to show that it is indeed the molecular picture which saves the day.

Let us begin with a set of N molecules in volume V so that $\rho = N/V = 1/v$ is known. The configuration of each molecule, assuming cylindrical symmetry is specified by $\tau \equiv (\hat{r},\hat{\Omega})$, a 5-component quantity. Prescribe for the molecules a pairwise interaction $v(i,j)$. The partition function is then:

$$Z = \int e^{-\sum_{i<j} v(i,j)/kT} \, d\tau_1 \ldots d\tau_N ,$$

and the n-particle distribution function is given by

$$P^{(n)}(1,\ldots,n) = \frac{N\ldots(N-n+1)}{Z} \int e^{-\sum_{i<j} v(i,j)/kT} \, d\tau_{n+1} \ldots d\tau_N .$$

In particular, when only orientational order is of concern,

$$P^{(1)}(1) = \frac{N}{Z}\int e^{-\sum_{i<j} v(i,j)/kT}\, d\tau_2 \ldots d\tau_N \equiv \rho f(\theta_i), \qquad (10)$$

and

$$P^{(2)}(1,2) = \frac{N(N-1)}{Z}\int e^{-\sum_{i<j} v(i,j)/kT}\, d\tau_3 \ldots d\tau_N$$

$$\equiv P^{(1)}(1)P^{(1)}(2)g(1,2) = \rho^2 f(\theta_1)f(\theta_2)g(1,2). \qquad (11)$$

The function $f(\theta)$ in Eq. (10) is what we used earlier in Eq. (1) to define the order parameter. It measures the long range order in the system. $g(1,2)$, a pair correlation function, provides an additional measure of short range order. In terms of the distribution functions, the entropy

$$S = -Nk\int P^{(N)}(1,\ldots,N)\,\ell n P^{(N)}(1,\ldots,N)\,d\tau_1 \ldots d\tau_N \qquad (12)$$

can be expanded in a cluster series.

The only input is the interaction potential $v(i,j)$. It is rather unfortunate that we often speak of liquid crystal molecules as rods, thus placing too much emphasis on the steric aspect of the interaction. Temperature has no meaning when we deal with hard rods. Let me stress that only cylindrical symmetry will be assumed in this talk. The molecules can be thought of as point particles each carrying an imaginary two-headed arrow. For a pair of such molecules, Oseen pointed out that the interaction potential depends only on the 5 scalars: r_{12}, $\hat{\Omega}_1\cdot\hat{r}_{12}$, $\hat{\Omega}_2\cdot\hat{r}_{12}$, $\hat{\Omega}_1\cdot\hat{\Omega}_2$, and $\hat{\Omega}_1\times\hat{\Omega}_2\cdot\hat{r}_{12}$. The last one offers a selection between two signs. It is a chiral term. We showed elsewhere that it is responsible for the cholestric phase. The scalars $\hat{\Omega}_1\cdot\hat{r}_{12}$ and $\hat{\Omega}_2\cdot\hat{r}_{12}$ break the symmetry between splay 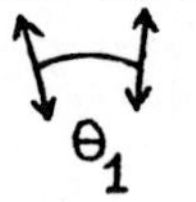 and bend 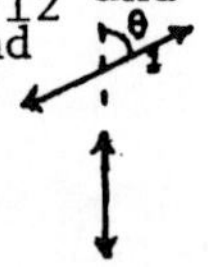. We have recently shown that functions with these arguments can be constructed to account for anisotropic repulsions, thus giving rise to macroscopic anisotropic properties. Also, they will help in classifying smectics. But for our present interest, and for the sake of simplicity, let us use only the variables r_{12} and $\hat{\Omega}_1\cdot\hat{\Omega}_2$. In fact, following Kobayashi and McMillan, let us write

$$v(1,2) = v_o(r_{12}) + v_2(r_{12})P_2(\hat{\Omega}_1\cdot\hat{\Omega}_2) + v_4(r_{12})P_4(\hat{\Omega}_1\cdot\hat{\Omega}_2). \qquad (13)$$

In the Maier-Saupe form of the mean field theory, the free energy functional appears as

$$\mathcal{F}[f(\theta)] = \mathcal{F}_o + \tfrac{1}{2} N\rho\gamma_2\sigma_2^2 + \tfrac{1}{2} N\rho\gamma_4\sigma_4^2 + NkT\int f(\theta)\ln f(\theta)d\hat{\Omega} \ , \qquad (14)$$

where

$$\gamma_\ell = \int v_\ell(r)d\vec{r}, \qquad (15)$$

$$\sigma_\ell = \int f(\theta)P_\ell(\cos\theta)d\hat{\Omega}, \qquad (16)$$

and the integral term represents the leading approximation of -TS. Possible phases of the system emerge from finding the extrema of $\mathcal{F}$ with respect to $f(\theta)$, the results being

$$f(\theta) = \frac{1}{4\pi Z} e^{-B_2P_2(\cos\theta)-B_4P_4(\cos\theta)} \ , \qquad (17)$$

with

$$B_2 = \frac{\rho\gamma_2}{kT}\sigma_2 \text{ and } B_4 = \frac{\rho\gamma_4}{kT}\sigma_4 \ . \qquad (18)$$

Equations (16)-(18) form a set of self-consistent equations. Among the solutions, that which yields the absolute minimum for $\mathcal{F}$ describes the stable phase at given ρ and T. Such a theory is very successful in predicting phase diagrams for homologous series of liquid crystals and order parameters, but very poor in predicting volume change $\Delta V_{IN}/V_I$, latent heat ΔH_{IN}, and the temperature T_ξ - often off by factors of 3-5 or even an order of magnitude, as pointed out by deGennes.

The reason for its shortcomings should be quite clear. It can best be seen by deriving the mean-field theory using an integrodifferential equation approach. Differentiate Eq. (10), the definition of $f(\theta)$, with respect to θ. Using Eq. (11), one obtains in two lines what is apparently the first of a hierarchy of (generalized BBGKY) equations:

$$\frac{\partial f(\theta_1)}{\partial\theta_1} = \frac{-\rho}{kT}\int f(\theta_1)f(\theta_2)g(1,2)\frac{\partial v(1,2)}{\partial\theta_1}d\tau_2 \qquad (19)$$

Setting g(1,2) to unity and integrating, one quickly recovers Eqs. (16)-(18). Thus, the mean field theory simply means ignoring short range or pair correlations, as the name "mean field" implies. But this is where things go wrong. One should not ake the mean field approximation so literally. In the Ising model, the spins are fixed on lattice sites. "Mean field" refers to the way in which magnetic interactions are handled. Spatially, short range correlations are already in effect: The particles have been kept apart by edict. Here, for liquid crystals, short range correlations have not been built into the model at the outset. Unless one wishes to resort to an unrealistic lattice model, spatial correlations will have to be accounted for in a natural way by the theory. We believe that the mean field approximation would work well for the orientational part of the distribution, but spatial short range correlations must not be ignored. After all, liquid crystals are at liquid densities. Nobody in his right mind expects the mean field approximation to work for ordinary liquids. So why should it work here? At least one should let g(1,2) be approximated by some radial function that contains a volume exclusion effect.

This then is the basis of our molecular model. Differentiating Eq. (11), the definition of $P^{(2)}(1,2)$, we obtain the second of the hierarchy of integrodifferential equations:

$$\nabla_1 P^{(2)}(1,2) = P^{(2)}(1,2)\nabla_1\left[\frac{-v(1,2)}{kT}\right]+\int P^{(3)}(1,2,3)\nabla_1\left[\frac{-v(1,3)}{kT}\right]d\tau_3 \ . \tag{20}$$

Letting $g(1,2) \approx g_e(r_{12})$ and averaging Eq. (20) over $\hat{\Omega}_1$ and $\hat{\Omega}_2$, we find

$$\nabla_1 g_e(r_{12}) = g_e(r_{12})\nabla_1\left[\frac{-\tilde{v}(r_{12})}{kT}\right]+\rho\int g_e^{(3)}(r_{12},r_{23},r_{31}) \times \nabla_1\left[\frac{-\tilde{v}(r_{13})}{kT}\right] d\vec{r}_3 \ , \tag{21}$$

$$\text{with } \tilde{v}(r) = v_o(r) + v_2(r)\sigma_2{}^2 + v_4(r)\sigma_4{}^2. \tag{22}$$

Eq. (21) is familiar from the theory of classical liquids. We know how it can be solved with the use of some superposition approximation. Placing the result in Eq. (1), we recover (under the realistic conditions that v_2 and v_4 are small compared to v_0) Eqs. (16)-(18), with the exception that γ_2 and γ_4 now pick up volume and temperature dependences through a modified Eq. (15).

$$\gamma_\ell = \int [g_\ell(r)v_o(r) + g_o(r)v_\ell(r)]d\vec{r} + kT\int g_\ell(r)[1+\ln g_o(r)]d\vec{r}, \tag{23}$$

with $g_\ell(r)$ defined by writing $g_e(r)$ as $g_o(r) + g_2(r)\ {}_2^{\,2} + g_4(r)\ {}_4^{\,2}$.

What we wind up with is a double-nested self-consistency made up of Eqs. (16)-(18) and (21)-(23).

In the empirical analysis there is no need to solve these equations. The various thermodynamic functions appear in combinations related to measurable quantities. For example, McColl and Shih, and Keyes and Daniels found that the transition curve can be represented quite well by

$$\rho = bT^{\frac{1}{\Gamma}}, \text{ with } \Gamma = 4\text{-}5 \tag{24}$$

for most nematogens. Since it turns out that Eqs. (16)-(18) require $B_\ell/\sigma_\ell = \rho\gamma_\ell/kT = (b/k)T^{1/\Gamma-1}\ \gamma_\ell$ to be a constant along the transition curve, γ_ℓ depends on T through the relation $\gamma_\ell = \gamma_{\ell 0}T^\alpha$, with

$$\alpha = 1 - 1/\Gamma = 0.75\text{-}0.80. \tag{25}$$

γ_ℓ is then a "measurable" quantity. To obtain γ_ℓ, detailed knowledge of $v_\ell(r)$ and $g_\ell(r)$ is not needed. [That γ_ℓ is chosen to depend only on T, rather than T and ρ, is somewhat arbitrary. Over a narrow range of T, the ρ dependence, which is secondary through Eq. (23), is weak. There is very little change by allowing some ρ dependence.]

In addition to Γ, we need empirical information on T_{IN}, σ_2, v_I, the compressibility κ, and the thermal expansion coefficient β, all at T_{IN}. Through calculating F and its thermodynamic derivatives and the help of the following Maxwell construction (Fig. 4), we find:

$$\frac{\Delta v_{IN}}{v_I} \equiv \frac{v_I - v_N}{v_I} \approx \kappa(P_c - P_A) \ , \tag{26}$$

and
$$\Delta H_{IN} = T(S_c - S_A) + T[(S_{c'} - S_c) + (S_A \cdot S_{A'})] \ , \tag{27}$$

with
$$S_{c'} - S_c = (v_{c'} - v_c)\left(\frac{\partial S}{\partial v}\right)_T \approx \frac{1}{2}\Delta v_{IN}\left(\frac{\partial P}{\partial T}\right)_v = \frac{\beta}{K} \ ; \tag{28}$$

we then obtain the volume change and the latent heat.

Furthermore, we obtain T_ξ as follows. Assume nematic embryos of volume V in an isotropic background. The fluctuation energy $\rho V_\xi[F(\sigma_N) - F(0)]$ is, by the equipartition theorem, 1/2 kT, Thus,

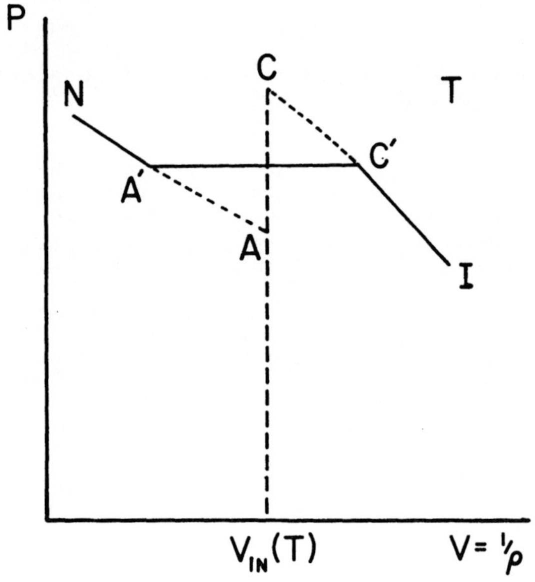

Figure 4

$$V_\xi \sim \frac{kT}{2\ [F(\sigma_N)-F(0)]}\ ,$$

or

$$I \propto \int_\infty <\sigma(0)\sigma(r)>d\vec{r} \approx \int_{V_\xi} <\sigma(0)\sigma(r)>d\vec{r} \approx V_\xi \sigma_N^{\ 2} = \frac{kT\ \ \sigma_N 2}{2\rho[F(\sigma_N)-F(0)]}\ . \quad (29)$$

[deGennes' result, Eq. (8), can be recovered immediately by taking $F(\sigma_N) \approx F(0) + a(T-T^*)\sigma_N^{\ 2}$.] The light scattering intensity thus diverges when $F(\sigma_N) = F(0)$. This temperature can be determined easily from constant-volume isotherms, (Note Fig. 5), thus:

$$T_\xi - T_{IN} \approx \frac{1}{2}\Delta v_{IN}/(\frac{dv}{dT})_{T_{IN}}\ . \quad (30)$$

Note that T_ξ by construction is also what we in a Van der Waals theory would refer to as the maximum supercooling temperature.

In contrast with the Landau-deGennes theory, we use G (through the Maxwell construction) instead of F. We do not define a

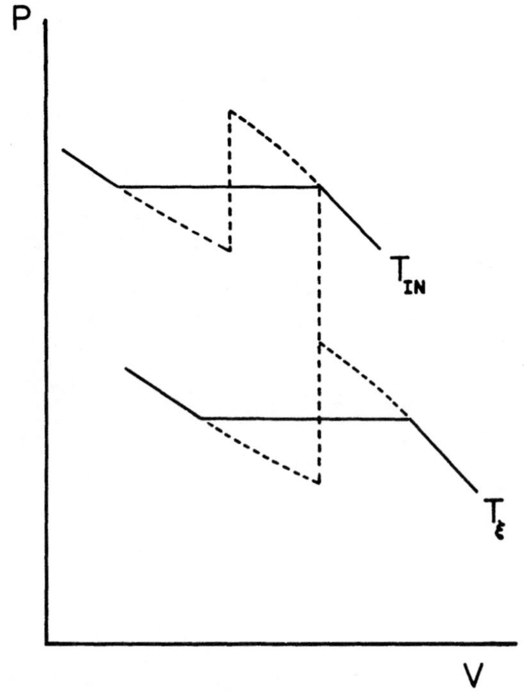

Figure 5

temperature T*. Our F(σ) diagram appears as in Fig. 6, while Fig. 3 should have its vertical axis relabeled G. Our results are shown in the table below for the two most popular nematogens and compared to experiment.

	PAA		MBBA	
	Calc.	Expt.	Calc.	Expt.
$\Delta v_{IN}/v_I$ (%)	0.32-0.41	0.30-0.36	0.11-0.14	0.11-0.14
ΔH_{IN} (J/mole)	635-821	574-760	280-360	284-381
$\|T_\xi - T_{IN}\|$ (K)	2.6-3.4	3.3	0.8-1.0	0.8

The agreement is excellent.

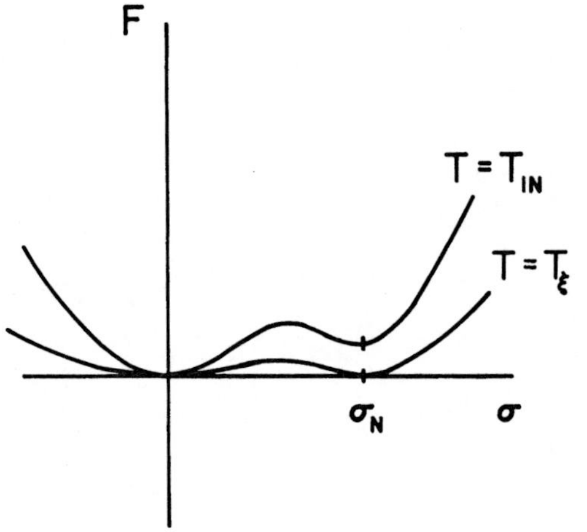

Figure 6

What I have given here is just one example of what a careful analysis based on a molecular model can do for nematogens. Similar analyses can be done for other liquid crystalline and plastic crystalline phases; some are at present being carried out at Northwestern.

The major contributors to the work described in this talk are Y. R. Lin-Liu and Y. M Shih.

References

In my opinion, the best way to approach the subject is to begin with E. B. Priestley, P. J. Wajtowicz, and P. Sheng's Introduction to Liquid Crystals (Plenum 1975), and then move on to P. G. deGennes' Physics of Liquid Crystals (Oxford 1974). M. J. Stephen and J. P. Straley's Rev. Mod. Phys. 46, 618 (1974) contains a wealth of useful material, while G. H. Brown, J. W. Doane and V. D. Neff's A Review of the Structure and Physical Properties of Liquid Crystals (CRC, 1971) contains a large number of older references. G. W. Gray's Molecular Structure and the Properties of Liquid Crystals (Academic, 1962) is by now out of date, but does have quite a bit of chemistry and structural analysis.

Papers from our group have appeared mainly in the last three years' Phys. Rev. A., Phys. Rev. Letters, Mol. Cryst. Liq. Cryst., and Phys. Letter A. A list will be provided upon request. From these papers one can obtain other references.

POSSIBLE "NEW" QUANTUM SYSTEMS - A REVIEW*

L. H. Nosanow

Division of Materials Research
National Science Foundation
Washington, D.C. 20550

Abstract: The current status of the theory of the possible "new" quantum systems, helium-six and spin-aligned hydrogen, is given. Results of ground-state calculations are combined with finite temperature experimental data to give an over-all view of quantum systems which obey both Bose-Einstein and Fermi-Dirac statistics. An essential feature of this work is the use of the quantum theorem of corresponding states, which allows one to treat the quantum parameter $\eta = \hbar^2/m\epsilon\sigma^2$ as a "Conceptual" thermodynamic variable. In this context, the liquid-to-crystal and liquid-to-gas transitions are discussed for quantum systems and shown to be affected profoundly by the statistics. Further, the construction of the phase diagram for ^{6}He and ^{6}He-^{4}He mixtures is reviewed and the possibility of detecting superfluid ^{6}He is discussed. Finally, the predicted properties of spin-aligned hydrogen are reviewed and a brief discussion of the experimental problem of preparing it is given.

I. INTRODUCTION AND SUMMARY

Recently there have been a number of theoretical papers,[1-5] which have considered the properties of possible "new" quantum

---

*This research was performed as part of the NSF Independent Research Program. However, the opinions expressed herein are those of the author and do not necessarily reflect the views of the NSF.

systems. The two systems which have been discussed are (1) spin-aligned hydrogen[1-4] (the isotopes of which we shall denote by H↑, D↑ and T↑), and (2) helium-six[5] (^{6}He). By "spin-aligned hydrogen," we mean hydrogen, suitably constrained (i.e., in large magnetic fields and at low temperatures) so that two hydrogen atoms can interact only in the b $^3\Sigma_u^+$ state (which has no bound state), instead of the usual X $^1\Sigma_g^+$ state (which has a bound state, the usual H_2 molecule). Helium-six is an unstable isotope which undergoes β-decay to lithium-six with a half-life $\tau = 0.32$ sec. Although this time is too short to make measurements at thermodynamic equilibrium, it is long enough, in principle, to observe superfluidity in ^{6}He. Of course, both spin-aligned hydrogen[6-8] and helium-six[9,10] have been discussed in the past. However, there are both theoretical and experimental reasons why these systems have been reconsidered recently. The purpose of the present work is mainly to discuss the current status of the theory of these "new" quantum systems and touch briefly upon possible experiments which may be done to prepare and study them. A more complete discussion of the current experimental situation for spin-aligned hydrogen has been given by Stwalley.[11]

One of the main reasons that these systems have been reconsidered recently is that they can be studied with techniques and results[12-14,2,4,5] which have been developed to treat quantum systems within the context of the quantum theorem of corresponding states (QTCS), originally proposed by de Boer[15] and co-workers.[16] The most important parameter of this theorem is the quantum parameter

$$\eta \equiv \hbar^2/m\varepsilon\sigma^2 = (\Lambda^*/2\pi)^2 , \qquad (1.1)$$

where Λ^* was introduced by de Boer;[15] the rest of the symbols are defined in Section II. One of the main points made in this recent work was that it is most illuminating to view the quantum parameter as a "conceptual" thermodynamic variable. In this way, it is possible to analyze quantum systems and to show clearly the effect of "statistics" on the solidification pressure,[12] to understand the liquid-to-gas phase transition at zero temperature for both Bose and Fermi systems,[13] to construct the phase diagram for ^{6}He and a somewhat speculative phase diagram for ^{6}He-^{4}He solutions,[5] and to construct the "generalized phase diagram" for quantum systems.[14]

Another reason for reconsidering spin-aligned hydrogen at this time is the calculation of very accurate H-H pair potentials in the X $^1\Sigma_g^+$ and b $^3\Sigma_u^+$ states by Kolos and Wolniewicz.[17] These potentials are pictured, tabulated and compared with the He-He interaction and fit with a Lennard-Jones potential in Ref. 4. The relevant parameters are given in Section II. These K-W potentials have been shown to be very accurate; thus, spin-aligned hydrogen is an unusual many-body system in that the pair potential is not one of the uncertainties of the problem.

The basic idea of the work we will review is to use a combination of theoretical results and experimental data to deduce the general behavior of quantum systems which obey both Bose-Einstein and Fermi-Dirac statistics. Then the properties of the aforementioned "new" quantum systems can be obtained within this context. In Section II, we give a brief review of the QTCS. In Section III, we summarize the results of variational calculations of the ground state of quantum systems in solid, liquid and gaseous phases. In Section IV, we summarize various finite temperature experimental data. We then utilize these data along with the results of Section III and the QTCS to construct the "generalized" phase diagrams for Bose-Einstein and Fermi-Dirac quantum systems in $P^* - T^* - \eta$ space. Finally, brief discussions of helium-six and spin-aligned hydrogen are given in Sections V and VI, respectively.

II. QUANTUM THEORY OF CORRESPONDING STATES

The QTCS[15,14] applies to a class of systems, each member of which interacts via a pair potential of the form

$$v(r) = \varepsilon v^*(r/\sigma), \tag{2.1}$$

where ε is the coupling constant (with dimensions of energy), σ is a range parameter (with dimensions of length), and $v^*(x)$ is the same dimensionless function of its argument for each member of this class of systems. In this paper we shall be interested in the class of quantum systems that interact via a "noble-gas-like" pair potential. A simple, phenomenological form for this potential is due to Lennard-Jones; i.e.,

$$v^*(x) = 4(x^{-12} - x^{-6}). \tag{2.2}$$

To state the QTCS, it is convenient to first introduce the dimensionless quantum parameter η defined by Eq. (1.1). We have found it notationally more convenient to use η than the de Boer parameter Λ^*. Values of η along with values of ε, σ and other useful quantities are given for various substances in Table I. It is further convenient to introduce several dimensionless or "reduced" variables as follows:

Table I. The quantum parameter η for various substances. Also given are the masses m, coupling constants ε, "core diameters" σ, ε/σ^3 and $N_0\sigma^3$. We used $\hbar = 1.05430 \times 10^{-27}$ erg sec, $N_0 = 6.02252 \times 10^{23}$ particles/mole, 1 amu = 1.66024×10^{-24}g and $k_B = 1.38054 \times 10^{-16}$ erg/particle deg.

Substance	m amu	ε deg	σ Å	ε/σ^3 atm	$N_0\sigma^3$ cm^3/mol	η
H	1.008	6.46	3.69	17.5	30.2	0.547
D	2.014	6.46	3.69	17.5	30.2	0.274
^{3}He	3.016	10.22	2.556	83.39	10.06	0.2409
T	3.016	6.46	3.69	17.5	30.2	0.183
^{4}He	4.003	10.22	2.556	83.39	10.06	0.1815
^{6}He	6.019	10.22	2.556	83.39	10.06	0.1207
H_2	2.016	37.0	2.92	202.5	15.0	0.0763
D_2	4.028	37.0	2.92	202.5	15.0	0.0382
Ne	20.18	35.6	2.74	235.8	12.4	0.0085
Ar	39.95	120.	3.41	412.3	23.9	0.00088

$$T^* \equiv k_B T/\varepsilon \tag{2.3a}$$

$$V^* \equiv V/N\sigma^3 \equiv 1/\rho^* \tag{2.3b}$$

$$P^* \equiv P\sigma^3/\varepsilon \tag{2.3c}$$

$$F^* \equiv F/N\varepsilon \ , \tag{2.3d}$$

where T is the temperature, V is the volume, N is the number of particles, ρ is the number density, P is the pressure and F is the Helmholtz free energy. The QTCS states that, for a one-component system,

$$F^* = F^*(T^*,V^*,\eta) \ , \tag{2.4}$$

where F* depends only on the form of v*(x) and on whether the particles obey Bose-Einstein or Fermi-Dirac statistics. A more complete discussion of the QTCS as it relates to the present work can be found in Refs. 12 and 14.

III. GROUND-STATE RESULTS

The ground-state properties of Boson and Fermion quantum systems have been studied extensively in NPP and MNP. They found that quantum systems can have solid, liquid or gaseous ground states depending on the value of η. In addition, they found that both the liquid-to-crystal and liquid-to-gas transitions are profoundly affected by whether the system obeys Bose-Einstein or Fermi-Dirac statistics. Their results for various special values of η are given in Table II.

In this work we shall give only an intuitive discussion of the liquid-to-crystal transition and refer the interested reader to MNP for a full discussion. The essential point of the physics is that, for a given value of η, the ground-state energy of the Fermi liquid is much higher than that of the corresponding Bose liquid. This energy difference is due to the existence of the fermi sea and is approximately given by the well-known kinetic energy term equal to $3\varepsilon_F/5$, where ε_F is the fermi energy. On the contrary, the energies of Bose and Fermi crystals are very nearly equal. This is because, in the crystalline phase, the atoms are so well localized that the exchange integrals are small - of the order of 1 mK. Thus, for a given value of η, the crystal-liquid energy difference is larger for bosons than for fermions. The reduced solidification pressures for both Bose and Fermi systems as functions of η are given in Fig. 1 of Ref. 14. The value of η for which the triple-point temperature and pressure vanish is called η_3; its values for bosons and fermions (η_{3B} and η_{3F}, respectively) are given in Table II.

We shall now give an intuitive discussion of the liquid-to-gas transition and refer the reader to MNP for a full discussion. In this case there are several points that make up the essential physics. In the first place, as η increases, the equilibrium density of the liquid phase decreases; therefore, it is possible to expand the energy in powers of the density in the neighborhood

Table II. Values of η at which various transitions occur in the ground states of quantum systems. The value deemed closest to experiment is given. The original discussions may be found in Refs. 12, 13, 15, 14 and 4.

TRANSITION	STATISTICS	η
Liquid-Solid	Bose	0.136
	Fermi	0.177
Liquid-Gas	Bose	0.456
	Fermi (2)	0.290
	Fermi (1)	0.274
Critical-Point	Fermi (2)	0.33
	Fermi (1)	0.35

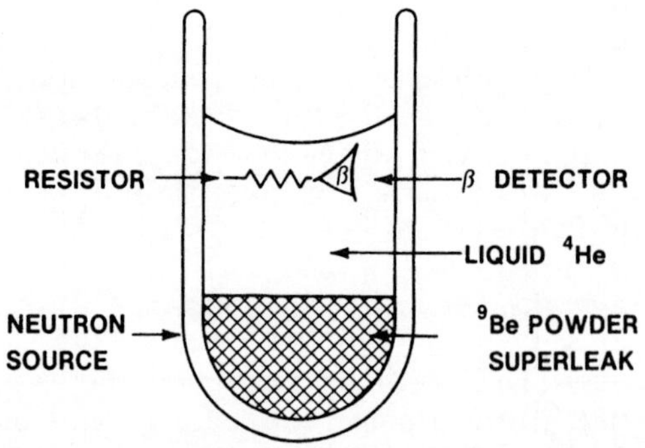

Figure 1. Schematic experiment to detect ^{6}He superfluidity in ^{6}He-^{4}He mixtures.

of the liquid-to-gas transition for both bosons and fermions. In the Bose case, the leading term in this expansion is proportional to the density. As shown in MNP, this leads to the fact that at zero temperature, Bose systems undergo a liquid-to-gas transition that is second order in η; it occurs at $\eta = \eta_{CB}$. At zero temperature the liquid and gaseous phases of a Bose system cannot coexist. On the contrary, the behavior of a Fermi system at zero temperature is quite different from that of a Bose system. This is because the leading term in the density expansion of the energy is proportional to $\rho^{2/3}$. Again, as explained in great detail by MNP, this leads to a van der Waals-like liquid-to-gas transition for fermions at zero temperature. At the liquid-to-gas transition, $\eta = \eta_{LF}$; at the critical point, $\eta = \eta_{CF}$. The values are given in Table II for Fermi systems with both one and two allowed nuclear spin states. The effect of the number of allowed nuclear spin states is discussed in Ref. 4. The full zero-temperature $P^* - \eta$ phase diagram is given in Fig. 5 of MNP.

IV. FINITE TEMPERATURE RESULTS AND "GENERALIZED" PHASE DIAGRAMS

The finite-temperature properties of Boson and Fermion quantum systems have been studied extensively in the original work of de Boer[15] and coworkers[16] and in the recent work of Nosanow.[5,14] The way to utilize the QTCS is to plot known experimental data for a given reduced thermodynamic quantity as a function of η. Then the value of this quantity can then be inferred for another substance by interpolation or extrapolation. In this way de Boer and Lunbeck[16] were able to predict some of the properties of ^{3}He before they were measured.

Experimental data of various thermodynamic quantities for systems with "helium-like" pair interactions are tabulated and plotted in Refs. 5 and 14 for both Bose and Fermi systems. In particular, the temperature, pressure and molar volume for both the triple point and the critical point are studied. In general, as η increases these temperatures and pressures decrease; whereas, the molar volume increases. These trends can all be understood straightforwardly, since an increase in η means that the kinetic energy per particle is increasing relative to the magnitude of the potential energy per particle.[18]

In both Refs. 5 and 14 the theoretical gound-state results (which were summarized in Section III) are combined with the experimental finite-temperature data This procedure allows better extrapolations to be made and also permits one to differentiate more clearly between the behavior of Bose and Fermi systems. In this way the "generalized" $P^* - T^* - \eta$ phase diagrams given in Figs. 4 and 5 of Ref. 14 were constructed. This method was also used in Ref. 4 to estimate the properties of spin-aligned deuterium.

These results will be summarized in Section VI. The finite temperature behavior of quantum systems and their critical behavior are extensively discussed in Ref. 14.

V. HELIUM-SIX

The question of the possible superfluidity of ^{6}He has been considered for many years.[9,10] The only experiments were those of Guttman and Arnold,[9] which did not show superfluidity in ^{6}He. Further, they obtained an atom fraction of ^{6}He of only 10^{-19} and it is questionable whether superfluidity can be observed with such a low concentration.[19,20] Since there has been substantial development in technique since the experiment of Guttman and Arnold, it was felt that a reexamination of this question was in order and this was the main subject of Ref. 5.

In that reference, the phase diagrams of pure ^{6}He and of ^{6}He-^{4}He mixtures were worked out using a combination of experimental data, theoretical results and some intuitive physics. Some of the results are summarized in Table III. It is believed that the phase diagram of pure ^{6}He is quite well founded and that there is a small, but definite, region where the pure liquid would be superfluid. The phase diagram for the ^{6}He-^{4}He mixtures is not so well grounded. Firstly, it is not known[19,20] what minimum concentration of ^{6}He would be necessary for it to exhibit superfluidity. Secondly, the existence of a λ-eutectic point, which appears quite naturally on grounds of continuity, must be regarded as a speculation given the present state of our understanding of the λ-transition itself.

Nevertheless, we believe that another experiment to detect superfluidity in ^{6}He would be worthwhile and feasible. A schematic diagram of a possible experiment is given in Fig. 1. Since ^{6}He is produced in the reaction ^{9}Be(n,α)^{6}He, the idea is to use the ^{9}Be powder not only to produce ^{6}He, but also as a superleak. Then neutrons can be allowed to impinge on the ^{9}Be superleak as the temperature is lowered. If β's are detected when a small current is applied to the resistor, it is clear that superfluid ^{6}He will have escaped from the superleak. There are certainly many other experiments which could be envisioned to detect superfluidity in ^{6}He and many experiments which could use ^{6}He as a probe to study other problems. Thus, we believe that there is much physics to be learned from the study of ^{6}He.

Table III. Estimated values for several properties of ^{6}He (taken from Ref. 5)

Property	Value
Critical Temperature	7.5 K
λ-Temperature	1.6 K
Triple-point temperature	1.2 K
λ-eutectic temperature	1.2 K

VI. SPIN-ALIGNED HYDROGEN

Spin-aligned hydrogen has also been considered for many years.[6-8] In the past, it has mainly been of interest because it is the simplest free radical.[6,7] However, recently it has been studied as a quantum system.[1-4] It is expected that, to a first approximation, H↑ and T↑ will obey Bose-Einstein statistics; whereas, D↑ will obey Fermi-Dirac statistics. It can be seen readily from the values of η for H↑ and D↑ given in Table I that, if these systems can be prepared, they should exhibit even more "extreme" quantum behavior than the helium isotopes. In this Section, we shall briefly review the properties of spin-aligned hydrogen which have been estimated theoretically and briefly consider a schematic experiment to prepare it.

In Refs. 2 and 4 it is shown that the Kolos-Wolniewicz $b^3\ \Sigma_u^+$ potential can be well approximated by a Lennard-Jones potential. Thus, the results of Sections III and IV can be applied to infer the properties of H↑ and D↑. Since $\eta(\mathrm{H}\uparrow) = 0.55 > \eta_{CB} = 0.45$, it is clear that H↑ cannot form a liquid phase. Thus, it is expected that H↑ will undergo a Bose-Einstein condensation. In Fig. 2 of Ref. 2, the Bose-Einstein condensation temperature is given as a function of pressure. This curve was estimated using the ideal gas expression, since H↑ is expected to exist at very low densities. In our opinion, the study of the Bose-Einstein condensation is one of the most important experiments that could be done with H↑. After all, at the present time there is really no fundamental

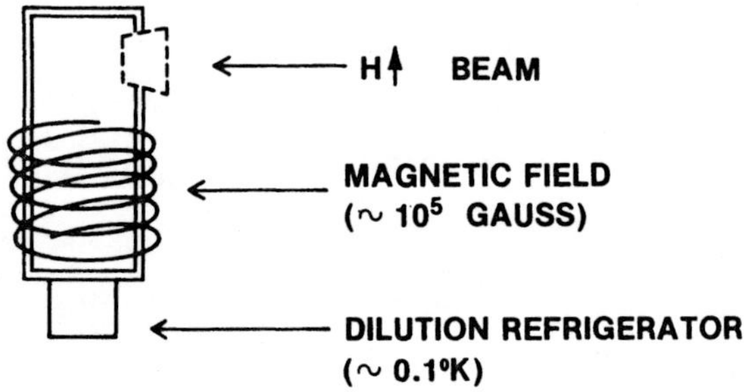

Figure 2. Schematic experiment to prepare spin-aligned hydrogen.

understanding of the λ-transition in ^{4}He. It is not even possible to state categorically that the Bose gas will be superfluid below T_{BE}. There could conceivably be a superfluid transition at a lower temperature. If the gas does become superfluid at T_{BE}, the study of this transition will undoubtedly be extremely important for our understanding of Bose systems and their superfluidity.

The properties of D↑ are studied extensively in Ref. 4. Two cases are considered: (1) $D{\uparrow}_1$, in which only one nuclear spin state is allowed, and (2) $D{\uparrow}_2$, in which two nuclear spin states are allowed. The ground-state energies for these systems are plotted versus density in Figs. 4 and 5 of Ref. 4. It turns out that the K-W potential yields a gaseous ground state in each case; whereas, the L-J potential yields a liquid. Unfortunately, the energies involved are so small that it is not possible to say with certainty whether real $D{\uparrow}_1$ or $D{\uparrow}_2$, will be bound or not. If they do have a gaseous ground state, then it would be possible to liquefy them at zero temperature by applying pressure. The finite temperature properties of $D{\uparrow}_1$ and $D{\uparrow}_2$ are also studied in Ref. 4 utilizing the same approach as described in Section IV. The results are given in Table IV along with those for other "noble-gas-like" systems for comparison. On this table are also shown values of the thermal de Broglie wave length evaluated at the critical point. Since these lengths are small compared to the correlation length for critical fluctuations near the critical

Table IV. The critical temperatures and thermal de Broglie wave lengths at the critical temperature for helium, spin-aligned deuterium, and other "noble-gas-like" systems.

SUBSTANCE	T_C (°K)	λ_C (Å)
$D\uparrow_2$	1.29	10.8
$D\uparrow_1$	1.68	9.5
^3He	3.32	5.47
^4He	5.20	3.82
H_2	33.3	2.13
Ne	44.5	0.57
Ar	151.	0.23

point, it is to be expected that quantum effects will not be important near the critical point.[21]

In conclusion, we shall briefly discuss the schematic experiments to prepare spin-aligned hydrogen shown on Fig. 2. The idea is to make a beam of H↑ atoms and shoot it into a "bottle" which is in a magnetic field of the order of 10^5 gauss at a temperature of the order of 0.1°K. When enough H↑ has been put into the "bottle," one would close it with a "cork" and do experiments on H↑. The major experimental problem, in our opinion, is the interaction with the walls. It is not known, at present, to what extent spin-flip collisions with the walls will occur. The successful operation of a hydrogen maser, however, does give one hope that this problem either will not be too severe or will be overcome by clever experimentation. In any case, spin-aligned hydrogen is clearly a most interesting system and merits a reasonable effort to try to prepare it. These questions are discussed more fully in the paper by Stwalley.[11]

References

(1) J. V. Dugan, Jr. and R. D. Etters, J. Chem. Phys. 59, 6171 (1973); R. D. Etters, J. V. Dugan, Jr. and R. W. Palmer, J. Chem. Phys. 62, 313 (1975); R. D. Etters, Phys. Letters 42A, 439 (1973); R. L. Danilowicz, J. V. Dugan, and R. D. Etters, J. Chem. Phys. 65, 498 (1976).

We shall refer to these references collectively as DEP.

(2) W. C. Stwalley and L. H. Nosanow, Phys. Rev. Letters 36, 910 (1976).

(3) W. C. Stwalley, Phys. Rev. Letters 37, 1628 (1976).

(4) M. D. Miller and L. H. Nosanow, Phys. Rev. B, to be published.

(5) L. H. Nosanow, J. Low Temp. Phys., 23, 605 (1976).

(6) J. T. Jones, M. H. Johnson, H. L. Mayer, S. Katz and R. S. Wright, Aeronutronics Systems, Inc. Publ. No. U-216 (1958), unpublished.

(7) M. W. Windsor, in "Formation and Trapping of Free Radicals," edited by A. M. Bass and H. P. Broida (Academic Press, New York, 1960), p. 400f.

(8) R. Hess, Adv. Cryog. Eng. 18, 427 (1973); R. Hess, doctoral dissertation, University of Stuttgart, 1973 (unpublished); and Deutsche Luft-und Raumfahrt, Forschungsbericht 73-74: Atomarer Wasserstaff (Institut fur Energie Wandlung und Elektrische Antriebe, Stuttgart/Braunschweig, 1973); W. Perhka, private communication.

(9) L. Guttman and J. R. Arnold, Phys. Rev. 92, 547 (1953).

(10) R. Packard, private communication.

(11) W. C. Stwalley, following paper.

(12) L. H. Nosanow, L. J. Parish and F. J. Pinski, Phys. Rev. B 11, 191 (1975); we refer to this paper as NPP.

(13) M. D. Miller, L. H. Nosanow and L. J. Parish, Phys. Rev. Letters 35, 581 (1975); Phys. Rev. B 15, 214 (1977); we refer to this work as MNP.

(14) L. H. Nosanow, J. Low Temp. Phys. 26, 613 (1977).

(15) J. de Boer, Physica 14, 139 (1948).

(16) J. de Boer and B. S. Blaisse, Physica 14, 149 (1948); J. de Boer and R. J. Lunbeck, Physica 14, 520 (1948).

(17) W. Kolos and L. Wolniewicz, Chem. Phys. Letters 24, 457 (1974); J. Mol. Spectrosc. 54, 303 (1975).

(18) This behavior is exhibited clearly on Fig. 7 of MNP.

(19) A. Fetter, private communication.

(20) J. Ruvalds, private communication.

(21) M. Suzuki, Prog. Theor. Phys. 56, 77 (1976).

PREDICTED STABILITY CONDITIONS FOR GASEOUS SPIN-ALIGNED HYDROGEN (H↑).*

William C. Stwalley

Department of Chemistry, University of Iowa
Iowa City, Iowa 52242

Abstract: It is predicted that spin-aligned hydrogen (H↑) can probably be maintained under the following conditions:

1. $\mathcal{H}/T \gtrsim 10^6$ G/K.

2. Thickly ^{4}He-coated container.

3. No species other than ^{4}He and H↑ present.

INTRODUCTION

Spin-aligned hydrogen (H↑) is an extremely interesting (but thus far hypothetical) form of hydrogen where all electronic spin projections (M_s) are parallel so that chemical recombination to H_2 cannot occur. Here only the lowest energy ($M_s = -1/2$) form is considered. Because it is the most quantal atomic system possible,[1] H↑ is predicted[1-3] to remain a gas to absolute zero (for pressures less than the solidification pressure[1] of $\sim$ 50 atm). It should show the extreme quantum behavior[1] of a nearly ideal Bose gas, including Bose-Einstein condensation and superfluidity. The purpose of this paper is to briefly review and update arguments[1,2,4-6] concerning the stability of gaseous H↑. Table 1 lists the processes to be discussed here. It is assumed here that all impurities (especially D↑, O_2; see Ref. 6) can be completely eliminated.

---

* Acknowledgement is made to the Donors of the Petroleum Research Fund, administered by the American Chemical Society, for support of this work.

Table 1. Possible Destruction Mechanisms[a] for $H\uparrow$ in a Thickly ^{4}He-Coated[b] Container with $T \leq 1K$ and $\mathcal{H} \geq 50$ kG. S refers to the ^{4}He Surface.

I. Gas Collision	1.	$H\uparrow + H\uparrow \rightarrow H\uparrow + H\downarrow$
" "	2.	$H\uparrow + H\uparrow + H\uparrow \rightarrow H\uparrow + H\uparrow + H\downarrow$
" "	3.	$H\uparrow + H\uparrow + H\uparrow \rightarrow H\uparrow + H_2(v=14)$
II. Wall Collision	4.	$H\uparrow + H\uparrow + S \rightarrow H\uparrow + H\downarrow + S$
" "	5.	$H\uparrow + H\uparrow + S \rightarrow H_2(v=14) + S$

a In processes 1-5, ^{4}He could be substituted for a $H\uparrow$ with a comparable to greatly diminished (e.g. for 1) rate constant. However, since $[^4He] << [H]$, the rates will be negligible compared to those given above.

b It is felt that H will not penetrate the bulk of a thick ^{4}He film (see text). Whether the $H\uparrow$ are free or very weakly physisorbed (binding $\sim$ 1.5 K) should not change the above processes significantly.

GENERAL STABILITY CONSIDERATIONS

It is obvious[1,2,4-6] that a necessary condition for the stabilization of $H\uparrow$ is that its Zeeman energy be large compared to kT, so that processes requiring energy for a spin-flip will be greatly attenuated. For example, the criterion[6] $\mathcal{H}/T = 10^6$ G/K puts a factor of $e^{-134} = 6/4 \times 10^{-59}$ into the rate of processes requiring this activation. Once, however, a single spin flips, a tremendous heat release is possible via reactions such as $H\uparrow + H\uparrow + H\downarrow \rightarrow H\uparrow + H_2$ since the binding energy of the H_2 molecule is 52000 K! The heating should then allow further spin-flips to occur, which in a chain reaction would destroy the system. Alternately, the spin-flip can occur within a molecule, producing $X^1\Sigma_g^+ H_2$ from $b^3\Sigma_u^+(H\uparrow)_2$. To minimize the energy deficits, it seems most likely that any singlet H_2 form will be initially highly vibrationally excited, most probably in v = 14 (bound by 210 K in the J = 0 state).

GAS-PHASE COLLISIONS

Let us now examine the gas-phase processes of Table 1 in order.

Endothermic process 1 was previously examined in some detail.[6] The essential results were that

(a) there was no possibility for process 1 below the threshold energy (equal to the Zeeman energy).

(b) above threshold, the electronic spin selection rule $\Delta S = 0$ broke down slightly because of hyperfine interaction, giving a small cross section but an unacceptably rapid rate for H↑ destruction when $\mathcal{H}/T \ll 10^6$ G/K.

(c) To eliminate process 1, use $\mathcal{H}/T \gtrsim 10^6$ G/K.

One way of looking at this is that there is essentially no singlet density of states for energies between the last bound level (v = 14 at -210 K) and the dissociation continuum at 0K. (In HD and D_2 this is not the case because of resonances discussed in Ref. 6).

Similarly, endothermic process 2 is expected to have a small but finite cross section which can be completely eliminated by using $\mathcal{H}/T \gtrsim 10^6$ G/K. Here one is concerned with the intersection of doublet and quartet potential surfaces which are approximately the sum of pairwise interactions at the large distances involved in the quartet collisions. The density of states for the H_3 doublet molecule could conceivably include a resonance between the 3H↑ asymptote and the 2H↑ + 1H↓ asymptote (the dissociation limit for H_3 doublet molecules), but this possibility is difficult to determine theoretically. In any case, the resonance effects should be avoidable by varying the magnetic field.[6]

Process 3 (and similar processes to produce $v \leq 13$) is unfortunately exothermic. The process does, however, appear to be exceedingly unlikely. The same doublet and quartet surfaces applicable to process 2 are applicable here, but a much larger change of energy (vibrational to translational) is necessary to produce H_2 (v = 14) and H with relative kinetic energy of $\sim$ 200 K. A roughly exponential decrease with increasing energy mismatch is frequently postulated. Moreover, the Franck-Condon factors for this transition should be quite small. Further work is needed, however, to more quantitatively evaluate this process.

Processes involving 4 or more atoms could only become important as the density approaches that needed for solidification ($\sim 10^{22}/cm^3$); they will not be considered here, but should show similar features to process 1-3.

SURFACE COLLISIONS

The collisions of $H_\uparrow$ with surfaces provide an additional mechanism for possible spin-flips. Previously, I have argued[6] that solid Ne or Ar coatings might minimize the probability of spin flip. Such arguments are somewhat weakened by the slight possibility[7] that the adsorption energy for H on solid Ne or Ar could assist in overcoming the energy barrier to spin-flip. It is known[8] that H can be trapped in Ne or Ar matrices and hence H should bind to Ne and Ar solid surfaces as well.

However, an alternative surface of even higher promise is available--liquid ^{4}He, particularly at temperatures where the ^{4}He vapor pressure is small (e.g. $T < 0.5K$ implies $p < 1.6 \times 10^{-5}$ torr[9]). It appears clear that there is a "quantum insolubility" effect[10] for H in ^{4}He; i.e. the zero-point motion is far too great and the H-He interaction too weak[11] to permit any concentration of H in bulk ^{4}He below say 0.5 K (i.e. less than 1 H/mole ^{4}He).

It is well known[12] that there is excess binding for ^{3}He on the surface of ^{4}He. Similarly, it is more likely that H would be bound to the surface of ^{4}He than found within the bulk; my estimates[13] suggest that there is a single bound state (bound by $\sim$ 1.5 K) for H on a ^{4}He surface. Thus this adsorption energy cannot contribute significantly to overcoming the spin-flip energy barrier.

It is also interesting that ^{4}He scattering from ^{4}He surfaces is very highly absorptive;[14] hence it is likely that H scattering from ^{4}He surfaces will also be highly absorptive, so that cooling of H↑ (prepared say in a discharge at $\sim$ 4K and accelerated into the high field region to $\sim$ 10-20K by the Zeeman force) will probably not require many wall collisions. Once H↑ covers the ^{4}He surface, however, absorptive collisions should no longer be possible.

Process 4 is endothermic and similar to process 1 (and also 2); the ^{4}He surface is not expected to be very important in breaking the $\Delta S = 0$ selection rule or in providing activation energy. Thus $\mathcal{H}/T \gtrsim 10^6$ G/K should also eliminate it.

Process 5 is exothermic and similar to process 3 except that a spinless surface is involved in place of H↑. Again the large magnitude of the energy change and the smallness of the Franck-Condon factors suggest an extremely small rate.

Under high density conditions where the mean free path is small compared to the container size, the gas phase three-body processes (2 and 3) will, of course, dominate if their probabilities per collision are comparable to the surface plus two-body collisions (processes 4 and 5). For H↑ below 1K, this means the number of gas-phase three-atom collisions in say a 1 cm container will be comparable to the number of wall plus two-atom collisions for number densities of $\sim 10^{19}/cm^3$; for higher densities, the gas-phase three-body collisions will dominate.

SUMMARY

Under the conditions

1. $\mathcal{H}/T \gtrsim 10^6$ G/K

2. Thickly ^{4}He-coated container

3. No species other than ^{4}He and H↑ present,

it is likely that gaseous spin-aligned hydrogen (H↑) is stable. Process 3, being exothermic, is the most questionable and may possibly limit the highest density achievable.

ACKNOWLEDGMENTS

Helpful discussions of Ref. 6 with L. H. Nosanow, D. Kleppner, P. A. Pincus and T. D. Holstein are gratefully acknowledged.

References

(1) (a) W. C. Stwalley and L. H. Nosanow, Phys. Rev. Letters 36, 910 (1976).

(b) L. H. Nosanow, Preceding Paper.

(2) (a) J. V. Dugan and R. D. Etters, J. Chem. Phys. 59, 6171 (1973).

(b) R. D. Etters, J. V. Dugan, and R. W. Palmer, J. Chem. Phys. 62, 313 (1975).

(c) R. L. Danilowicz, J. V. Dugan and R. D. Etters, J. Chem. Phys. 65, 499 (1976).

(3) L. W. Bruch, Phys. Rev. B 13, 2873 (1976).

(4) J. T. Jones, M. H. Johnson, H. L. Mayer, S. Katz, and R. S. Wright, "Characterizations of Hydrogen Atom Systems", Aeronutronics Systems, Inc. Publication U-216 (1958).

(5) M. W. Windsor, in "Formation and Trapping of Free Radicals", edited by A. M. Bass and H. P. Broida, (Academic Press, New York, 1962), p. 400f.

(6) W. C. Stwalley, Phys. Rev. Letters 37, 1628 (1976).

(7) L. H. Nosanow, private communication.

(8) S. N. Foner, E. L. Cochran, V. A. Bowers and C. K. Jen, J. Chem. Phys. 32, 963 (1960).

(9) H. Van Dijk and M. Durieux, Physica 24, 920 (1958).

(10) W. C. Stwalley, "Quantum Insolubility: H in ^{4}He", to be submitted.

(11) J. P. Toennies, W. Welz and G. Wolf, Chem. Phys. Letters 44, 5 (1976).

(12) C.-W. Woo, in "Quantum Statistics and the Many-Body Problem", edited by S. B. Trickey, W. P. Kirk and J. W. Dufty, (Plenum, New York, 1975), p. 175.

(13) W. C. Stwalley, "The Predicted Adsorption of H on a ^{4}He Surface", to be submitted.

(14) D. O. Edwards et al., in "Quantum Statistics and the Many-Body Problem", edited by S. B. Trickey, W. P. Kirk and J. W. Dufty, (Plenum, New York, 1975), p. 195.

SURVEY OF THE CENTRAL ISSUES CONCERNING SUPERFLUID SOLIDS

W. M. Saslow

Department of Physics, Texas A&M University
College Station, Texas 77843

I. INTRODUCTION

If the subject of superfluidity in solids were a body of water, it would probably be described as murky, turbid, or perhaps even turbulent. In this survey I shall try to give the reader at least an understanding of what is going on at the surface. Although it will not be an exhaustive survey, it should serve as a useful introduction to the subject, and should present some benchmarks for those whose interest carries them to study the literature first-hand.

Let me begin by presenting a simple-minded characterization of what is meant by superfluidity in solids, and of how to calculate the superfluid fraction, following that in Sect. II, with provisos of a more sophisticated nature.

Consider a macroscopically large rotating annulus of height h, inner radius R, and thickness $d \ll R$ made up of a (slightly strained) crystal of atoms of mass m (e.g., ^{4}He), and at T=0. The ground-state wavefunction Ψ_o for the N particles in the lattice has a series of quantum numbers specifying, in part, the crystalline sites, whose number M need not equal N. Due to zero-point motion, the density profile $\rho(\vec{r})$ (which reflects the site positions) is somewhat delocalized. In a conventional rotation experiment, the lattice (and thus the density profile) rotates. In superfluid rotation the lattice moves and Ψ_o changes in such a way that, in the rotating frame, matter flows yet the density profile does not change.

If we let all the single-particle states develop the same phase factor ϕ ("universal phase factor"), then the excess energy in the rotating frame is given by[1]

$$E - E_o = \frac{m}{2} \int \rho(\vec{r})\, \vec{v}_s^{\,2}(\vec{r}) d\vec{r} \tag{1}$$

where $\vec{v}_s(\vec{r}) = (h/m)\vec{\nabla}\phi$ varies on an atomic scale. From this one may define the superfluid number density ρ_s by[1]

$$E - E_o = \frac{m}{2} \rho_s v_o^2 V \tag{2}$$

where v_o is the average flow velocity about the annulus ($\Delta\phi = 2\pi m v_o R$ is, in the rotating frame, the phase change on going around the annulus; in this frame the wavefunction is not single-valued - see Sect. II), and V is the crystal volume. Using the condition[2]

$$\vec{\nabla} \cdot \vec{j}(\vec{r}) = 0 \tag{3}$$

where $\vec{j}(\vec{r}) = \rho(\vec{r})\vec{v}_s(\vec{r})$, it is possible to obtain the flow pattern $\vec{v}_s(\vec{r})$ for a known $\vec{v}_o$, and therefore to compute ρ_s by equating (1) and (2). Explicit calculations show that ρ_s/ρ_o (where $\rho_o = N/V$) is near unity for a reasonably delocalized $\rho(\vec{r})$, whereas it is near zero when (r) is rather localized, the transition between the two regimes being fairly abrupt.[2] For solid ^{4}He, most likely $0.05 \lesssim \rho_s/\rho_o \lesssim 0.20$, given our present uncertain knowledge of atomic localization for this substance.

However, a universal phase factor is not the only way to obtain a phase change $\Delta\phi$ on going around the annulus. The problem is much more subtle than that, so that the system could well find a way to make $\rho_s/\rho_o \approx 0$. In short, the calculation just described only gives an upper bound on ρ_s/ρ_o.

II. THE CONDITIONS FOR SUPERFLUIDITY IN SOLIDS

The fundamental paper on this subject is by Leggett,[1] although the possibility of a different sort of superfluidity in solids (to be discussed in Sect. III) was considered by Andreev and Lifshitz somewhat earlier.[3] Leggett, in turn, acknowledges a debt to the work of Yang,[4] Kohn,[5] and Bloch[6] (as well as to a seminal conversation with Valatin).

Leggett's considerations on superfluidity in solids center about the possibility of a nonclassical moment of inertia (NCMI), an idea which Kohn and Sherrington[7] believe to have been first used by Blatt and Butler.[8] Leggett considers the specific case of a large rotating annulus, as we did in the previous section. He shows that, in the frame rotating with the container walls,

$$\Psi(\ldots\theta_i+2\pi\ldots) = \exp(-2\pi i m R^2\omega/\hbar)\Psi(\ldots\theta_i\ldots) \tag{4}$$

on moving the ith particle once around the annulus, returning to its starting place. [Note that $-\omega R$ corresponds to v_o of Eq. (2)]. From this it follows that $E_{min}(\omega) = E_{min}(\omega + \frac{\hbar}{mR^2})$, so $= 0(\frac{\hbar}{mR^2})$. (Such periodicity is reminiscent of Bloch's work on persistent currents.[6])

He also deduces, under the provision that the internal energy does not change, and with I_o the classical moment of inertia, that

$$\rho_s/\rho_o = \lim_{\omega\to o} I_o^{-1}\, \partial^2 E_{min}(\omega)/\partial^2\omega^2 \quad , \tag{5}$$

where $E_{min}(\omega)$ is the energy of the ground state in the rotating frame. Since $I_o \simeq NmR^2$, if $|E_{min}(\omega)-E_{min}(0)|$ is of order $\hbar^2/mR^2$ or less, then $\rho_s/\rho_o = 0(N^{-1})$ and hence there is no superfluidity.

Leggett next considers how one might keep $|E_{min}(\omega)-E_{min}(0)|$ of order $\hbar^2/mR^2$ or less. One possibility, he notes, appropriate to electrons in a metal, is for the system to make jumps between mutually orthogonal states (taken with respect to the non-rotating frame). Because we think of atoms in a solid as being in localized states, it is hard to imagine this possibility occurring, although in a footnote[1] Leggett recognizes that once one admits the possibility of superfluidity in solids, then one has to consider delocalized atomic states deriving from coherent atomic diffusion, and these states should also engage in (ordinary) incoherent atomic diffusion which "might well give rise to a certain amount of 'normal metal' type behavior".[1] Most likely, if these states satisfy the Landau criterion for superfluidity,[9] then we may safely ignore this possibility. Fermi systems, like ^{3}He, might more easily violate this criterion than would Bose systems, like ^{4}He, so we consider the latter to be a better class of candidates for exhibiting a NCMI.

The second possibility that Leggett considers is that the system is sufficiently localized to maintain itself in a wavefunction which satisfies the boundary condition of Eq. (4) by developing phase factors which contribute only negligibly to the energy of the system. (His discussion of this point owes much to some of Kohn's work on the theory of the insulating state.[5]) It is of considerable interest to expand upon this discussion, so the reader can see some concrete examples.

Therefore let us begin by considering a wavefunction similar to that employed by Nosanow[10] to describe solid ^4He, except that for simplicity we will neglect the Jastrow functions, which introduce particle correlations (i.e. multiple configurations). That is, each particle s in a state localized about a single lattice site (e.g. employ Wannier functions). By letting each state develop a phase factor which accrues its phase change $-2\pi m R^2 \omega/\hbar$ when the coordinate is far away from its site of localization, the energy of each state, and therefore of $E_{min}(\omega)$, changes by a negligible amount, so $\rho_s/\rho_o \approx 0$ by Eq. (4).

If the Jastrow functions are now included, so that more than one configuration is considered, our conclusions remain quite unchanged, provided that the configurations only involve localized states. However, note that, with more than one configuration, phase changes can cause the density $\rho(\vec{r})$ to change because of interference between configurations. This can increase the potential energy, but for localized configurations this increase should be negligible.

Now consider the case where a macroscopic fraction of the single-particle states in the various configurations are delocalized, but there is no Bose-Einstein condensation (BEC). These states cannot satisfy Eq. (4) without there being a significant increase in the kinetic energy of the system, so that $\rho_s/\rho_o > 0$. However, the changes in the wavefunction might cause $\rho(\vec{r})$ to change appreciably, thus causing an appreciable increase in the potential energy (which there is no simple way to calculate) and which would tend to increase ρ_s/ρ_o. Eventually, when the potential energy has increased too much, the system might go into a superfluid state which corresponds to the case of a universal phase factor, thus leaving $\rho(\vec{r})$ unchanged.

Finally, consider the case where there is BEC into a delocalized state, so that the system possesses off-diagonal-long-range-order (ODLRO[4]). The wavefunction of Lowy and Woo,[11] in the absence of Jastrow factors, would have this form. Either by changes in the internal phase factors, or by developing a universal phase factor (in which case the procedure of Sect. I would apply), one would expect the system to develop a $\rho_s/\rho_o > 1$ here.

It turns out that Kohn and Sherrington[7] have given a very general proof that ODLRO is a necessary and sufficient condition for superfluidity. Applying this theorem to the above two cases, the case without BEC (but with a macroscopic fraction of delocalized states) has $\rho_s/\rho_o = 0$, whereas the case with BEC has $\rho_s/\rho_o > 0$.

Unfortunately, their proof gives no specifics (e.g. is the Landau criterion violated when there is no BEC? how does one compute ρ_s/ρ_o when it is non-zero?), and therefore it provides little to aid one's physical intuition. As an existence proof it is very important, but on the value of ρ_s/ρ_o it is mute. the interested reader is encouraged to read Sec. 6 of Ref. 7.

Another paper on the subject, by Chester,[12] employs a proof by Reatto[13] on Bose-Einstein condensation (BEC) in quantum fluids, and a conjecture that the probability distribution arising from a Jastrow-type fluid wavefunction would also produce a solid phase, for a certain class of two-body interactions. From this Chester deduces that BEC can occur in solids. However, on formation of the solid phase, the wavefunction leading to the probability distribution would necessarily differ from the quantum liquid wavefunction considered by Reatto, and only this latter was shown by him to lead to BEC. Therefore, until Reatto's proof is generalized to quantum solids, I think it is not consistent to employ Chester's argument. Recall that BEC into a delocalized state is considered to be equivalent to ODLRO, so that if BEC obtains, ODLRO obtains, and therefore superfluidity may occur.

I will conclude this section by expressing the opinion that superfluidity may occur in solid ^{4}He if the Lowy-Woo wavefunction[11] gives an accurate representation of the true ground-state wavefunction of that system, for then ODLRO occurs.

III. PROPERTIES OF SUPERFLUID SOLIDS, IF THEY EXIST

In 1969 Andreev and Lifshitz considered the effect of impurities, interstitials, and vacancies on quantum solids at very low temperatures.[3] By permitting these defects to have hopping matrix elements, they showed that "defecton" bands would form (see also Guyer and Zane[14] in this regard), and they argued that such defectons could even be present in the ground state (most likely "vacancions", at least, are not present in the ground state[15]). From a model calculation they showed that the defection excitations would not violate the Landau criterion for superfluidity, and therefore they proceeded to develop a macroscopic hydrodynamic theory of superfluid quantum crystals. Although I believe there are difficulties with their paper, it presents the first hydrodynamic theory which allows for crystal defects, and this theory is certainly correct in the absence of superfluidity.

The superfluid velocity introduced by Andreev and Lifshitz (AL) has the property that it is a Galilean velocity, and therefore it differs from $v_s = \frac{\hbar}{m} \vec{\nabla}\phi$, where ϕ is the phase (assumed to be the same for all single-particle states). This is because ϕ

is tied to the lattice, and has variations on a microscopic scale. The $\vec{v}_s^{(AL)}$ refers to an average motion of matter (but not of excitations), whereas $\vec{u}$ refers to the average motion of the lattice ($\vec{u}$ is a displacement of the lattice from its equilibrium position). In equilibrium AL find that $\vec{v}_n = \dot{\vec{u}}$, so that the excitations move with the lattice (Umklapp processes associated with the lattice can cause such equilibration). They present no theory for ρ_s, and thus it is not possible to attribute any precise meaning to this quantity. They do not say that $\rho_s/\rho_o = 1$ at $T = 0$, and they are presumably of the opinion that $\rho_s/\rho_o = 0$ at $T = 0$ in the absence of zero point defectons.[15] They find that the propagating hydrodynamic modes of this system are: elastic waves, loaded only by the normal fluid density; and fourth sound, loaded only by the superfluid density.

More recently I have derived the hydrodynamics for the case of a non-Galilean $\vec{v}_s$.[16] Assuming a universal phase factor, an explicit expression for $\rho_s(T)/\rho_o$ is derived, with $\rho_s(0)/\rho_o < 1$. In most ways other than this, my results agree with those of AL. However, the fact that $\vec{v}_s$ is not a Galilean velocity has some important consequences which are worthy of discussion. First, the system should have no anomalous thermal properties because there will be no two-fluid counterflow in the presence of a heat current (in contrast to the work of Andreev and Lifshitz). Second, it may be difficult to bring the superfluid into rotation. This is because the total angular momentum is given by $L = I_o\omega_n + I_s\omega_s$ (where ω_n is the rotation frequency of the lattice, ω_s is the average frequency of superfluid rotation with respect to the lattice, and $I_s = (\rho_s/\rho)I_o$ is the superfluid moment of inertia), and it may be difficult to cause a $\omega_s \neq 0$ to develop on passing from above to below T_c because L, I_o, and ω_n do not change. However, if rapid acceleration or deceleration takes place, nonequilibrium processes might occur which would bring the superfluid into rotation.

Limitations of time and space would be sufficient to prevent me from summarizing what is known about T_c and what is known about thermal anomalies in the vicinity of T_c. Fortunately or unfortunately, these limitations coincide very well with the amount of space required for such a summary, since very little is known about either subject. A theoretical expression exists for calculating $\rho_s(T)$, assuming a universal phase factor, and $\rho_s(T_c) = 0$ should give T_c; however, to my knowledge not enough is known about solid ^{4}He to perform such a calculation. On the experimental side, there is no evidence to indicate that solid ^{4}He undergoes a phase transition, although to my knowledge there has been no concerted attempt to detect superfluidity in such a system, either by looking for a NCMI, for fourth sound, or for some other such signature.

References

(1) A. J. Leggett, Phys. Rev. Lett. 25, 1543 (1970).

(2) W. M. Saslow, Phys. Rev. Lett. 36, 1151 (1976).

(3) A. F. Andreev and I. M. Lifshitz, Zh. Eksp. Teor. Fiz. 56, 2057 (1969). [Sov. Phys. JETP 29, 1107 (1969)].

(4) C. N. Yang, Rev. Mod. Phys. 34, 644 (1962).

(5) W. Kohn, Phys. Rev. 133, A171 (1964).

(6) F. Bloch, Phys. Rev. 137, A787 (1965).

(7) W. Kohn and D. Sherrington, Rev. Mod. Phys. 42, 1 (1970).

(8) J. M. Blatt and S. T. Butler, Phys. Rev. 100, 481 (1955).

(9) L. D. Landau, J. Phys. U.S.S.R. 5, 71 (1941).

(10) L. H. Nosanow, Phys. Rev. 146, 120 (1966).

(11) D. N. Lowy and C.-W. Woo, Phys. Rev. B 13, 3790 (1976).

(12) G. V. Chester, Phys. Rev. A 2, 256 (1970).

(13) L. Reatto, Phys. Rev. 183, 334 (1969).

(14) R. A. Guyer and L. I. Zane, Phys. Rev. Lett. 24, 660 (1970).

(15) S. M. Heald, D. R. Baer, and R. O. Simmons, Bull. Am. Phys. Soc. 21, 352 (1976).

(16) W. M. Saslow, Phys. Rev. B 15 (1977) (to be published).

NUMBER FLUCTUATIONS IN A MACROSCOPIC SUBVOLUME OF QUANTUM FLUIDS:

A POSSIBLE EXPERIMENT

T. P. Bernat

Department of Physics and Astronomy
Louisiana State University
Baton Rouge, La. 70893

Abstract: In this paper, the particle number fluctuations in a macroscopic subvolume of a quantum Bose fluid are considered, and the effects of the existence of a Bose-Einstein condensate on the magnitude of number fluctuations is discussed. The feasibility of measuring such particle number fluctuations is analyzed and the possibility is offered that such measurements could provide information on the existence of a Bose-Einstein condensate in liquid ^{4}He as well as the recently discussed "exotic" quantum fluid, polarized atomic hydrogen.

I. EFFECT OF A BOSE-EINSTEIN CONDENSATE ON NUMBER FLUCTUATIONS IN MACROSCOPIC SUBVOLUMES

In the following discussion it is assumed that the number of particles in a small but macroscopic subvolume of a completely isolated many-particle system can be measured as a function of time, $n(t)$. The results of such a measurement would ideally be compared to the results of an exact calculation of $n(t)$. Such a calculation is not possible, and at best one could calculate the spectrum of the random fluctuation. Such a calculation should be microscopic in nature if it is to reveal the effect of the single-particle momentum distribution function on $n(t)$ or $\tilde{n}(\omega)$.

A simpler experiment and calculation is involved in examination of the r.m.s. number fluctuation, defined by

$$<(\Delta n)^2> = <[n(t) - \bar{n}]>_t = <n^2>_t - <n>_t^2 \tag{1}$$

i.e., a time average of the squared density fluctuations. This is a time-independent quantity and in principle can be evaluated from a knowledge of the N-body state for the particles enclosed in the volume V of which v is a subvolume by calculating

$$<(\Delta n)^2> = <N|\hat{N}^2|N>_v - [<N|\hat{N}|N>_v]^2 \tag{2}$$

where the subscript indicates that integrals are to be performed only within the subvolume. Equation (2) has been evaluated for an ideal Bose-Einstein gas by Johnston,[1] with the result being that in the presence of a B-E (Bose-Einstein) condensate of n_o molecules in the volume v,

$$<(\Delta n)^2> \simeq \underline{n} + 2 \sum_{k\neq o} |(k,0)_v|^2 n_k n_o. \tag{3}$$

In this expression, n_ℓ is the occupation number for the single particle state with wavevector k_ℓ and the quantities $(k,0)_v$ are overlap integrals for single particle wavefunctions of wavevector k_k and k_o, evaluated in the subvolume. $<(\Delta n)^2>$ can be evaluated for an ideal gas of ^{4}He atoms by using standing wave single particle states and London's expression for n_o and $n_{\ell\neq 0}$.[2] In carrying out the calculation, the ground state wavefunction is taken to be $\phi_{n=0} = (V)^{-1/2}$.

V and v are assumed to be cubes of volumes L^3 and ℓ^3 respectively, with v symmetrically located in V. The wavevectors for single particle states have the form $\vec{k} = \pi/L(m\hat{x} + n\hat{y} + p\hat{z})$ with m, n,p positive integers. If L/ℓ is taken as 10π, the sum in Eq. (3) can be cut off for $m,n,p > 10$ since $(k,0) \approx 0$ for larger wavevectors. Since those values of k contributing to the sum have energies $\varepsilon << kT$ for $T > 10^{-14}$ K a high temperature approximation can be made for the occupation numbers. If L is 1 cm, the resulting relative rms number fluctuation for ideal ^{4}He is found to be

$$\left(\frac{<(\Delta n)^2>}{n^2}\right)^{1/2} \simeq \left[\frac{1}{n} + 1.67 \times 10^{-7} \left(\frac{T}{T_c}\left[1-\left(\frac{T}{T_c}\right)^{3/2}\right]\right)\right]^{1/2} \tag{4}$$

where T_c is the B-E transition temperature, and the density of real liquid ^{4}He has been used. $n^{-1} \simeq 10^{-18}$ is negligible at most temperatures. Equation (4) is shown graphically in Fig. 1, and indicates an anomalously large number fluctuation in a subvolume of

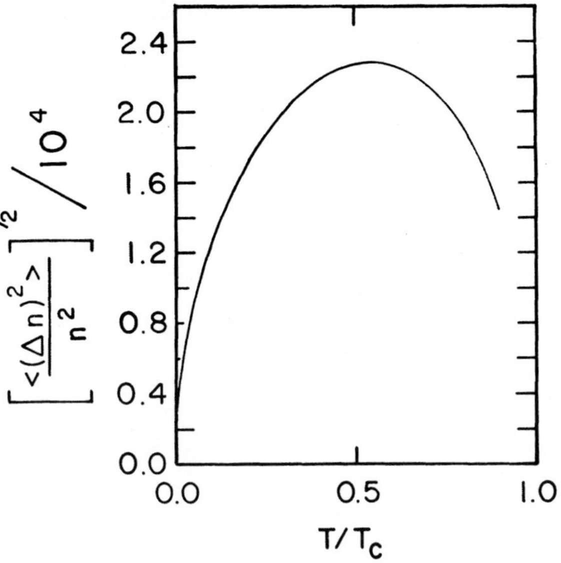

Figure 1. Anomalous number fluctuatuions in a small subvolume of ideal liquid ^{4}He, according to calculation in text.

an ideal Bose gas. This anomaly is dependent on the existence of a macroscopically occupied single-particle ground state, and its experimental observation would indicate the presence of this (B-E) condensate in, for instance, spin aligned atomic hydrogen which should be an excellent approximation to an ideal Bose gas.[3]

To estimate the effect of a B-E condensate on the number fluctuations in a subvolume of a system of interacting bosons, the interaction would have to be included in calculating the system wavefunction as well as evaluating $<(\Delta n^2)>$. An exact solution is probably not possible, and in order to get a sort of "0th order" approximation of the effects of interaction on the previous ideal gas calculation of the fluctuation, Johnston's evaluation of $<(\Delta n)^2>$ for the ideal gas will be assumed valid, and the wavefunction $|N>$ will still be considered an eigenfunction of the occupation number operators $\hat{n}_\ell$. However, the eigenvalues n_ℓ will now be replaced with the quantum mechanical average $<n_\ell>$ as calculated in, for instance, the Bogoliubov approximation.[4] The dispersion in n_ℓ introduced by the interaction will be neglected. The depletion of the B-E condensate then leads to a reduction in the subvolume number fluctuations, which is easily seen if the Bogoliubov ground state momentum distribution function is used.[4] To the same approximations as for the ideal gas calculation, and again making a high temperature approximation for the distribution function, by summing over the same terms as before and using the same volume and sub-volume, at T = 0°K,

$$\left[\frac{\langle(\Delta n)^2\rangle}{n^2}\right]^{1/2} \simeq \frac{1}{n}\left[1 + 2.6 \times 10^{-3} \left(\frac{a\rho}{\pi}\right)^{1/2} \alpha\right] \tag{5}$$

where α is the fractional Bose-Einstein condensate:

$$\alpha = \frac{n_o}{n} = 1 - \frac{8}{3}\left(\frac{\rho a^3}{\pi}\right)^{1/2} .$$

For spin aligned atomic hydrogen[3] at 0°K and molar volume 200 cm^3, the relative fluctuation calculated from Eq. (5) is 80 $n^{-1/2}$, or eighty times larger than the classical value or the ideal Bose gas value (Eq. 4) at T = 0°K. The relative fluctuation for liquid ^4He in this approximation with $\rho = 2.2 \times 10^{22}/cm^3$ and a = 2.56 Å is 81 $n^{-1/2}$.

The temperature dependence of the relative fluctuations can be calculated in the above approximation if thermodynamic averages of momentum distribution functions are used. As temperature increases above absolute zero the Bose-Einstein condensate is reduced, but the populations of the lowest momenta single-particle states increase by a relatively larger amount.[5] Thus the relative density fluctuations initially increase. In the Bogoliubov approximation, the initial increase is linear in T.

While the Gogoliubov Bose gas model has been employed in these calculations, the anomalous number fluctuations would be present for any single particle momentum distribution function with a macroscopic condensate and large population of low momentum states. For example, McMillan's momentum distribution function, derived from a variational technique,[6] leads to an rms relative fluctuation about twice as large as has been calculated for the Bogoliubov distribution at 0°K.

The calculation for interacting bosons presented here is far from rigorous or complete. However, the magnitude of the fluctuation anomaly present at this somewhat naive level of approximation compels an attempt at a more exact treatment including time dependence. If an anomaly persists in such a calculation, it can be used to search for the Bose-Einstein condensate in quantum fluid systems, or to set an upper limit to α.

II. MEASUREMENT OF SMALL NUMBER FLUCTUATIONS

Very high Q superconducting resonant circuits can be employed to detect very small changes in any parameter of a system that is capable of modulating either the capacitance or the inductance of the circuit. The density of ^{4}He atoms in a subvolume is one such parameter. The properties of superconducting parametric transducers have been discussed in the literature.[7-9] The general operating principle is that any small change in a system parameter, δp can produce a shift in the resonant frequency of the circuit by an amount $\delta\omega \propto \delta p$. If the capacitance is linear in p and the inductance is independent of p, then to first order

$$\frac{\delta\omega}{\omega_o} = -\frac{1}{2}\frac{\delta p}{p} \tag{6}$$

where ω_o is the unshifted resonant frequency. In principle, $\delta\omega/\omega_o$ can be measured to parts in 10^{14} using existing technology.[7] The measurement requires a circuit Q of 10^6 to 10^8, but the sensitivity is mainly limited by frequency noise in the driving oscillator, and to a lesser extent by electronic noise voltages in mixers, amplifiers, etc.

Recently this measurement technique has been used by Berthold et al.,[10] to measure the temperature-dependent density of liquid ^{4}He. In that work, the density was coupled to the frequency of a superconducting microwave cavity through the ^{4}He dielectric constant, and the shifting resonant frequency was measured directly. Resonant frequencies, and hence densities, could be measured to parts in 10^9.

Superconducting circuit technology can readily be adapted to the measurements of number fluctuations in macroscopic subvolumes of fluid systems. Fig. 2a is a schematic of a re-entrant rf superconducting cavity whose resonant frequency is strongly modulated by the number of atoms, n, in the volume v between the "capacitor plates." As in the work of Berthold et al.,[8] the coupling is via the dielectric constant and the Clausius-Mossotti relation. The number fluctuations would be indicated by a fluctuating phase shift in the rf across the cavity (Fig. 2b). Using state-of-the-art technology, phase shifts of parts in 10^{14} should be resolvable.[9] Since the dielectric constant of liquid ^{4}He is about 1.05, this would reflect relative fluctuations in n of parts in 10^{12}. On the other hand, if $v = 10^{-4}$ cm^3, then for a liquid ^{4}He density of 2.2×10^{22}/cm^3, the classical rms relative fluctuation in n is $n^{-1/2} \simeq 10^{-9}$, which is three orders of magnitude greater than the estimated sensitivity.

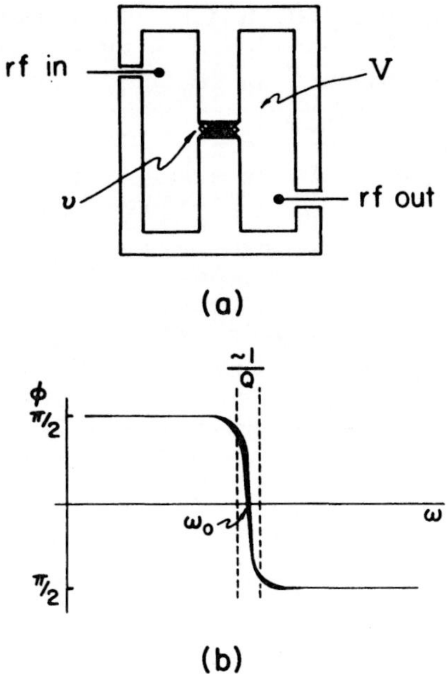

Figure 2. (a) Schematic of superconducting re-entrant cavity. The cavity is cylindrically symmetric about its long axis. (b) Relative phase between input and output of a cavity with high Q and resonant frequency ω_o.

References

(1) James R. Johnston, Am. J. Phys. 38, 516, April 1970.

(2) Fritz London, Superfluids, Vol. II, Dover Publications, Inc., New York, 1964.

(3) William C. Stwalley and L. H. Nosanow, Phys. Rev. Lett. 36, 910, April 1976.

(4) Alexander L. Fetter and John Dirk Walecka, Quantum Theory of Many-Particle Systems, McGraw-Hill Book Co., New York, 1971, p. 314.

(5) Ibid., p. 494.

(6) W. L. McMillan, Phys. Rev. 138, 442, April 1965.

(7) G. J. Dick and H. C. Yen in Proceedings of the IEEE 1972 Applied Superconductivity Conference, p. 684.

(8) D. G. Blair, T. P. Bernat and W. O. Hamilton in Proc. 14th Int. Conf. Low Temp. Phys., (North-Holland, Amsterdam, 1975), p. 254.

(9) W. C. Oelfke and W. O. Hamilton, "Superconducting Accelerometers for the Study of Gravitation and Gravitational Radiation," in Proc. XXVII Congress Astronautical Federation, 1976.

(10) J. E. Berthold, H. N. Hanson, H. J. Maris, and G. M. Seidel, Phys. Rev. B 14, 1902, (1976).

SOLID o-p-HYDROGEN: A COMPARISON OF THEORIES OF STATISTICALLY DISORDERED CRYSTALS

W. Biem, H.-W. Diehl, and K. Menn

Institut für Theoretische Physik, Universität Giessen
West Germany

1. INTRODUCTION

The solid hydrogens (p-H_2, o-H_2; o-D_2, p-D_2) are the simplest molecular crystals with well known intermolcular forces and they are quantum crystals. The anisotropic forces between the molecules are of purely quadrupolar origin apart from very small additional terms (overlap and van der Waals). In these solids p-H_2 and o-D_2 molecules occupy their lattice sites with spherical rotational states ($J = 0$) and the anisotropic forces do not influence them. But o-H_2 and p-D_2 molecules are in $J = 1$ rotational states at low temperatures. Pure crystals of these modifications are known to exhibit a phase transition into an ordered phase when cooled down to 3 K. The structure of this phase is the cubic Pa3. The crystal can be compared with an antiferromagnet. While in antiferromagnets magnons are elementary excitations of the magnetic ordering, in o-H_2 resp. p-D_2 the comparable excitations of the rotational ordering are called librons. Their existence is well established by experiment and theory including anharmonic effects.[1]

Mixtures of o-H_2 and p-H_2 or p-D_2 and o-D_2 resp. are model substances for theories and experiments about disordered crystals. They can be considered in parallel to dilute antiferromagnets. The spectra of librons as a function of the concentration c of the "nonmagnetic" species (p-H_2; o-D_2) can be measured and calculated. Especially optical experiments with Raman scattering have already been performed.[2] The $k = 0$ spectra measured in these experiments are the basis of a discussion of different approximations where the same quantity can be calculated.

We have calculated the average T-matrix approximation (ATA) and the coherent potential approximation (CPA) for different concentrations c of $p\text{-}H_2$ resp. $o\text{-}D_2$.

2. THEORY

For the theory we refer to several beautiful survey articles[3] and restrict ourselves to some definitions. In the Hamiltonian of a disordered crystal in a fixed configuration

$$H = H_o + V = (H_o+\bar{V}) + (V-\bar{V}) = \tilde{H} + \tilde{V}, \qquad (1)$$

H_o represents the ideal crystal (pure $o\text{-}H_2$), V the disorder produced by the admixed $p\text{-}H_2$. $\bar{V}$ is a periodic potential which can be conveniently choosen. With $\tilde{G}$, the Green's function of $\tilde{H}$, the Green's function of H is

$$G(z) = (\tilde{G}(z)^{-1}-\tilde{V})^{-1} = \tilde{G} + \tilde{G}\,\tilde{V}\,G = \tilde{G} + \tilde{G}\,J\,\tilde{G}. \qquad (2)$$

From this equation for a single configuration we have to go to configurationally averaged quantities $\langle G\rangle$, $\langle J\rangle = T$ and arrive at

$$\langle G(z)\rangle = [\tilde{G}(z)^{-1} - M(z)]^{-1} = \tilde{G}(z) + \tilde{G}(z)T(z)\tilde{G}(z) \qquad (3)$$

where we have introduced the mass operator

$$M(z) = T(1+\tilde{G}T)^{-1}\,. \qquad (4)$$

There are in principle two possibilities to solve the problem of the disordered crystal: (A) Starting from the ideal crystal $\tilde{H}$ we calculate the T-matrix $T[\tilde{V}]$. This can be done only approximately. The results are "average t-matrix approximations" (ATA). (B) We vary $\bar{V}$ and look for a solution where $\langle G(z)\rangle = \tilde{G}(z)$, that is we solve

$$T[\tilde{V}(z)] = 0\quad . \qquad (5)$$

With this method we reach self-consistency. We determine the ideal Green's functions $\tilde{G}$, which have the same dynamics as the configurationally averaged Green's functions $\langle G \rangle$. Also in this case one has to make approximations, and the result is the famous "coherent potential approximation" (CPA).

Th simplest 'single-site forms' of ATA and CPA can be obtained with two approximations: (A) We assume $\tilde{V}$ to consist of single site contributions

$$V = \sum_j V^{(j)} n_j \tag{6}$$

with $n_j = 1(0)$ if there is a p(o)-molecule on site j. This is a severe approximation if V consists of perturbation potentials $V^{(j)}$ which extend to the neighbours of the defect site as is inevitable in our problem. (B) We express the T-matrix by the single defect T-matrices $t^{(j)}$. With them we get in the ATA a mass operator

$$M_{ATA}(z) = \sum_j M_{ATA}^{(j)}(z) \tag{7}$$

where

$$M_{ATA}^{(j)}(z) = \bar{V}^{(j)} + (1 + \langle t^{(j)} \rangle \tilde{G})^{-1} \langle t^{(j)} \rangle \tag{8}$$

$$= c\, \tilde{V}^{(j)} (1 - (1-c)\tilde{G}\, \tilde{V}^{(j)})^{-1} .$$

Within the CPA instead of (5) the simple relation

$$\langle t^{(j)} [\tilde{V}(z)] \rangle = 0 \tag{9}$$

has to be solved self-consistently.

We have used two choices of perturbation potentials $V^{(j)}$: $V^{(j)}_{rigid}$ is the potential of one p-H_2 surrounded by o-neighbours in their rigid unperturbed orientation, and $V^{(j)}_{tip}$ includes the tipping of the neighbours since the defect changes the orienting field.

3. COMPARISON OF THE RESULTS

We have calculated the total density of states

$$Z(\omega) = - \frac{2\omega}{\pi} \frac{1}{2N} \text{Im Tr } G(\omega+i0) \tag{10}$$

and the density of states of the wave vector $\underline{k} = 0$

$$Z(\underline{k}=0, \omega) = - \frac{2\omega}{\pi} \text{Im Tr } G(\underline{k}=0, \omega+i0) \tag{11}$$

which has been measured by Raman scattering. The moments of the spectra $\int \omega^{2n} Z(\omega)\, d\omega$ have been used to study the quality of the theories. In the CPA with a nondiagonal random scattering matrix only the first three moments are exact (instead of 6 for diagonal scattering). In the higher moments errors of the magnitude 1 are to be expected instead of Z^{-2} for diagonal scattering (Z number of nearest neighbours). All the same the CPA results show positive aspects if compared with the ATA on the one hand, and both ATA and CPA if compared with the experiments.

Fig. 1 shows spectra calculated with $V^{(j)}_{rigid}$, Fig. 2 the same with $V^{(j)}_{tip}$ (including the spectrum of pure o-H_2) both for c = 20%. $Z_{ATA}(\omega)$ shows structure especially at the band edges compared with $Z_{CPA}(\omega)$ which is smooth. This structure is fictitious and a result of the ATA procedure; modes sifted outside the band of the ideal crystal lose their damping because they can no longer interact with the continuum of states. Contrary to this, the CPA spectrum is always self-consistent with its modes. Therefore these modes show a steadily growing width with growing c^4 (see Fig. 3 and 4).

The mean energies of the $\underline{k} = 0$ modes as functions of c can be compared with the Raman measurements. Both ATA and CPA show the correct order of magnitude but there remain discrepancies. They may result from using a harmonic model fitted to the strongly anharmonic librons. Up to now all theories of statistically disordered crystals are based on harmonic models. Here we see that it is highly desirable to include anharmonicities in the treatment. Work in this direction is underway now.

The peak frequencies of the $\underline{k} = 0$ modes are functions of the concentration c: $\omega(c) = \omega(0)(1-Kc...)$ where K has experimental values between 0.9 and 1.3 for the different modes. The CPA with tipping yields K_{CPA} between 0 and 1.3. In the ATA only an average mode frequency can be defined with K_{ATA} between 0.4 and 1.3.

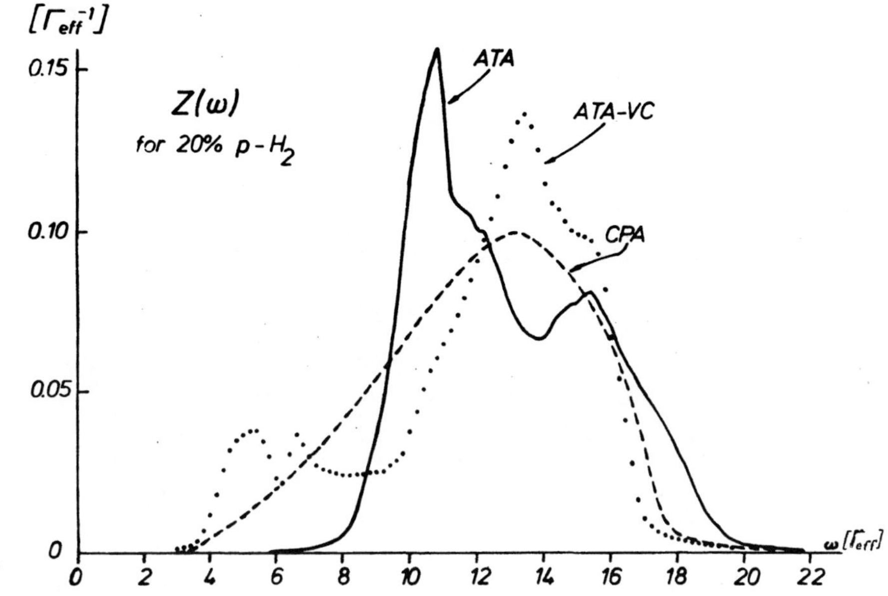

Figure 1. Z(ω) calculated with $V_{rigid}^{(j)}$. ---- CPA, —— ATA with $H = H_o$, ···· ATA with H of a virtual crystal of concentration c = 20%.

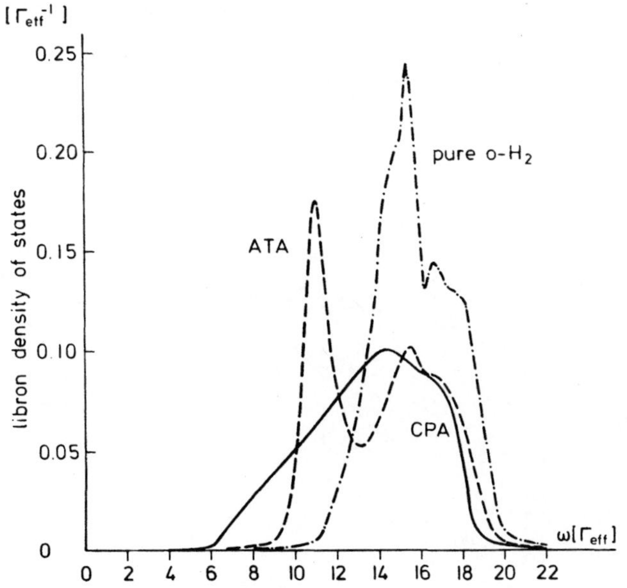

Figure 2. Z(ω) calculated with $V_{tip}^{(j)}$ for c = 20%. —— CPA, ---- ATA with $H = H_o$. For comparison Z(ω) of the pure o-H_2 crystal.

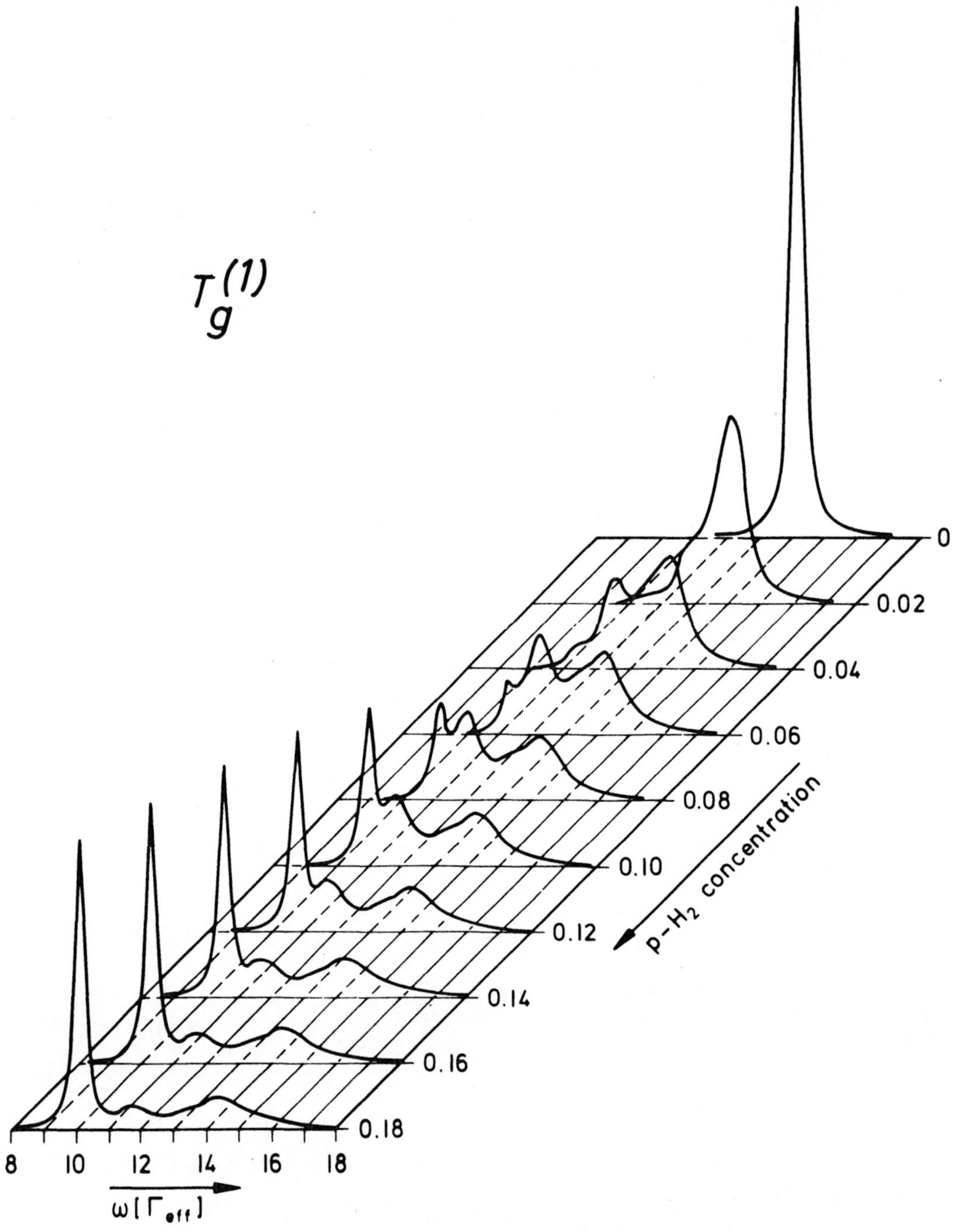

Figure 3. One of the three $\underline{k} = 0$ modes in ATA as a function of c.

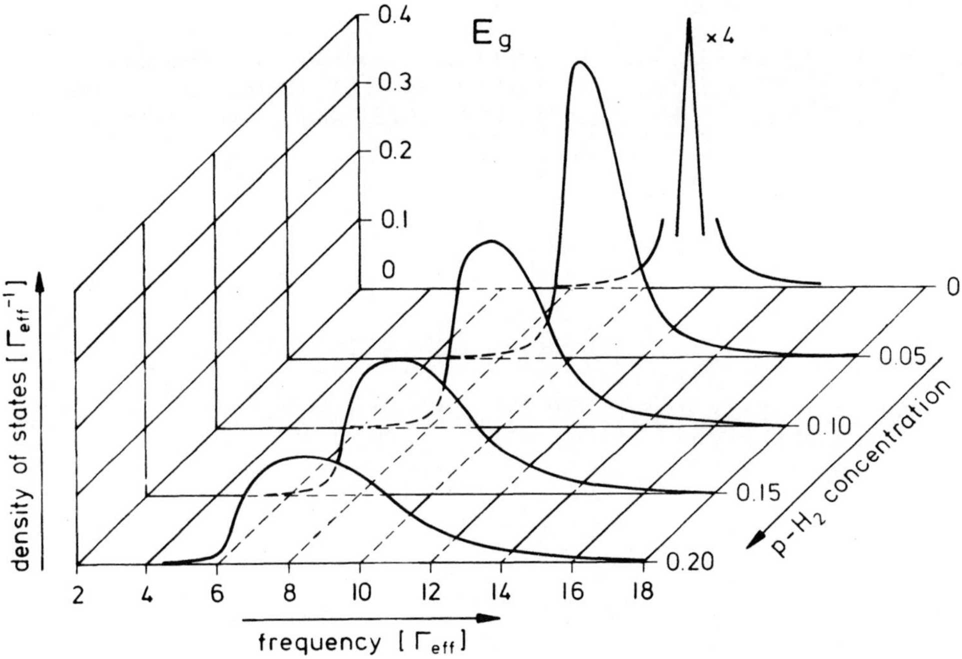

Figure 4. One of the three $\underline{k} = 0$ modes in CPA as a function of c.

References

(1) For solid hydrogen see e.g. C. F. Coll and A. B. Harris, Phys. Rev. B 4, 2808 (1971); H. W. Diehl, Z. Physik 271, 235 (1974) and references therein.

(2) W. N. Hardy, I. F. Silvera and J. P. McTague, Phys. Rev. B 12, 753 (1975).

(3) R. J. Elliott, J. A. Krumhansl and P. L. Leath, Rev. Mod. Phys. 46, 465 (1974); H. Ehrenreich and L. M. Schwartz, Solid State Physics 31, 149 (1976).

(4) Some more results and a complete discussion of them can be found in H. W. Diehl, Z. Physik B24, 161 (1976); H. W. Diehl and W. Biem, Z. Physik B25, 197 (1976) and to be published.

SPECIFIC-HEAT MEASUREMENTS ON SOLID ^{3}He

D. S. Greywall

Bell Laboratories
Murray Hill, N. J. 07974

Abstract: The specific heat of bcc ^{3}He has been measured at five molar volumes between 21.5 and 24.5 cm^3 and for temperatures between 50 mK and the melting curve. The data below 0.5K show no evidence of the large anomalous contribution to the specific heat which has been observed in all of the previous specific heat measurements on ^{3}He, and thus indicate that this long-standing anomaly is not due to an intrinsic property of this quantum solid.

Over the past fifteen years the specific heat at constant volume C_V of bcc ^{3}He has been measured by several groups[1-4] in the temperature range extending from below 0.3 K to the melting curve. All of these measurements have shown, in addition to the lattice specific heat proportional to T^3, an anomalous excess contribution for temperatures $T \lesssim 0.5$ K. In each experiment the anomalous component was considered to be a real effect and not due to experimental difficulties. In addition to the lattice, the only other known contribution to the specific heat at low temperatures is due to the ordering of the nuclear spins. Although this term has a T^{-2} temperature dependence and thus becomes increasingly significant as the temperature is lowered, its size is too small to explain the observations. Several other possible sources for the excess specific heat have been suggested.[4-12]

One of the conjectures, which for a time seemed very promising, attributed the observed phenomena to an upward-curving phonon dispersion implying an extremely large elastic anisotropy. Although subsequent sound velocity measurements[13] in single ^{3}He crystals of known orientation did indicate a large elastic anisotropy, the anisotropy was not large enough, in fact, to explain any of the excess specific heat. With no other proposal seeming particularly convincing, the problem was returned to its original state of uncertainty.

The present high-precision specific heat measurements were undertaken in an attempt to help clarify the situation. It was anticipated that a precise determination of the temperature dependence of the excess specific heat would finally yield meaningful indications as to possible sources of this long-standing anomaly. Instead, the data[14] show no evidence of the anomalous contribution at all, and thus demonstrate that the anomaly is not due to an intrinsic property of this quantum solid.

A cross-sectional drawing of the cylindrical calorimeter is shown in Fig. 1. It was constructed of copper and had a nominal sample-chamber volume of 9 cm^3. The calorimeter was positioned firmly below the mixing chamber of a dilution refrigerator using three graphite tubes which were attached to the sample cell via a copper flange machined as part of the cell body. This flange was also the platform on which the germanium thermometers were mounted. There were twelve 0.15 cm diameter copper rods passing through the sample chamber. Although the presence of these rods did increase

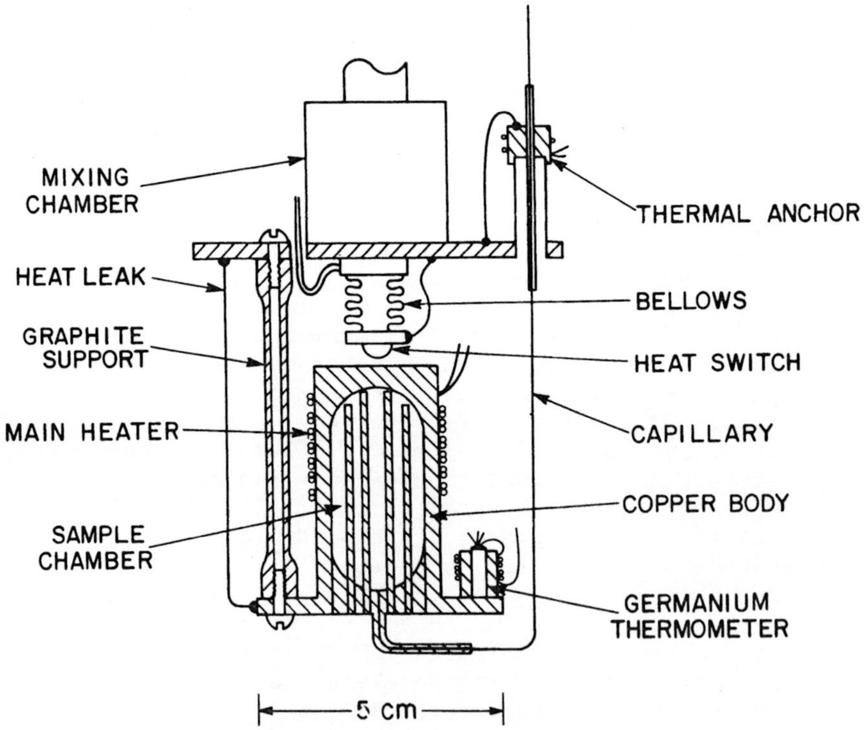

Figure 1. Calorimeter.

the surface area by a factor of two, their main purpose was to reduce the average distance of the sample from the nearly isothermal surface of the calorimeter and thereby reduce the time necessary for thermal equilibrium to be established. The stainless-steel filling capillary between the mixing chamber and the calorimeter was 8 cm long and had an i.d. of 0.005 cm and an o.d. of 0.015 cm. To counterbalance the residual heat leak of 2 nW into the cell, a fine copper wire, providing a conduction of 10^{-7}W/K at 0.1K linked the calorimeter to the mixing chamber. The mechanical heat switch was used to cool the sample chamber initially to less than 50 mK. The calorimeter was completely surrounded by a copper radiation shield at the mixing chamber temperature.

The heat capacity measurements were obtained using the standard heat pulse technique with temperature increments equal to approximately 5% of the temperature. Figure 2 shows the temperature recorded before and after the application of a heat pulse to the calorimeter which contained a ^{3}He sample at 24.454 cm^3/mole. At this molar volume the thermal equilibrium time constant was very short; however, the time constant at low temperature ($T \lesssim 0.1$K) increased rapidly with increasing density. This is demonstrated by

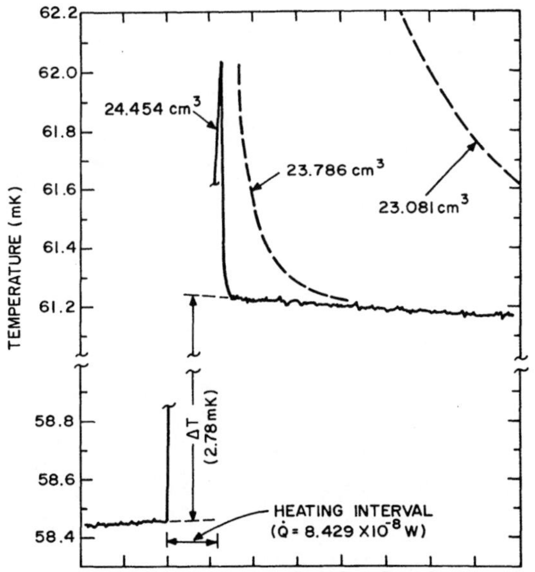

Figure 2. Temperature recorded before and after the application of a heat pulse. The time axis has an arbitrary origin. The molar volume of the sample was 24.454 cm^3. At the higher densities, the longer thermal relaxation times encountered at these low temperatures are indicated by the dashed curves.

the dashed curves in Fig. 2. At a molar volume of 23.081 cm^3 and for T $\lesssim$ 0.1K it was necessary to monitor the temperature for at least 40 min after the heat pulse in order to determine the steady-state after-drift rate. As a result the low temperature data at this volume are much less precise than those obtained for the two lowest densities. At even higher densities, the extremely-long spin-lattice relaxation times prevented any data from being obtained below 0.1K. For all of the densities considered, the thermal relaxation times above 0.1K were of the order of a few seconds.

Shown in Fig. 3 are the specific heat results for T $\lesssim$ 0.6K plotted as a function of temperature on log-log scales. For T $\lesssim$ 0.1K the specific heat is due mainly to the nuclear-spin contribution proportional to T^{-2} while at higher temperatures the phonon contribution proportional to T^3 dominates. The solid curve in the figure shows the smoothed results of Castles and Adams[4] at 24.40 cm^3/mole which is nearly equal to our largest molar volume. The two sets of measurements are in agreement at T $\approx$ 0.065K and for T $\gtrsim$ 0.3K; however, in the region of the minimum in the curves there is a large discrepancy of about 30%. Although not indicated in the figure, a similar difference exists at the other molar volumes. Measurements of the specific heat by Sample and Swenson[2] and by Pandorf and Edwards[3] made prior to the work of Castles and

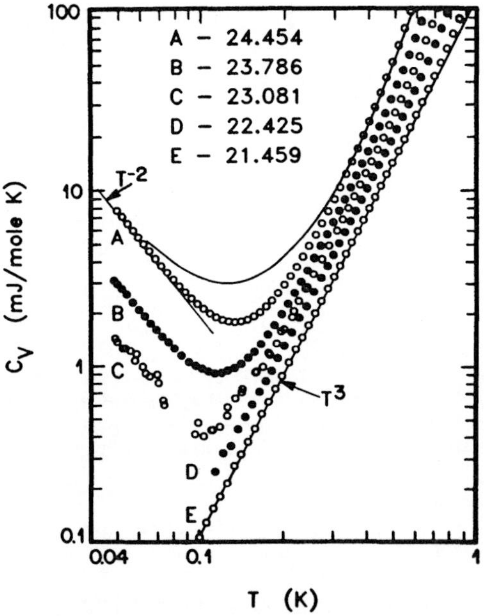

Figure 3. Specific heat of bcc ^{3}He versus temperature. The numbers are the molar volumes in cubic centimeters. The smooth curve corresponds to the work of Castles and Adams (Ref. 4) at 24.40 cm^3/mole.

Adams[4] do not extend in temperature to the region of the specific heat minimum and so it is not possible to make as detailed a comparison with these results. However, in the region of temperature overlap these three earlier sets of data are qualitatively very similar.

Another representation of the present specific heat results is given in Fig. 4. Plotted here is the temperature dependent Debye temperature $\Theta(T)$ calculated from the data at each of the five densities using the equation

$$C_V - C_V^{spin} = (12\pi^4/5)R[T/\Theta(T)]^3 \qquad (11)$$

where

$$C_V^{spin} = 3R(J/k_BT)^2 . \qquad (12)$$

At the three lowest densities the values of $|J|$ were determined using the present specific heat data. At the two highest densities the values of $|J|$ used were taken from the work of Panczyk and Adams.[15] If what remains after subtracting the spin-ordering contribution from the total measured specific heat of the sample is due to the lattice alone, then $\Theta(T)$ should be nearly constant for sufficiently low temperatures ($T/\Theta_0 \lesssim 0.02$). This is the behavior observed in most crystalline solids including hcp ^{4}He.[16] The present data at each of the densities are consistent with this expectation, but contrary to the rapid decrease in $\Theta(T)$ with decreasing temperature that has been reported on the basis of all of the previous C_V measurements[1-4] on bcc ^{3}He. The smoothed results of Castles and Adams are also indicated. The small decrease in $\Theta(T)$ in the present data at the largest molar volume for $T/\Theta_0 \lesssim 0.008$ could be due to an error in the determination of $|J|$ by as little as 0.1%. The small increase in $\Theta(T)$ at the smallest molar volume for $T/\Theta_0 \lesssim 0.008$ could be due to an error in the empty calorimeter specific heat of less than 1%.

Castles and Adams were able to describe their results, at each of the several molar volumes considered by the function

$$C_V = \alpha T^{-2} + \beta T + \gamma T^3 \qquad (1)$$

with β equal to 9, 8, and 7 mJ/mole K^2 at molar volumes of 24.40, 23.81 and 22.91 cm^3, respectively. A weighted fit of the present data at the three largest volumes and for $T < 0.3$K with the same function yielded values of 0.02 ± 0.05, 0.14 ± 0.05, and -0.9 ± 0.3 mJ/mole K^2 for nearly corresponding molar volumes of 24.454, 23.786, and 23.081 cm^3, respectively. The two sets of

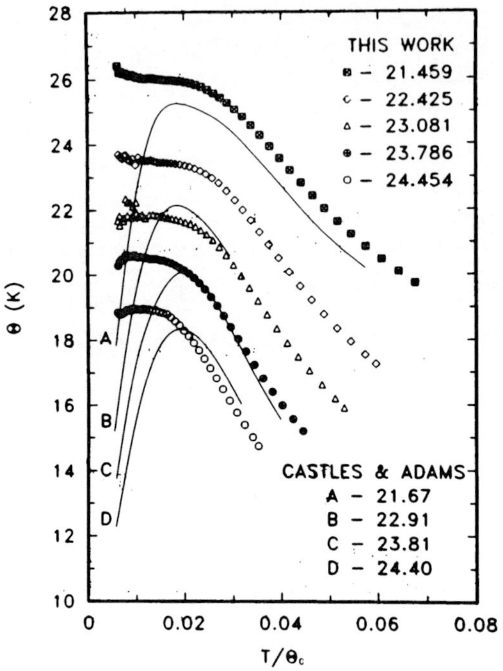

Figure 4. The Debye thetas for bcc ^{3}He versus reduced temperature. The numbers are the molar volumes in cubic centimeters. The smooth curves were determined using the data of Castles and Adams tabulated in Ref. 4 with Θ_0 set equal to the maximum value of $\Theta(T)$.

parameters differ by an order of magnitude in absolute value. Although the standard error bars on the present balues of β do not include zero for each of the molar volumes, the deviation from zero is not considered to be significant. Small systematic errors in the measurements or the fact that other terms due to the lattice and proportional to higher powers of T should have been included in the fitting function could account for this difference. The present low temperature results thus show no evidence for any contribution to the specific heat other than that attributable to the nuclear spins or to the phonons. The fact that the large anomalous contribution to the specific heat, reported in all of the previous specific heat work on bcc ^{3}He, was not also observed in this work demonstrates that after all the long standing specific-heat anomaly in ^{3}He is not due to some intrinsic property of this quantum solid. The origin of the anomaly in the previous work remains however an interesting matter.

At one time a similar anomaly had been reported in the specific-heat measurements[17-19] on hcp ^{4}He. But now it is generally believed that this excess contribution was due to various experimental difficulties such as an improper accounting of the large heat capacity of the stainless-steel filling capillary, errors in the low temperature thermometer calibration, or errors in the measurement of the heating interval. It is possible though that crystal defects may have played a role in these measurements.[19] The more recent measurements of the specific heat of hcp ^{4}He,[16,20,21] excepting the work of Castles and Adams,[4] show no low temperature anomaly. Castles and Adams do see an unexplained contribution to the specific heat, but state that is nearly within their experimental uncertainty. The absence of an anomaly in ^{4}He is consistent with the recent results of Hanson et al.[22] They observed no anomalous contribution to the entropy of the solid determined using high-resolution measurements of the density of liquid ^{4}He along the melting curve.

In the case of the most recent specific-heat results on bcc ^{3}He by Castles and Adams care was taken to guard against some of the now-obvious pitfalls and still the anomaly remained. Since the present experiment differed in several technical respects from that of Castles and Adams and also from the other previous work it is not possible to say definitely which one of these differences explain why the anomalous specific heat was not also observed in the present work.

Table I is a comparison of some of the perhaps - significant experimental parameters in the present and in previous work on bcc ^{3}He. The differences which stand out are in the size of the calorimeter, in the ratio of surface area to volume and also in the time used to solidify the sample. That the samples were grown extremely slowly and in a large calorimeter with a very open geometry suggests that the solid samples were of higher crystal quality than in the previous experiments. This leads to the speculation that crystal defects may have been responsible for the anomaly reported in the earlier works. It should also be noted that although Eq. 1 can be used to describe the data of Ref. 4, this does not imply that the term βT is the anomalous contribution to the specific heat. It has been shown[23] that if γ is required to be consistent with sound velocity measurements (and consequently also with the present results), then the excess specific heat of Ref. 4 has a Schottky-like temperature dependence. This again suggests crystal defects. To be kept in mind, however, is the possibility that the excess specific heat observed previously may not in each experiment have been due to the same effect. If this is the case, then an intercomparison of the various experimental parameters can be both confusing and misleading. Another point to be considered is the fact that the PVT measurements of Henriksen et al.,[24] which can be related simply to the specific heat, are consistent with the existence of the anomaly. Since these strain-gauge pressure measurements involve completely different experimental techniques and since there is no equivalent to the empty-cell heat capacity to be subtracted from these data, it would appear that the specific heat anomaly can not be dismissed as having been due to some trivial experimental difficulty.

TABLE I

	This work	Castles and Adams (Ref. 4)	Pandorf and Edwards (Ref. 3)	Sample and Swenson (Ref. 2)
Principle cell construction materials.	Copper	BeCu and Copper	Copper	BeCu
Nominal cell volume (cm^3)	9	1	0.3	1
Surface area/ volume (cm^{-1})	6	275	10^4	10
$C_{empty}/(C_{empty} + C_{sample})$ at 0.3K and at 24 cm^3/mole	0.2	0.4	0.6	0.4
Size of stainless-steel capillary	0.05mm I.D. 0.15mm O.D. by 8cm long	0.10mm I.D. 0.20mm O.D. by 10cm long	0.13mm I.D. 0.25mm O.D. filled with 0.08mm dia. niobium wire	0.25mm I.D. 0.46mm O.D. filled with 0.2mm dia. steel wire
$C_{capillary}/C_{empty}$ at 0.1K	0.002	0.009	-	-
Transfer gas used in cool down	none	^4He	^3He	H_2
Heat switches	mechanical	supercon-ducting (tin)	supercon-ducting (indium)	mechanical and super-conduting (lead)
^4He concentration (ppm)	2.4	2	300	1800
Freezing Method	blocked capillary	blocked capillary	blocked capillary	blocked capillary
Solidification time (hours)	13-42	1-2	< 0.5	0.3
Annealing of sample after completion of solidification	no	yes (2 hours)	yes	yes (0.3 hour)
$\Delta T/T$	0.05	0.1-0.5	0.05	0.05
Temperature drift rate at 65mK (μK/min.)	10	400	-	-

References

(1) E. C. Heltemes and C. A. Swenson, Phys. Rev. 128, 1512 (1962).

(2) H. H. Sample and C. A. Swenson, Phys. Rev. 158, 188 (1967).

(3) R. C. Pandorf and D. O. Edwards, Phys. Rev. 169, 222 (1968).

(4) S. H. Castles and E. D. Adams, Phys. Rev. Lett. 30, 1125 (1973), and J. Low Temp. Phys. 19, 397 (1975).

(5) C. M. Varma, Phys. Rev. Lett. 24, 203, 970 (E) (1970).

(6) H. Horner, J. Low Temp. Phys. 8, 511 (1972).

(7) R. A. Guyer, J. Low Temp. Phys. 6, 251 (1972).

(8) I. E. Dzyaloshinskii, P. S. Kondratenko, and V. S. Levchenkov, Zh. Eksp. Teor. Fiz. 62, 1574 (1972) [Sov. Phys. JETP 35, 823 (1972].

(9) D. S. Greywall, Phys. Rev. B11, 4717 (1975).

(10) J. P. Franck and I. D. Calder, in Proceedings of the Fourteenth International Conference on Low-Temperature Physics, Otaniemi, Finland, 1975, edited by M. Krusius and M. Vuorio (North-Holland, Amsterdam, 1975), Vol. 1, p. 495.

(11) A. Widom and J. B. Sokoloff, J. Low Temp. Phys. 21, 463 (1975).

(12) J. B. Sokoloff and A. Widom, Phys. Rev. Lett. 35, 673 (1975).

(13) D. S. Greywall, Phys. Rev. B, 11, 1070 (1975).

(14) D. S. Greywall, Phys. Rev. Lett. 37, 105 (1976) and Phys. Rev. B 15, 2604 (1977).

(15) M. F. Panczyk and E. D. Adams, Phys. Rev. 187, 321 (1969).

(16) W. R. Gardner, J. K. Hoffer and N. E. Phillips, Phys. Rev. A7, 1029 (1973).

(17) F. J. Webb, K. R. Wilkinson and J. Wilks, Proc. Roy. Soc. (London) A214, 546 (1952).

(18) E. C. Heltemes and C. A. Swenson, Phys. Rev. 128, 1512 (1962).

(19) J. P. Franck, Physics Lett. 11, 208 (1964).

(20) D. O. Edwards and R. C. Pandorf, Phys. Rev. 140, A816 (1965).

(21) G. Ahlers, Phys. Rev. A 2, 1505 (1970); Physics Lett. 22, 404 (1966).

(22) H. N. Hanson, J. E. Berthold, G. M. Seidel, H. J. Maris, Phys. Rev. B, 14, 1911 (1976).

(23) D. S. Greywall, Phys. Rev. B11, 4717 (1975).

(24) P. N. Henriksen, M. F. Panczyk, S. B. Trickey, and E. D. Adams, Phys. Rev. Lett., 23, 518 (1969).

METASTABLE STATES OF SOLID ^{4}He NEAR THE HCP-BCC PHASE TRANSITION

V. B. Yefimov and L. P. Mezhov-Deglin

Solid State Physics - Acad. of Sciences of USSR
Chernogolovka, Moscow District, USSR

SUMMARY

Observations of metastable hcp and bcc phases occurring near the crystallographic phase transitions in ^{4}He are reported. It has been established that both the supercooled bcc phase and super heated hcp phase are observable at temperatures differing from the equilibrium transition temperature by 0.8 to 3%. The metastable phase lifetime is found to exceed at least 10^2 sec. At high rates of motion of the hcp-bcc interface ($v \gtrsim 1$ cm/sec.), the crystal developed visually discernible cracks and flaws (the maximum stress at the interface being 30 Nt-cm^{-2}), whereas at low rates ($v \lesssim 0.1$ cm/sec.), the stresses arising at the interface were able to relax on the crystal surface.

The original report of this research may be found (in Russian) in Physics of Low Temperatures (Acad. of Sciences of the Ukranian SSR, 5, 652 (1976).

KAPITZA RESISTANCE AT A SOLID HELIUM-COPPER INTERFACE UNDER HEAVY THERMAL LOADS

L. P. Mezhov-Deglin

Solid State Physics Institute, Acad. of Sciences of USSR
Chernogolovka, Moscow District, USSR

SUMMARY

Results were presented from three sets of measurements of the heat conduction between solid helium and a cold copper conductor in the 0.45 to 1.5K temperature range. The heat flux through the boundary could be described satisfactorily by $Q = A(T^4_{He} - T^4_{Cu})$, with the parameter A weakly dependent on the temperature. That is, the temperature dependence of the Kapitza resistance $R_K = (1/4AT^3)$ is close to the cubic dependence predicted by the acoustic detuning theory. However, the magnitude of R_K is one to two orders smaller than that which is predicted and, in contrast with that theory, it does not depend on the density or sound velocity in He.

A comparison of the resistances at the boundaries between He and a Cu single crystal or a polycrystalline Cu sample shows that the quality of the crystal structure of the Cu sample does not affect R_K. The resistance at the He-Cu boundary is determined primarily by the properties of the surface layers of the Cu sample. This result follows unambiguously from the fact that R_K is independent of the He impedance (which increases five times on transition from He II to solid He at 185 atm.) and independent of the degree of perfection of the bulk structure of the Cu sample and He crysals; it also follows from the considerable change (up to five times) of R_K upon substitution of the cold conductor.

The heat conductivity data for ^{4}He and ^{3}He crystals prepared from technically pure gases at pressures between 40 and 150 atm. were presented.

The original report of this research may be found (in Russian) in Zhurnal Eksperimentalnoi i Teoreticheskoi Fiziki 71, 1453 (1976).

THERMODYNAMIC PRESSURE OF SOLID ^{3}He IN HIGH MAGNETIC FIELDS*

E. B. Flint, E. D. Adams, and C. V. Britton

Physics Department - University of Florida
Gainesville, Florida 32611

Abstract: New measurements have been made of the thermodynamic pressure of solid ^{3}He in high magnetic fields. The previous departure from the Heisenberg nearest neighbor model, observed by Kirk and Adams, has been confirmed. Careful measurements in several fields establish a B^2 dependence of the pressure change with field at a fixed temperature.

The thermodynamic pressure, $P(T,B)_V$, provides a convenient quantity for studying the effects of nuclear spin ordering in solid ^{3}He at temperatures well above the expected magnetic transition. Although it is no longer thought to be adequate, it is convenient to introduce the Heisenberg Hamiltonian with only nearest neighbor interactions (HNN)

$$H = -2J \sum_{i<j} I_i \cdot I_j - \sum_i \mu_i \cdot B, \qquad (1)$$

where J is the exchange energy, I is the nuclear spin operator, and μ_i the nuclear magnetic moment. With this Hamiltonian, the pressure is given by

$$P(T,B)_V = \frac{RT}{V} \frac{\partial \ln|J|}{\partial \ln V} \left[3 \frac{J}{kT}^2 + \ldots + \left(\frac{\mu B}{kT}\right)^2 \left(\frac{2J}{kT} + \ldots\right)\right]. \qquad (2)$$

---

*Work supported by the National Science Foundation.

From measurements of the pressure in zero magnetic field, Panczyk et al.[1] obtained the exchange energy as a function of molar volume. The first difficulty with the HNN model was seen from high magnetic field $P(T,B)_V$ data of Kirk and Adams (KA).[2] Since the work of KA, a number of theoretical models have been introduced to try to account for the various experimental results.[3-6] Generally, it has not been possible to fit all of the results with any of the models and the thermodynamic consistency of the data has been questioned.[7]

With this in mind, we have undertaken new measurements of $P(T,B)_V$ with two major objectives: (1) to verify the departure from the HNN model observed by KA and (2) to take careful data at several fields to establish the dependence of $P(T,0)_V - P(T,B)_V$ on B. The experimental arrangement was very similar to that of KA with only slight differences. Temperatures were measured with a carbon resistor which was calibrated against the ^{3}He melting pressure[8] during the same run. The resistor was located outside the main field and was shielded from the fringing field by a Nb_3Sn cylinder. Thermal contact to the ^{3}He was through a brush of 2×10^5 cu wires 0.025 mm in diam welded to the body of the cell. These had a filling factor of about 50%. In the present work it was possible to extend the measurements to 15 mK.

The results for a sample of molar volume v = 23.96 cm^3/mole in several magnetic fields are shown in Fig. 1. These are in the form $P(T,B)_V - P_o$, where P_o is a constant chosen so that the results extrapolate through zero at T = ∞. Because of slight changes in either the capacitance gauge or the bulk pressure on changing the field, the constant P_o was slightly different for each field. The solid lines through the points are simply to guide the eye. The dashed curve with no points is the calculated behavior for B = 4.9 T with the HNN model using a value of J determined from the B = 0 data. Comparing this curve with the data for B = 4.9 T (inverted triangles) we again find, as did KA, that $P-P_o$ is about a factor of two smaller than expected from the HNN model.

Following the work of KA, there have been several theoretical efforts to account for these high-field pressure results. When only the high-field results are considered, a good fit can be achieved using triple exchange, which gives an effective next-nearest neighbor ferromagnetic interaction.[3] However, the difficulty with this approach is that it would apparently lead to a transition higher than $T_N \approx 2$ mK of the HNN model,[9] whereas the observed transition is near 1 mK.[10] So far, there is no suitable theoretical explanation for the high-field measurements. We know of no experimental difficulties which could affect the data and consider the present confirmation of the KA results as further evidence of their correctness. Thus we believe that additional

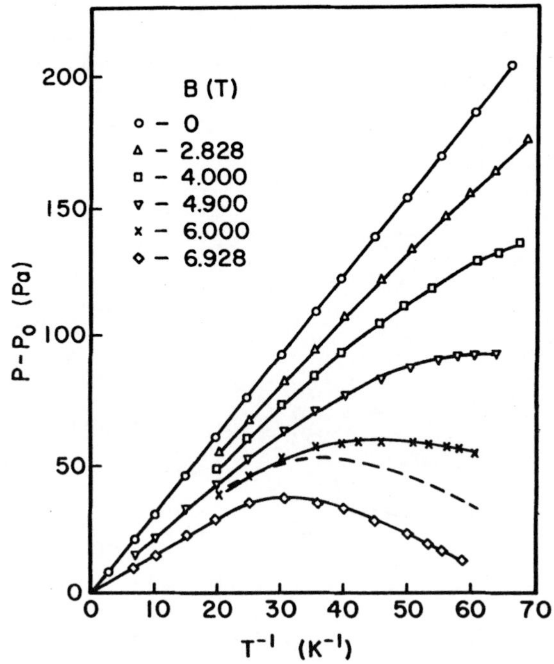

Figure 1. The thermodynamic pressure versus T^{-1} for various magnetic fields. The dashed curve is the behavior calculated from the HNN model.

theoretical effort is required.

As previously stated, one of the objectives of the present study was to establish the field dependence of $P(T,0)_V - P(T,B)_V$ at a fixed temperature. Experimentally, it is impractical to maintain a constant temperature while varying the field. However, this information can be gotten from Fig. 1. The results, in the form $P(0,T)_V - P(B,T)_V$ versus B^2 for various values of $1/T$, are shown in Fig. 2. It is seen that for each of the values of $1/T$, the required B^2 dependence is followed very closely. Observation of the B^2 dependence gives further confidence in the correctness of the data.

Returning to Fig. 1, we note one major difference between these results and those of Panczyk and Adams[1] in zero field. In the present case, the slope is about a factor of three smaller (this smaller slope was used to obtain J from Eq. 2 for comparison with the HNN model). At the present time, we do not know the reason for this behavior. One possibility is that there is a surface effect produced by the large number of small wires in the brush, which lowers the exchange interaction. However, this seems unlikely since the space between wires is more than 10^4 interatomic

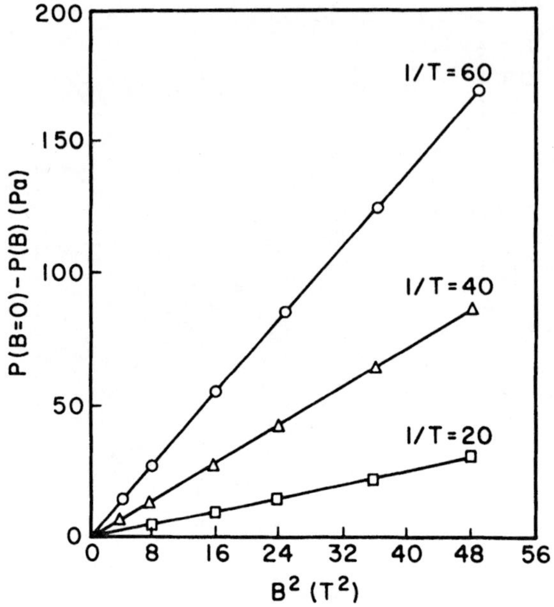

Figure 2. The magnetic field dependence of $P(T,0)_V - P(T,B)_V$ for various temperatures.

spacings of the helium.

Note Added - It was pointed out[11] at the symposium that the solid helium in the brush might be "frozen" to it and not free to move with pressure changes. The approximation that the helium is at constant volume might then be invalid. The connection between measured slopes dP/dT and the desired constant-volume slopes $(\partial P/\partial T)_V$ is[12]

$$\left(\frac{\partial P}{\partial T}\right)_V = \frac{dP}{dT}\left[1 + \frac{1}{k_T V}\frac{dV}{dP}\right] , \qquad (3)$$

where k_T is the compressibility of the helium and $V^{-1}dV/dP$ is the fractional change in the volume of the cell with pressure. If the entire volume of helium in the cell contributed to changes in pressure, the last term above would be a small correction. However, if only the small amount below the brush is active, the effective volume is so small that this term might be ~ 2.

Subsequently, we have made measurements with a thicker diaphragm for which dV/dP is reduced by a factor of 8, with the cell geometry otherwise unchanged. In this case, the pressure changes were as expected from the Panczyk and Adams results, indicating that the above "correction" term was the source of the small slopes. Thus the pressure scales of Figs. 1 and 2 should be increased by a factor of $\sim$3. The conclusions discussed above concerning the departure from calculations using the HNN model (recall that a J appropriate to the smaller slope was used in the calculation) and the B^2 dependence of the results are still valid since all pressure changes should be scaled by the same factor.

References

(1) M. F. Panczyk and E. D. Adams, Phys. Rev. 187, 321 (1969).

(2) W. P. Kirk and E. D. Adams, Phys. Rev. Lett. 27, 392 (1971).

(3) L. I. Zane, J. Low Temp. Phys. 9, 219 (1972).

(4) A. K. McMahan and R. A. Guyer, Phys. Rev. A. 7, 1105 (1973).

(5) L. I. Zane and J. R. Sites, J. Low Temp. Phys. 17, 159 (1974).

(6) A. K. McMahan and J. W. Wilkins, Phys. Rev. Lett. 35, 376(1975).

(7) R. A. Guyer, Phys. Rev. A 9, 1452 (1974).

(8) R. A. Scribner and E. D. Adams, in Temperature, Its Measurement and Control in Science and Industry (Instrument Society of America, Pittsburgh, 1972) Vol. 4.

(9) L. I. Zane, Phys. Lett. 41A, 421 (1972); E. D. Adams and L. H. Nosanow, J. Low Temp. Phys. 11, 11 (1973).

(10) W. P. Halperin, C. N. Archie, F. B. Rasmussen, R. A. Buhrman, and R. C. Richardson, Phys. Rev. Lett. 32, 927 (1974); R. B. Kummer, E. D. Adams, W. P. Kirk, A. S. Greenberg, R. M. Mueller, C. V. Britton, and D. M. Lee, Phys. Rev. Lett. 34, 517 (1975).

(11) W. P. Halperin, D. D. Osheroff, H. Meyer, private communication.

(12) G. C. Straty and E. D. Adams, Rev. Sci. Instr. 40, 1393 (1969).

TEMPERATURE AND FREQUENCY DEPENDENCE OF LONGITUDINAL SOUND IN HCP ^{4}He

I. D. Calder and J. P. Franck

Department of Physics, University of Alberta
Edmonton, Canada

We present here data on the temperature- and frequency-dependence of the longitudinal sound velocity of single crystals of hcp ^{4}He. The data show the following: the longitudinal sound velocity has very large dispersion at all temperatures from the melting point on down; the anomalous behavior is not always present; in crystals that show an anomaly (the majority of those grown), the anomaly moves to lower temperatures with increasing frequency.

Measurements of the absolute sound velocity were made using the pulse-echo-overlap method.[1-4] We used McScimin's method[5] to obtain the correct overlap of the rf oscillations; the results obtained using this criterion always agreed well with the velocity obtained from the average time between echoes along the entire echo train. For the measurement of the temperature-dependence the same set of rf oscillations were followed over the entire temperature range. Sensitivities for deteching changes in the velocity were about 1×10^{-5} at the fundamental (5MHz), 2×10^{-5} at the third harmonic (15MHz), and 1×10^{-4} at the fifth harmonic (25MHz). For the measurement of dispersion, McScimin's method had to be used separately at each frequency. We found that this criterion always led to the smallest possible shift of velocity with frequency. Measurements at different frequency also introduce errors due to finite pulse-width[6] and to diffraction.[7] Both effects can be estimated to be below 5×10^{-5} when going from 5 to 15 MHz, or from 15 to 25 MHz. These estimates were borne out by observations on the liquid above the melting point where observed changes in the measured velocity were -6×10^{-5} between 5 and 15 MHz and -1×10^{-5} between 15 and 25 MHz.

The experiment was cooled by a SHE minifridge. Temperatures were determined from the resistance of a calibrated 220 ohm, ½ watt Speer resistor mounted on the pressure cell.

In Figs. 1 and 2 we show results for two crystals that exhibit the anomaly. At high temperature the sound velocity is well described by an expression of the form

$$v(T) = v_o - aT^4 + bT^6. \quad (1)$$

This behavior is expected in the adiabatic range. At some temperature near 1.8K the attenuation starts to rise sharply with falling temperature, as judged by the decrease in the amplitude of the echoes. A crystal that, near melting, showed in excess of 100 echoes may go down to about 3-5 visible echoes at the lowest temperature for constant excitation levels. We are refraining from quoting absolute values of attenuation because of the difficulty of separating out purely geometric effects. The attenuation

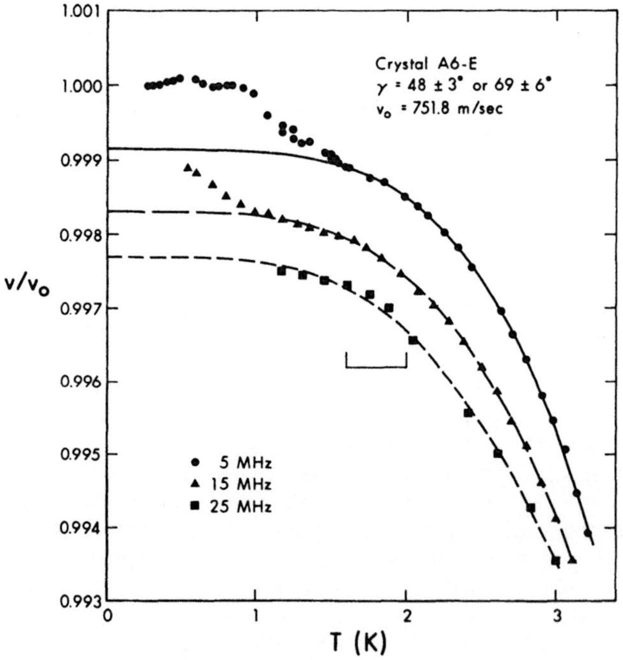

Figure 1. Reduced velocity vs. temperature for crystal A6-E at a most probable orientation of 69° from the c-axis. The curves are fits to equation (1). The horizontal bracket indicates the range within which the attenuation began to rise sharply.

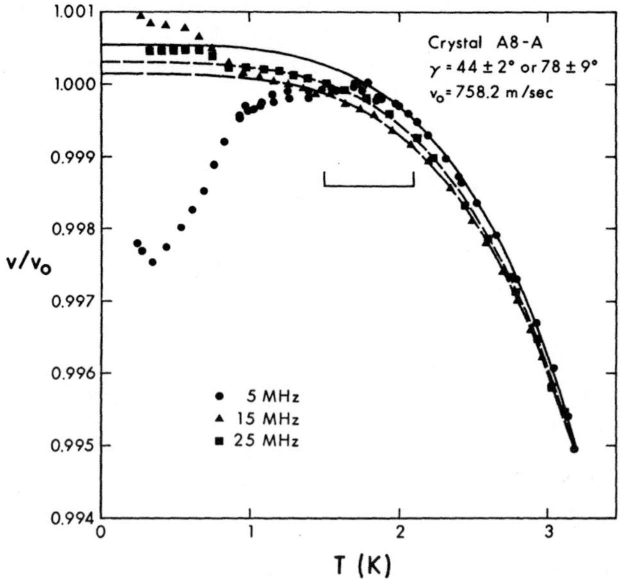

Figure 2. Reduced velocity vs. temperature for crystal A8-A at a most probable orientation of 78° from the c-axis. The curves are fits to equation (1). Horizontal bracket as in Fig. 1.

in the adiabatic range is, however, certainly very low.

At or below the temperature at which the attenuation rises steeply, the phase velocity also shows strongly anomalous behavior. In most crystals the phase velocity rises steeply above the adiabatic velocity, Fig. 1, reaching a plateau at temperatures about 0.2 to 0.5K lower. The relative change from the extrapolated adiabatic velocity can be as large as 3×10^{-3}. The onset of the velocity anomaly was observed to occur in the range 0.9 to 1.6K for different crystals. In two of the crystals we observed a decrease in phase velocity, of the order 2×10^{-3}; such a crystal is shown in Fig. 2. This decrease in velocity started at about 1.8K.

Measurements of the frequency-dependence of the sound velocity were done on each crystal by going to the third and fifth harmonics (15 and 25 MHz). The following were observed: all crystals show a remarkably large dispersion, at all temperatures from the melting point on down. The velocity decreases in most crystals with rising frequency (Normal dispersion) by between 2×10^{-4} and 8×10^{-4} for a frequency change of 10 MHz. This dispersion is exceptionally large; it obviously can not persist over too large a frequency interval. The large normal dispersion in the 5 to 25 MHz region will have to be accompanied almost necessarily by an equally large anomalous dispersion region, probably at higher frequency. In fact, the crystal in Fig. 2 shows anomalous dispersion already in the 15 to 25 MHz range. The velocity anomaly becomes generally smaller and moves to lower temperature as the frequency increases. The negative-going velocity anomaly at 5 MHz in the crystal of Fig. 2 becomes positive-going and also moves to lower temperature.

We have also, surprisingly, observed two crystals in which both anomalies are absent, Fig. 3. In those crystals we estimate that sound progagated at 43° and 72° respectively from the c-axis. The sound velocity follows equation (1) down to the lowest temperatures. These crystals show a slight decrease in echo height down to about 1.8K. Below this temperature there is little or no further change. At 0.25K we observed about 50 echoes in these crystals, as compared to 3-5 in the other crystals.

All crystals measured have always shown reproducible behavior, both for temperature-cycling and for changes in frequency. We have so far not found any discernible orientation dependence of the effect. We feel that the effect must be connected with some type of lattice imperfections that differ from crystal to crystal. Any orientation dependence that might exist is, at this stage, overshadowed by the influence of imperfections.

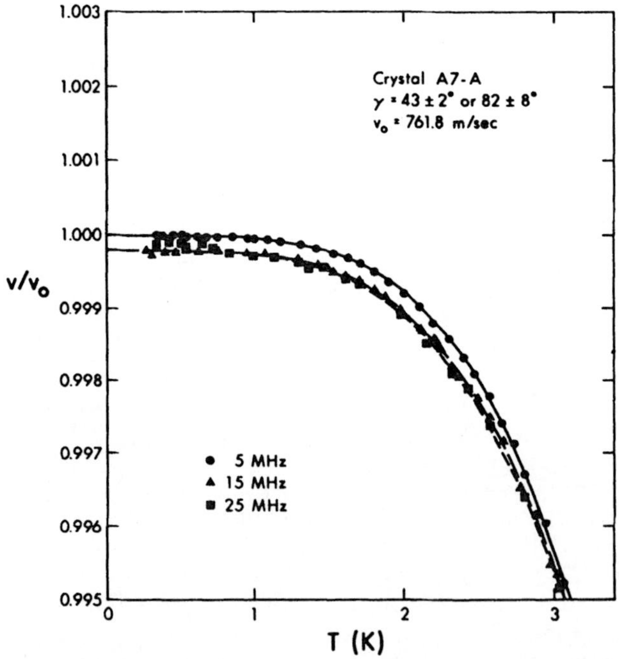

Figure 3. Reduced velocity vs. temperature for crystal A7-A at a most probable orientation of 43° from the c-axis. The attenuation in this crystal did not experience a sharp rise.

References

(1) J. E. May, Jr., IRE Nat. Conv. Rec. 6, Pt. 2, 134 (1958).

(2) E. P. Papadakis, J. Appl. Phys. 35, 1474 (1964); J. Acoust. Soc. Amer. 42, 1045 (1967).

(3) D. H. Chung, D. J. Silversmith and B. B. Chick, Rev. Sci. Instr. 40, 718 (1969).

(4) J. P. van Nest, T. S. Hutchinson and D. H. Rogers, Can. J. Physics 47, 1797 (1969).

(5) H. J. McScimin, J. Acoust. Soc. Amer. 33, 12 (1961).

(6) L. G. Merkulov and V. A. Tret'yakov, Soviet Physics Acoustics 20, 358 (1975).

(7) E. P. Papadakis, J. Acoust. Soc. Amer. 40, 863 (1966).

CRITICAL DYNAMICS IN $He^3 - He^4$ MIXTURES

Eric D. Siggia

Department of Physics, University of Pennsylvania
Philadelphia, Pennsylvania 19104

A phenomenological model for the critical dynamics of He^3 - He^4 mixtures is studied in the tricritical region and along the lambda line for $T > T_\lambda$. Calculations are done to first order in $\varepsilon = 4-d$ since our predictions begin to deviate from mean field theory below four dimensions. At the tricritical point, the mass diffusion constant, D, varies with the inverse correlation length as κ^x where $x = \varepsilon/3 + \theta(\varepsilon^2)$ in disagreement with previous mode coupling theories. The crossover scaling functions from the tricritical to critical regions are calculated for all the singular transport and kinetic coefficients in our model. A suitably defined effective exponent is found to vary non-monotonically in the crossover regime. Although our results are consistent with present experiments, at the tricritical point our model suffers from ambiguities similar to those discussed by Halperin, Hohenberg and Ma for their model C.

The critical properties of a number of transport coefficients are completely different in pure He and the mixtures. We have calculated the universal scaling functions that describe the crossover of the thermal conductivity and diffusion constant from their behavior in pure He to that characteristic of the lambda line to first order in ε along with leading correction terms. The crossover exponent is determined to all orders in ε which implies that the thermal conductivity at T_λ varies inversely with the concentration.

For details of this work, see References 1 and 2.

References

(1) Eric D. Siggia and David R. Nelson, Phys. Rev. B 15, 1427 (1977).

(2) Eric D. Siggia, Phys. Rev. B 15, 2830 (1977).

OPTICAL EXPERIMENTS IN ^{3}He-^{4}He MIXTURES NEAR THE TRICRITICAL POINT

P. Leiderer

Physik-Department E 10 der Technischen Universität München
8046 Garching, West Germany

In the vicinity of a critical point, a number of important properties can be determined by means of optical methods.[1] As an example let us consider a binary mixture near its consolute critical point where the observed "critical opalescence" is due to the increasing fluctuations in concentration.

From the intensity of the light scattered from these fluctuations the concentration susceptibility with its critical exponent γ can be obtained. The wave vector dependence of the scattered light provides insight into the size of the concentration fluctuations, i.e., the correlation length ξ, with its critical exponent ν. Spectral resolution of the scattered light shows the time scale of the concentration fluctuations, and in general a "critical slowing down" of these fluctuations is observed on approaching the critical point. The order parameter, which for a binary mixture is the concentration difference between the two coexisting phases, can also, among other methods, be determined optically by measuring the difference in the index of refraction. Finally, the interfacial tension in the phase separated mixture can be obtained using light scattering from interfacial waves.

In principle, helium is not an ideal candidate for light scattering experiments because its polarizability is quite small and therefore the scattered light intensities are very weak. Nevertheless, all the properties mentioned above have been measured in ^{3}He-^{4}He mixtures near the tricritical point (temperature $T_t \simeq 0.867$K, ^{3}He-concentration $x_t \simeq 0.675$[2]) demonstrating that even in this case optical methods can serve as a versatile tool in determining critical exponents.

The experimental set-up for all the measurements described here was essentially the same, with some modifications according to the particular quantity to be measured. It is shown schematically in Fig. 1. Because of the low light levels, a photon counting technique with the appropriate analyzing equipment was used for detection of the scattered intensities. In the following the main results of these measurements are described.

ORDER PARAMETER

Of the two order parameters near the tricritical point - the superfluid order parameter and the order parameter of the non-ordering density, which is the concentration difference $\Delta x=x_+-x_-$ between the coexisting normal and superfluid phases - only the latter is easily accessible optically. The limiting angle of total reflectivity of the interface in the phase separated mixture yields directly the discontinuity of the refractive index, $\Delta_n=n_--n_+$, across the interface, and by means of the relation $\partial n/\partial x = 0.0070$[3] the concentration difference Δx is obtained. This method determines the order parameter free of gravity perturbations, since only the interface itself is involved in the measurement, and in addition provides calibration for temperature differences from T_t to better than 0.1mK.

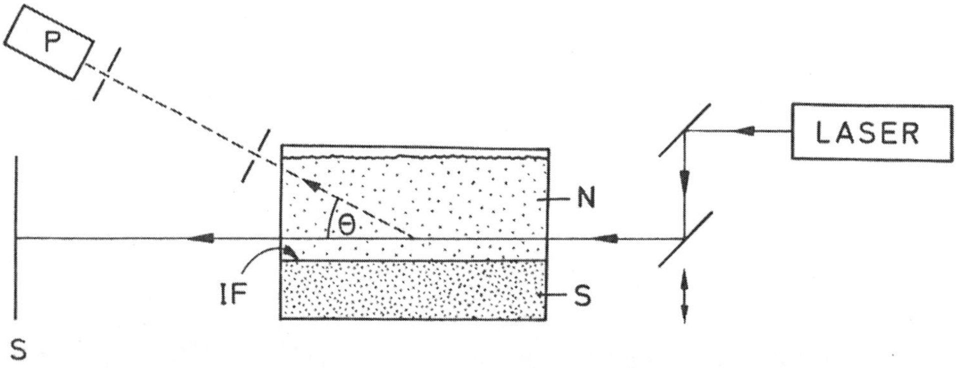

Figure 1. Schematic set-up for optical experiments in $^3He-^4He$-mixtures. When the laser beam is scanned vertically across the sample cell, the photomultiplier P registers the light scattered at an angle Θ from the uper normal (N) or lower superfluid (S) phase, respectively. When the laser beam traverses the cell exactly at the position of the interface IF, an additional spot appears on the screen S due to diffraction of part of the incident light at the limiting angle of total reflectivity.

Δn is found to decrease linearly on approaching T_t, as shown in Fig. 2, implying $\Delta x \propto t^\beta$ ($t = T_t - T/T_t$), where the tricritical exponent is $\beta = 1.0$ in the range $8 \times 10^{-4} < t < 10^{-2}$. This result confirms that the phase separation curve in $^3He-^4He$ mixtures, as opposed to conventional binary mixtures, has an angular top, as first observed by Graf et al.[4] This already indicates that $^3He-^4He$ mixtures near the tricritical point belong to a universality class different from that of simple fluids and conventional binary mixtures, with completely different, mean-field-like critical exponents.[5,6]

CONCENTRATION SUSCEPTIBILITY AND CORRELATION LENGTH

Light scattering intensities give a direct measure of the concentration susceptibility $(\partial x/\partial\Delta)_{T,P}$ near T_t in the small scattering vector limit ($q \to 0$), in contrast to experiments on specific heat[7] and saturated vapor pressure.[8] In the hydrodynamic regime, $q\xi \ll 1$, the scattering power is proportional to the extinction length

$$h = A(\lambda)\,\frac{k_B T}{v^2}\left[\frac{1}{v}\left(\frac{\partial v}{\partial x}\right)^2_{TP}\left(\frac{\partial x}{\partial \Delta}\right)_{TP} + \beta_{Tx}\right] \tag{1}$$

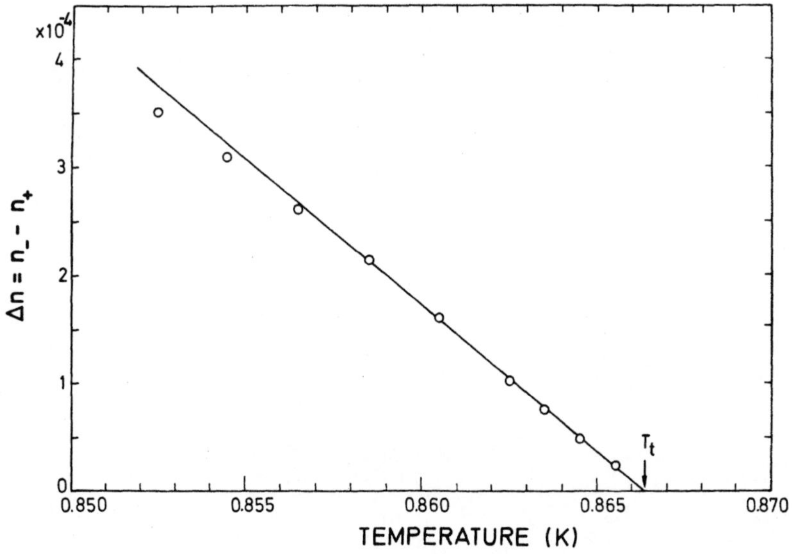

Figure 2. Discontinuity of the index of refraction, n, across the interface of the phase separated mixture, as determined from the limiting angle of total reflectivity. The accuracy of the absolute temperature scale is $\pm$ 1mK due to uncertainties in the ^3He vapor pressure calibration. Error bars for Δn are smaller than the size of the data symbols.

where $A(\lambda) = (2\pi/\lambda)^4 (4\pi\alpha N_o)^2/6\pi$ and $\beta_{Tx} = -(1/v)(\partial v/\partial p)_{Tx}$.

Here $\Delta = \mu_3 - \mu_4$ is the difference between the chemical potentials of ^{3}He and ^{4}He, v is the molar volume, αN_o is the molar polarizability of helium, and λ is the light wavelength. The isothermal compressibility β_{Tx} varies only weakly near the tricritical point, whereas the concentration susceptibility is expected to diverge. Since the atomic volumes of ^{3}He and ^{4}He differ, concentration fluctuations dominate the scattering and the critical opalescence is determined by the divergence of $(\partial x/\partial \Delta)_{TP}$.[9]

To keep gravitational effects on light scattering to a minimum while very close to T_t, the incident laser beam was focused and scanned vertically across the sample cell. In this way the scattered intensity was monitored <0.2 mm away from the interface, thus providing data essentially free of gravity effects when $T_t - T > 0.002$K. Apparent concentration susceptibility data $(\partial x/\partial \Delta)_{TP}$, shown in Fig. 3, were obtained from the scattered intensities by using Eq. (1), i.e. by heuristically assuming isotropic scattering for the $q \to 0$ limit. The results were fitted by the expected power-law divergence of the form $(\partial x/\partial \Delta)_{TP} = G \cdot t^{-\gamma}$.

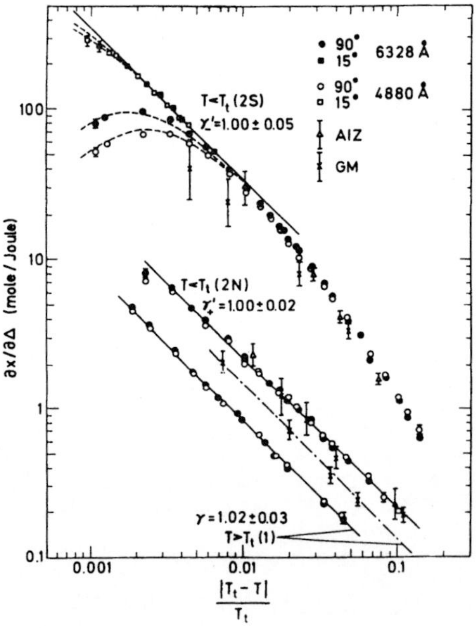

Figure 3. Log plot of $(\partial x/\partial \Delta)_{TP}$ vs. t for the single phase (1), and for the coexisting normal (2N) and superfluid (2S) phases for a mixture with x=0.6749. For comparison, data of Alvesalo et al. (AIZ) and Goellner and Meyer (GM) (Refs. 7 and 8) are shown. The dashed lines show the best fit to our data calculated according to Eq. (2). At $t \lesssim 10^{-3}$, the remaining small discrepancy is attributable to gravity effects.

For the single phase critical isochore and the normal coexisting phase, power laws were found to hold in the entire investigated temperature interval. In contrast, the tricritical region for the superfluid seems to be confined to $T_t - T < 10mK$, about the same temperature interval in which the phase diagram is linear (c.f. Fig. 2). The best-fit critical exponents are $\gamma = 1.02 \pm 0.03$, $\gamma_+' = 1.00 \pm 0.02$, and $\gamma_-' = 1.00 \pm 0.05$, where γ applies for $T<T_t$ along the critical isochore and γ_+' and γ_-' for $T<T_t$, the subscripts + and - referring to the normal and superfluid coexisting phases, respectively.

To obtain the correlation length ξ of the concentration fluctuations, the wave number dependence of the scattered intensity was determined by using two wavelengths, λ=6328Å and 4880Å, and two scattering angles, 90° and 15°, thus providing data at four wave numbers. The onset of q-dependence of the 90° scattering intensities in the coexisting superfluid is clearly evident in Fig. 3 for $t < 0.007$. A comparison with the modified Ornstein-Zernicke expression for the correlation function calculated by Furman and Blume[10] shows that the data can be well represented by the correlation function

$$C(\vec{q},t) \approx \frac{\text{const.}}{q} \left(\frac{q/t}{1.6\times10^{-16}\, q^2/t^2+1} \right) \tag{2}$$

If we define the critical exponents η and ν by writing $C(\vec{q},t) = 1/q^{2-\eta}\, F(q/t^{\nu})^6$, we find that the experimental results are consistent with the critical exponents $\nu_-'=1$ and $\eta_-'=1$. Thus the correlation length in the superfluid phase is given by $\xi_-' = 1.3\ t$ Å. Note that in the normal phases a much smaller, if any, deviation from the straight lines in Fig. 3 suggests a considerably smaller correlation length there - another indication of the large asymmetry in the amplitude of the power laws near T_t, as obvious already from the magnitudes of the concentration susceptibility in Fig. 3, and also from the different slopes of the phase separation curves.[2]

TRICRITICAL SLOWING DOWN

Similar to the correlation length, the relaxation time of the fluctuations also increases near a critical point. This is reflected in the spectrum of the scattered light as a narrowing of the central Rayleigh line (which is the one dominated by the scattering from concentration fluctuations) to typically less than 10^3 Hz. For a spectral analysis this requires resolutions $\Delta\nu/\nu < 10^{-13}$, nowadays attained by light beating spectroscopy. However, for this technique relatively high light intensities are necessary, and in $^3He-^4He$ mixtures only the superfluid phase has so far been accessible to measurement;[11] in the normal phase where the concentration susceptibility and consequently the scattering power is an order of magnitude smaller than in the superfluid phase, the light intensities were too small for such an analysis.

The resulting reduced Rayleigh linewidth Γ/q^2 for scattering at $\theta=15°$ is plotted in Fig. 4 for laser wavelengths $\lambda=4880$ and 6328Å, corresponding to scattering vectors $q=3.35\times10^4$ and $2.59\times10^4\ cm^{-1}$, respectively. The reduced linewidth is used because by comparison with Fig. 3 the data are expected to lie in the hydrodynamic regime, where the linewidth generally is given by $\Gamma=D_{eff}\cdot q^2$ with an effective diffusion coefficient D_{eff}. Indeed in this representation the data for different q collapse to a single line

$$\frac{\Gamma}{q^2} = D_{eff} = 1.5\times10^{-4}\cdot t^{0.95\pm0.07}\ cm^2/sec \tag{3}$$

Siggia and Nelson[12] have treated the problem of fluctuation lifetimes theoretically by means of renormalization group methods. In the hydrodynamic regime they predict the existence of two diffusive modes: one with eigen-frequency $\omega_1 \propto tq^2$, which seems to be observed here, and one with $\omega_2 \propto t^{-2/3}q^2$ which has been detected very recently in sound attenuation measurements.[13]

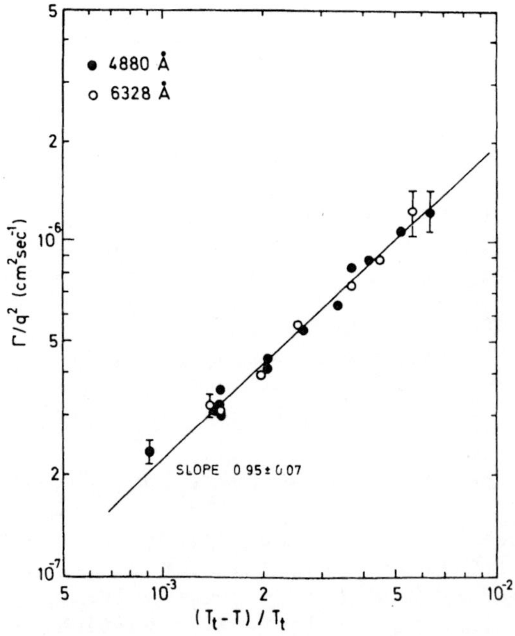

Figure 4. Reduced Rayleigh linewidths at 15° scattering angle.

LIGHT SCATTERING FROM THE INTERFACE

One can now use the above results for the critical exponents to predict the behavior of other properties near the tricritical point. One such property is the interfacial tension.

As a phase separated binary mixture approaches its critical point and the two phases become more and more similar, the interfacial tension - which is the free energy associated with unit interface area - vanishes as[14]

$$\sigma_i \propto (T_c - T)^{\mu} \tag{4}$$

Up to now no rigorous theory exists which links the interfacial tension to properties of the bulk system. However, a phenomenological approach which should be valid in the vicinity of the critical point has been developed by Fisk and Widom,[15] who obtain for the interfacial tension

$$\sigma_i = c \cdot \frac{L}{\chi} \cdot (\Delta x)^2 \tag{5}$$

where L is the thickness of the interface, χ the concentration susceptibility, Δx the concentration difference of the two phases, and c some numerical constant of the order 1/6, the exact value depending on the detailed shape of the interfacial profile. The interfacial thickness is generally assumed to be of the order of the bulk correlation length ξ. Hence (5) implies the scaling relation[15]

$$\mu = \gamma' - \nu' + 2 \quad . \tag{6}$$

At the consolute critical point of a binary mixture like cyclohexane-methanol Eq. (6) yields $\mu \approx 1.28$, in good agreement with experiment.[16] At the tricritical point of $^3He-^4He$, with the exponents given as discussed above, the prediction from scaling is $\mu=2$, as pointed out by Papoular.[17]

One way to check this prediction and to determine the interfacial tension experimentally is to study interfacial waves. If we neglect damping for the moment, the dispersion relation of interfacial waves (ripplons) is given by

$$\omega^2 = \frac{\rho_- - \rho_+}{\rho_- + \rho_+} gk + \frac{\sigma_i}{\rho_- + \rho_+} k^3 \tag{7}$$

where the restoring force in the first contribution is supplied by gravity and in the second term by the interfacial tension. Here and k are the angular frequency and wave number of the ripplon, ρ_+ and ρ_- the density of the ^{3}He-rich and ^{4}He-rich phase, respectively, and g the acceleration of gravity.

In our experiment, the interfacial waves were generated artificially by a coil of bifilar wires (cf. inset in Fig. 5), to which an a.c. voltage was applied, a set-up similar to that of Boldarev and Peshkov.[18] For the detection of the waves, the incident laser beam was reflected from the interface at a small angle. The interfacial waves act like a phase grating on the reflected beam and give rise to a Debye-Sears-like pattern of interference fringes, the fringe distance yielding the wavelength of the interfacial waves.

Knowing the wavelength and frequency of the ripplons, the interfacial tension can be obtained from the dispersion relation (7).[19] The resulting interfacial tension is plotted in Fig. 5. In the whole temperature range investigated, $20mK<T_t-T<370mK$, σ_i can be represented by

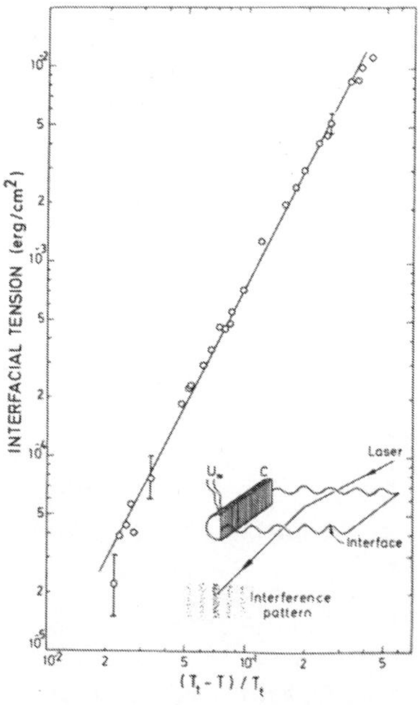

Figure 5. Interfacial tension of phase separated ^{3}He-^{4}He mixtures, calculated from the dispersion of interfacial waves. Damping of the waves has been taken into account.[19] The straight line is a fit according to Eq. (8). In the inset, the scattering geometry is shown. The interfacial waves are generated with the bifilar coil C.

$$\sigma_i = \sigma_{i0} t^{\mu} \tag{8}$$

where σ_{i0}=0.076 erg/cm^2, and μ=2.0, as expected from scaling.[17] This is a much stronger variation with temperature than observed near conventional critical points, where $\mu \simeq 1.28$, and indicates the influence of the tricritical point.

In most critical systems studied previously the borderline dimensionality where mean field theory should apply[20] is d=4 and is hence not accessible experimentally. From the experiments described here, a nearly complete set of critical exponents has been obtained for the first time for a system with borderline-dimensionality[21] d=3, where the critical exponents are mean-field-like. These exponents $\beta=\gamma'_{\pm}=\nu'_{-}=\eta'_{-}=1$, $\mu=2$ are consistent with scaling, *e.g.* $\gamma=(2-\eta)\cdot\nu$, $\mu=\gamma-\nu+2\beta$, etc. Much more accurate measurements are necessary to determine any logarithmic corrections to these pure power law dependences, as predicted by renormalization group theory.[21,22]

ACKNOWLEDGEMENTS

The work reported here was done in collaboration with W. W. Webb, D. R. Watts, D. R. Nelson, M. Wanner, and H. Poisel. Thanks are due to P. C. Hohenberg, B. Widom, M. Papoular and H. Kinder for many valuable discussions. Financial support by the Deutsche Forschungsgemeinschaft is gratefully acknowledged.

References

(1) For a review, see P. Heller, Rep. Progr. Phys. 30, 731 (1967).

(2) See for example G. Ahlers, in "The Physics of Liquid and Solid Helium," edited by K. D. Bennemann and J. B. Ketterson (Wiley, New York, 1976), Part I.

(3) H. A. Kierstead, J. Low Temp. Phys. 24, 497 (1976).

(4) E. H. Graf, D. M. Lee, and J. D. Reppy, Phys. Rev. Lett. 19, 417 (1967).

(5) R. B. Griffiths, Phys. Rev. Lett. 24, 715 (1970), and Phys. Rev. B7, 545 (1973), and references cited therein.

(6) E. K. Riedel, Phys. Rev. Lett. 28, 675 (1972).

(7) T. Alvesalo, P. Berglund, S. Islander, G. R. Pickett, and W. Zimmermann, Jr., Phys. Rev. Lett. 22, 1281 (1969), and Phys. Rev. A 4, 2354 (1971).

(8) G. Goellner and H. Meyer, Phys. Rev. Lett. 26, 1534 (1971); G. Goellner, R. Behringer, and H. Meyer, J. Low Temp. Phys. 13, 113 (1973).

(9) P. Leiderer, D. R. Watts, and W. W. Webb, Phys. Rev. Lett. 33, 483 (1974).

(10) D. Furman and M. Blume, Phys. Rev. B 10, 2068 (1974).

(11) P. Leiderer, D. R. Nelson, D. R. Watts, and W. W. Webb, Phys. Rev. Lett. 34, 1080 (1975).

(12) E. D. Siggia and D. R. Nelson, to be published.

(13) D. B. Roe, G. Ruppeiner, and J. Meyer, to be published.

(14) For a review, see B. Widom, in "Phase Transitions and Critical Phenomena," edited by C. Domb and M. S. Green (Academic, New York, 1972), Vol. 2, Chap. III.

(15) S. Fisk and B. Widom, J. Chem. Phys. 50, 3219 (1969).

(16) C. Warren and W. W. Webb, J. Chem. Phys. 50, 3694 (1969).

(17) M. Papoular, Phys. Fluids 17, 1038 (1974).

(18) S. T. Boldarev and V. P. Peshkov, JETP Lett. 17, 297 (1973).

(19) For a more detailed analysis, the damping of the interfacial waves has to be taken into account. See M. A. Bouchiat and J. Meunier, J. Phys. (Paris) 32, 561 (1971); J. Phys. (Paris) 33, C1-141 (1972); E. S. Wu and W. W. Webb, Phys. Rev. A8, 2077 (1973); P. Leiderer, H. Poisel, and M. Wanner, to be published.

(20) See, for example, B. Widom in "Fundamental Problems in Statistical Mechanics," Vol. 3, Edited by E. D. G. Cohen (North-Holland Publishing Company, Amsterdam, 1975).

(21) E. K. Riedel and F. J. Wegner, Phys. Rev. Lett. 29, 349 (1972).

(22) M. J. Stephen, E. Abrahams, and J. P. Straley, Phys. Rev. B 12, 256 (1975).

TRANSPORT PROPERTIES IN LIQUID ^3He-^4He MIXTURES NEAR THE TRICRITICAL POINT*

H. Meyer, D. Roe and G. Ruppeiner

Department of Physics, Duke University
Durham, N. C. 27706

In this talk, we present results of acoustic attenuation measurements in ^3He-^4He mixtures, carried out in the range between 1 and 45 MHz. Furthermore, an account is given of an analysis of transient effects occurring during sound velocity measurements, and which lead to some conclusions on the thermal diffusivity ratio K_T. Because the results have been described in detail in two papers,[1,2] the presentation here can be made brief.

Sound attenuation and velocity dispersion measurements can give valuable information on diffusion and relaxation processes near critical points. It is well known, for instance, that the relaxation time τ characterizing density fluctuations near a liquid-gas critical point T_c diverges strongly as T_c is approached.[3] The same is true for τ of concentration fluctuations in a binary mixture and for τ of superfluid density fluctuations near the lambda transition of pure ^4He and of ^3He-^4He mixtures with ^3He mole fraction X.[4,5] The mass diffusion D of the mixtures is also expected to show an anomalous behavior near the lambda point, and the predicted divergence has indeed been observed by Ahlers and Pobell.[6] The interesting problem is now to observe the passage of τ and of D from the critical behavior to the tricritical behavior for mixtures when the tricritical point (T_t=0.867K, X_t=0.674) is approached. The tricritical region is characterized by the competition of superfluid density ("order parameter") fluctuations and

*Supported by a grant from the National Science Foundation.

of concentration fluctuations.[6] The tricritical exponents characterizing the divergence of the order parameter relaxation time, τ, and of D can be expected to be quite different from those along the lambda ("2nd order") line.

Although a comprehensive theory for the sound attenuation and the dispersion $\Delta U = U_o(\omega) - U_o(0)$ for He^3-He^4 mixtures has not been worked out, a rather simple and crude analysis of our attenuation data is able to determine the exponents of D and τ and the respective amplitude variations as T_t is approached along various directions in the X-T plane.

Near a critical point, where anomalous behavior is observed in α, it is convenient to set

$$\alpha = \alpha_{reg} + \alpha_{sing.} \qquad (1)$$

where α_{reg} expresses the "background" behavior that is supposed to change only insignificantly with temperature throughout the critical region. It is the quantity α_{sing} that describes the behavior associated with the transition. It arises because the long internal relaxation times in the fluid bring about irreversible processes during the passage of the sound waves, causing an energy dissipation. In addition, the pressure gradient during the passage of the sound wave in a mixture gives rise to irreversible mass diffusion of the two fluid components between the trough and the crest of the wave. Briefly, when a ^3He-^4He mixture of a given composition is cooled through the lambda point, the attenuation and the dispersion pass through a peak situated slightly below T_λ, and the shape of peak is asymmetric on the temperature scale.[5,8] Theory[9] and experiments[1,5,8] show that the attenuation peak height and the width at half the maximum, $(\alpha_{sing}(T_\lambda)/2)$ strongly increase with frequency. It is therefore desirable to investigate the acoustic phenomena over a large frequency range.

A few words about the acoustic techniques might be useful. For the measurements on ^3He-^4He mixtures along the lambda line, it is necessary to work with flat cavities having a height of only a few mm to reduce gravity effects that produce concentration gradients. Flat acoustic resonators cover a range up to about 1 MHz, and this technique has been exploited very successfully by Rudnick and his associates[4] and still more recently by Buchal and Pobell.[5] The method using quartz transducers covers the range from about 0.5 MHz to frequencies of the order of several hundred MHz, and there can be, in principle, an overlap of the frequency ranges covered by both methods.

In the experiments we shall describe, a temperature-controlled copper cell containing an etalon with a spacing of 3 mm between the transducers was used. Two separate pairs of X-cut quartz plates were used, each having a diameter of 2.1 cm and with fundamental frequencies of 1 and 5 MHz respectively. For the attenuation meausrements, square pulses of 2 μsec were generated and sent to the first quartz transducer. The signal received in the second one was first amplified, then sent through a precision step attenuator to a tuned receiver. The peak of the received pulse was monitored with a box-car integrator.

Striking changes occur in the sound attenuation as pure ^4He is diluted with ^3He. At a given frequency, the peak height strongly decreases as X increases, passes through a broad minimum in the region of $X \simeq 0.5$ and through a maximum at the tricritical point. The half width at half maximum increases even more drastically. At a frequency of 15 MHz, for instance, this half width at $X=X_t$ is approximately 40 times larger than in pure ^4He. As the phase separation curve is approached, the attenuation again rises. In Fig. 1, we show the change of the shape of α_{sing} at a frequency of 15 MHz for several mixtures near the tricritical point. At X=0.553, the attenuation is clearly resolved into two peaks, one slightly below T_λ and the other becoming prominent as the phase separation T_σ is approached. As X is increased to X_t, these two peaks merge. Even in this relatively small range of X, there is a noticeable broadening of the peaks near T_λ as X is increased.

We now sketch briefly the analysis of the attenuation data and first consider the region far away from the transition. For practical considerations this is the regime where $\alpha_{sing} \lesssim \frac{1}{4}\alpha_{sing}(T_\lambda)$ and

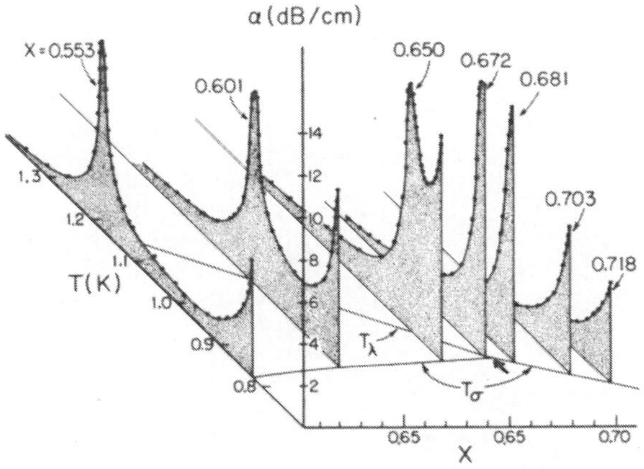

Figure 1. Attenuation at 15 MHz near the tricritical point for several mixtures. The lambda line and the phase separation lines are denoted respectively by T_λ and T_σ. They all join at the tricritical point, marked by an arrow.

here it is found that $\alpha_{sing} \propto \omega^2$, which corresponds to the hydrodynamic regime $\omega\tau \ll 1$. The attenuation is expressed in terms of hydrodynamic modes such as bulk and shear viscosities ζ and η and diffusion,[10]

$$\alpha_{sing} = \alpha_{visc.} + \alpha_D \tag{2a}$$

$$\simeq \omega^2 f(\eta,\zeta) + \frac{\omega^2 DM}{2\rho^2 U_o(0)} \left(\frac{\partial X}{\partial \Delta}\right)_T \left(\frac{\partial \rho}{\partial X}\right)^2 . \tag{2b}$$

Here $\partial X/\partial\Delta)_{T,P}$ is the concentration susceptibility, ρ is the mass density and M is the molar mass, which are known from other experiments. A calculation of α_D shows that its contribution to the singular attenuation increases as ^{4}He becomes diluted with ^{3}He, and for $X \simeq 0.55$ these estimates indicate that α_D accounts for practically the whole observed α_{sing}. Hence we shall assume that this is also the case for the more concentrated ^{3}He-^{4}He mixtures. Based on this assumption we calculate D, using the experimental data for α_{sing} and Eq. 2. The result is shown in Fig. 2. For dilute ^{3}He-^{4}He mixtures, it is found from a direct diffusion experiment[6] that $D \propto (T-T_\lambda)^{-z}$ with z approximately 0.3 to 0.4, and for X=0.55 this is confirmed roughly by the acoustic data analysis. But as more concentrated mixtures are used, the behavior of D changes drastically. Finally, along the path $X=X_t$, where[11] $(\partial X/\partial\Delta)_T \propto \varepsilon_t^{-1.05}$, one finds $D=D_o\varepsilon_t^z$ with $z=0.32 \pm 0.1$ and where $\varepsilon_t \equiv (T-T_t)/T_t$. This is in agreement with recent theoretical predictions by Siggia and Nelson,[12]

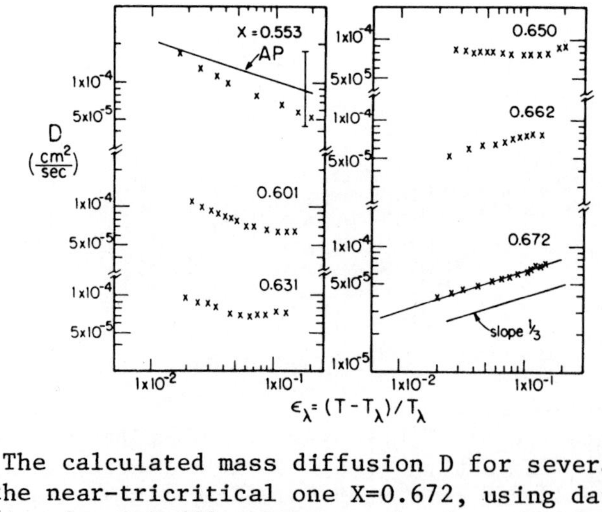

Figure 2. The calculated mass diffusion D for several mixtures, including the near-tricritical one X=0.672, using data at 15 MHz. The solid line for X=0.672 expresses the data derived from scaling all the acoustic data for this mixture.

who calculated the exponent z using the renormalization group approach, and who found z=1/3 from an epsilon expansion scheme.

In an effort to achieve an internally consistent and coherent analysis in terms of mass diffusion, we assume that (2b) constitutes the hydrodynamic approximation of a more general expression that describes α_{sing} in terms of the fluctuation-lifetime spectrum with a characteristic τ, namely

$$\alpha_{sing} = \omega^{1+y} f(\omega\tau) . \tag{3}$$

In the hydrodynamic regime, where $\omega\tau \ll 1$, $f(\omega\tau) \propto (\omega\tau)^{1-y}$ to agree with (2b), while for $\omega\tau \gg 1$, $f(\omega\tau) \to$ constant, because α_{sing} becomes finite. Along the path $X=X_t$, we can expect a power law for the lifetime

$$\tau = \tau_o \, \varepsilon_t^{-x} , \tag{4}$$

where x is a constant. The experimental data at the various frequencies can indeed be represented by this general expression (3) and from a fit one obtains the two exponents

$$y = 0.58 \pm 0.04; \; x = 1.7 \pm 0.15 . \tag{5}$$

Identification is made with the hydrodynamic regime (2b) and the exponents z, x and y are found to be internally consistent. From Siggia and Nelson's prediction[11] we can infer y=3/5 and x=5/3, and there is again good agreement with the experiment.

It is not possible at this time to determine the absolute amplitude τ_o, since the exact expression for $f(\omega\tau)$ is not known. However, we can get an estimate for the <u>variation</u> of τ_o as (T_t, X_t) is approached along various directions in the X-T plane. We shall assume that (3) expresses the data with the same exponents as before and that τ can be written in terms of polar coordinates, namely

$$\tau = \tau_o(\theta) r^{-x} , \tag{6}$$

where r is the "distance" from the tricritical point (see Fig. 3).

For a linear path in the reduced X-T plane, our analysis shows that the curves for $\tau_o(\theta)/\tau_o(0)$ taken for the various mixtures scale reasonably well on a common curve, as evidenced in Fig. 3. The amplitude $\tau_o(\theta)$ is hence extremely anisotropic. On the other hand $D_o(\theta)$, obtained from an analysis in the hydrodynamic regime, is nearly isotropic except in a direction nearly parallel to the lambda line, where D diverges and crossover effects can be expected.

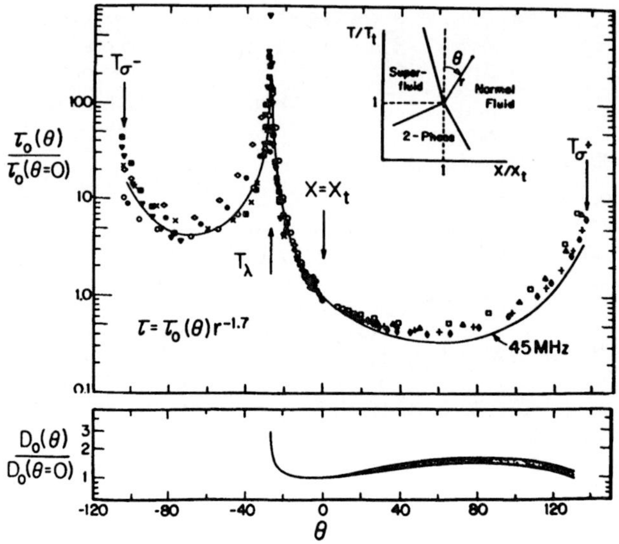

Figure 3. Top: The amplitude ratio $\tau_o(\Theta)/\tau_o(0)$ of the life-time amplitudes in the tricritical region. The experimental points show the data for 15 MHz, the symbols indicating various mixtures. The data for 25 and 35 MHz fall in between those for 15 and 45 MHz. Bottom: Variation of the diffusion amplitude $D_o(\Theta)$.

We now briefly describe an experiment from which we can obtain information on the thermal diffusion ratio K_T in ^3He-^4He mixtures near the tricritical point. In the course of detailed high-resolution sound velocity measurements, we observed that the measured velocity overshoots the eventual equilibrium value as the acoustic cell is cooled (or warmed) from one temperature to the next. The amount of overshoot $\delta U = U(max) - U(equilibrium)$ increases as the superfluid transition is approached. It is largest close to the tricritical point in the superfluid phase. We believe that this overshoot effect is caused by concentration gradients set up in the cell as its temperature is changed. Upon a programmed temperature decrease, the walls of the cell cool first and there will be a temperature gradient δT in the mixture between the wall and the center where the velocity is measured. This gives rise to a concentration gradient δX which causes the apparent velocity to differ from its true average value by $\delta U = (\partial U/\partial X)_T \delta X$. Hence the velocity overshoot is a direct measure of δX between the walls and the center. An analysis of the data as described in Ref. 2 leads to the ratio $(\delta X/\delta T)_{stat}$ under stationary conditions. This ratio in turn leads to the thermodiffusion ratio K_T via the relation, valid in the normal phase

$$\frac{K_T}{T} = \frac{M_4}{M_3}\frac{c^2}{X^2}\left(\frac{\delta X}{\delta T}\right)_{stat} \tag{7}$$

where c is the mass concentration. Our experiment determines $(\delta X/\delta T)_{stat}$ within a multiplicative constant which is the same for all mixtures and which is determined by a fit of the data to calculations using the relation

$$\left(\frac{\delta X}{\delta T}\right)_{stat} = X\left(\frac{\partial X}{\partial \Delta}\right)_{T,P}\left(\frac{\partial(s/X)}{\partial X}\right)_{T,P} \qquad (8)$$

valid in the superfluid phase. Here s is the molar entropy and $(\partial(s/X)/\partial X)_T$ is almost constant in the superfluid phase near T_t. Fig. 4 shows the experimental data for three mixtures and the calculations of (8) extended into the normal phase. The data for x=0.631, normalized to the corresponding curve, determine the free parameter. It can be seen that the fit in the superfluid phase is satisfactory, but there are deviations in the normal phase, where (8) might not be valid.

In the normal phase under stationary conditions, K_T is seen to diverge strongly for $X=X_t$ as the tricritical point is approached. This divergence is somewhat stronger than ε_t^{-1} which is predicted by theory. A more detailed discussion of the results is given in Ref. 2.

In conclusion, acoustic attenuation experiments and transient effects associated with high resolution sound velocity measurements near the tricritical point have led to information on the singularities of certain transport properties such as D and K_T and of the superfluid density fluctuation lifetime τ.

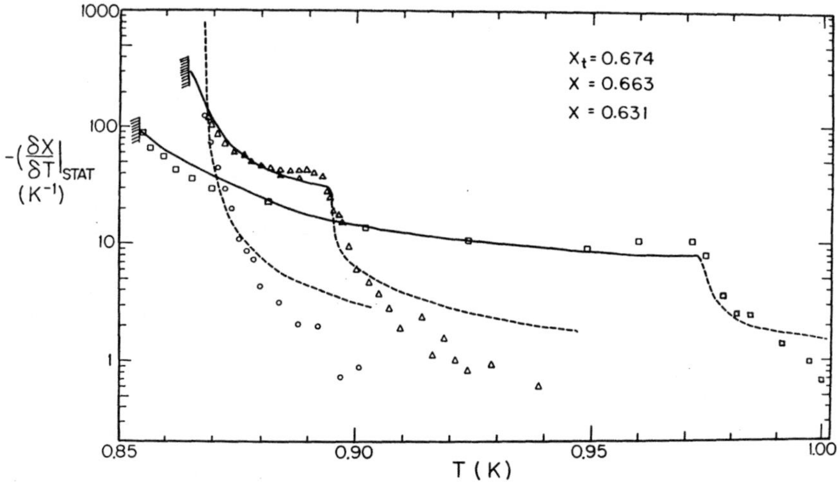

Figure 4. The ratio $(\delta X/\delta T)_{stat}$ for three mixtures, one of them being at the tricritical concentration $X_t=0.674$. The calculated curve is from Eq. 5, which might not be valid in the normal phase, as suggested by the dashes.

References

(1) D. B. Roe, G. Ruppeiner and H. Meyer, J. Low Temp. Phys. 27, (1977).

(2) D. B. Roe and H. Meyer, (to be published).

(3) See for instance, C. W. Garland in Physical Acoustics, edited by W. P. Mason and R. N. Thurston, (Academic Press, New York, 1970) vol. 7, p.

(4) R. D. Williams and I. Rudnick, Phys. Rev. Lett. 25, 276 (1970).

(5) C. Buchal and F. Pobell, Phys. Rev. B 14 (1976).

(6) G. Ahlers and F. Pobell, Phys. Rev. Lett. 32, 144 (1974).

(7) E. K. Riedel, Phys. Rev. Lett. 28, 675 (1972).

(8) A. Ikushima, D. Roe and H. Meyer, in Phonon Scattering in Solids, edited by L. J. Challis, V. W. Rampton and A. F. G. Wyatt (Plenum Publish. Co., New York, 1976) p. 255 and to be published.

(9) K. Kawasaki, Phys. Lett. 31A, 165 (1970).

(10) L. D. Landau and E. M. Lifshitz, Fluid Mechanics, Pergamon Press (1959), Chapter 8.

(11) G. Goellner, R. P. Behringer and H. Meyer, J. Low Temp. Phys. 13, 113 (1973); P. Leiderer, D. R. Watts and W. W. Webb, Phys. Rev. Lett. 33, 483 (1974).

(12) E. D. Siggia and D. R. Nelson, Phys. Rev. B, to appear.

(13) S. T. Islander and W. Zimmermann, Phys. Rev. A 7, 188 (1973).

PHONON-QUASIPARTICLE SCATTERING IN DILUTE $^3He/^4He$ MIXTURES

R. B. Kummer*, V. Narayanamurti and R. C. Dynes

Bell Laboratories
Murray Hill, New Jersey 07974

Abstract: Phonon-^3He quasiparticle scattering times in dilute mixtures of ^3He in liquid ^4He have been determined using the fast-heat-pulse technique. These scattering times were found to be in reasonably good numerical agreement with those theoretically calculated for Rayleigh-like scattering by Baym and Ebner. The scattering is shown to exhibit a strong pressure dependence.

The fast-heat-pulse technique has proven quite useful in the study of elementary thermal excitations in HeII[1] as well as in solid He.[2] From an analysis of the received heat pulse shapes, one can extract the velocities and, in some cases, the interaction times among the excitations. In the present work, this method is applied to dilute mixtures of ^3He in liquid ^4He, where, at low temperatures and pressures, thermal phonons are scattered primarily by ^3He quasiparticles. Measurements were made in the temperature range 0.05-1.5 K with ^3He concentrations from less than 10^{-3} up to about 10^{-2}. At three concentrations, data were taken at pressures of 5, 15, and 24 bar as well as at SVP.

The experimental apparatus and techniques are essentially the same as described in Ref. 1. A constantan film heater and magnetically biased indium film bolometer with extremely fast thermal response times were used. Both the heater and bolometer were $\sim$ 3.5 x 3.5 mm and were separated by a distance of $\sim$ 2.6 mm. This geometry virtually eliminated any boundary scattering effects.

---

*Postdoctoral Research Fellow

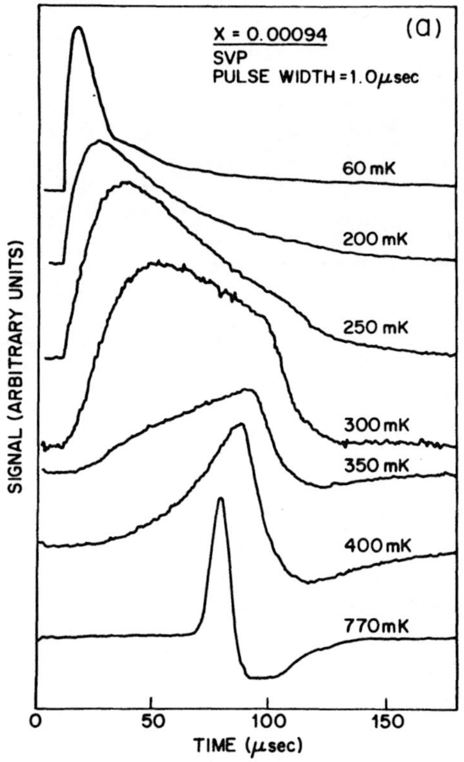

Figure 1 (a) Pulse shapes at SVP showing the transition from essentially ballistic phonon propagation to second sound. ^{3}He concentration X = 0.00094 = $n_3/(n_3+n_4)$, where n_3 and n_4 are the numbers of moles of ^{3}He and ^{4}He, respectively. Cell length = 2.596 mm. The 1.0 μsec input pulse width is five times larger than that used at low temperatures for the pulses which were analyzed quantitatively.

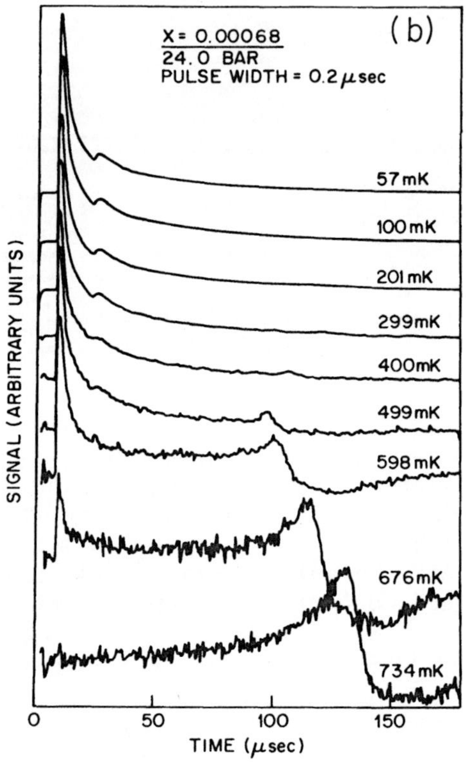

Figure 1 (b) Typical pulse shapes at 24.0 bar showing reduced phonon scattering at high pressures. Ballistic "echo" signal clearly evident up to T $\approx$ 500 mK. X = 0.00068; input pulse width = 0.2 μsec.

Very small heat pulse energies were required in order to avoid self-heating effects. The heater power was usually $\sim$ 1.6 mW/mm^2 and pulse width was typically 0.2 μsec. Voltage pulses from the bolometer were amplified and recorded by A Biomation 8100 transient recorder and then signal averaged with a Nicolet model 1074 signal averager.

Typical pulses at SVP and at 24.0 bar are shown in Figs. 1a and 1b, respectively. At the lowest temperatures in each case the phonon "cloud" produced at the heater travels essentially ballistically to the detector. As the temperature is increased at SVP (Fig. 1a) phonon-^{3}He quasiparticle (p-qp) scattering begins to occur, and the pulse is broadened. At $\sim$ 300 mk p-qp scattering is sufficiently strong that a collective second sound mode begins to develop in the phonon–quasiparticle "gas", and eventually at higher temperatures the heat is transported entirely as second sound. The situation is quite different at higher pressures as shown in Fig. 1b for 24.0 bar. The striking feature of these data is that the ballistic pulse retains its narrow width up to about 700 mK indicating that the p-qp scattering time in this case is approximately larger than the pulse propagation time ($\sim$ 10 μsec) even at this relatively high temperature.

The p-qp scattering times were derived from an analysis of the phonon pulse shapes in the ballistic-second sound transition region. This method utilizes a general dispersion relation for temperature waves derived by Rogers[3] (using hydrodynamic heat flow equations)

$$\frac{1}{3} k^2 v_1{}^2 = \omega^2 + \frac{i\omega}{\tau_R} + \frac{i\omega k^2 \tau (v_1{}^2 - v_2{}^2)}{1 - i\omega\tau} . \tag{1}$$

Here τ_v and τ_R are the normal- and resistive-process scattering times, respectively, and $\tau^{-1} = \tau_N{}^{-1} + \tau_R{}^{-1}$, v_2 is the velocity of phonon second sound, and v_1 is the first sound velocity, k is the wave number and ω the angular frequency of the temperature wave.

For the purposes of the present analysis, the ^{3}He quasiparticles in the mixture are treated as resistive phonon scattering centers. This picture is probably valid in the weak and intermediate scattering regime ($\omega\tau \gtrsim 10^{-1}$), but not in the region of very strong scattering ($\omega\tau \lesssim 10^{-2}$) where heat propagates as a collective mode (second sound) in the entire gas of phonon, quasiparticle, and (at higher temperatures) roton excitations. Thus Eq. 1 was solved in the limit $\tau_R \ll \tau_N$ for the phase velocity $v = \mathrm{Re}[\omega/k)$ as a function of $\omega\tau$. The product $\omega\tau$ was then determined from the peak velocity of each heat pulse in the transition region. The effective "frequency" characterizing each heat pulse was taken to be $\bar{\omega} = 2\pi/2t$, where t is the pulse propagation time.

The final results at SVP are shown in Fig. 2 where we have plotted $\tau(\omega\tau/\bar{\omega})$ multiplied by the ^{3}He concentration (X) versus temperature. (A quantitative analysis of the data at higher concentrations and/or pressures was not possible because the ballistic-second sound transition "window" was too narrow.) The line labeled "T^4" in the figure is the result of a theoretical calculation by Baym and Ebner[4] of the Rayleigh-like scattering of phonons by ^{3}He quasiparticles.[5] While our data fall slightly below the theoretical curve and show a somewhat weaker temperature dependence, the overall numerical agreement is considered good, particularly in light of the fact that there are no adjustable parameters in the theory.

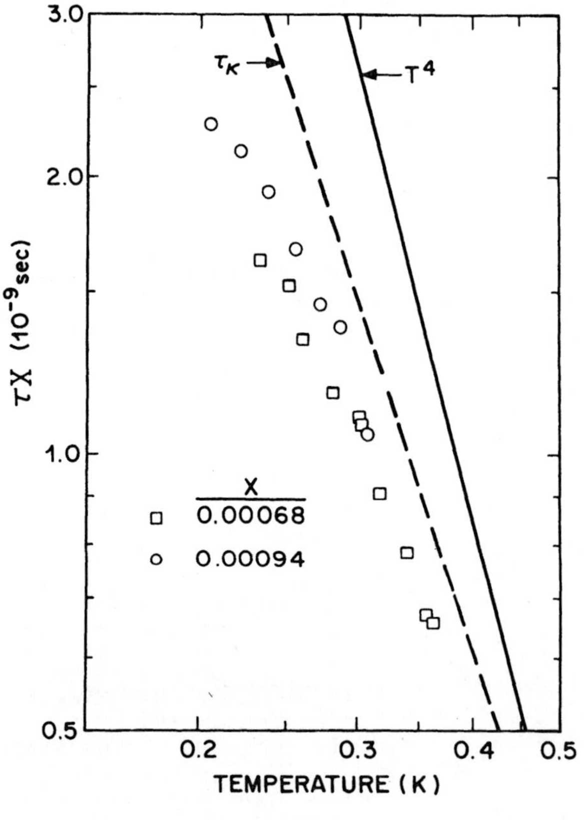

Figure 2. Phonon-quasiparticle scattering time (τ) multiplied by ^{3}He concentration (X) versus temperature. The dashed line labeled "τ_K" represents a least-squares fit to the effective scattering time obtained from the thermal conductivity measurements of Rosenbaum et al.[5] The solid line labeled "T^4" is from the theoretical calculation of the Rayleigh-like scattering time in the dominant phonon approximation by Baym and Ebner.[4]

The dashed line labeled "τ_κ" in Fig. 2 represents the effective phonon scattering time inferred from recent thermal conductivity (κ) measurements in dilute mixtures[6] through the relation $\kappa = 1/3\ c_V v_1{}^2 \tau_p$, where c_V is the phonon specific heat, and τ_p^{-1} is equal to the effective p-qp scattering rate τ_κ^{-1} plus that due to boundaries τ_b^{-1}. Again the agreement with our data, as shown in the figure, is reasonably good.

We will briefly mention a second aspect of the present experiment which is more fully dealt with elsewhere.[7] The second sound velocity in the mixtures was measured up to $\sim$ 1.5 K and compared with calculated velocity curves assuming various models for the ^{3}He quasiparticle excitation spectrum. It was found that neither the Landau-Pomeranchuk spectrum nor the recently proposed "^{3}He roton" model[8] yield acceptable agreement with the experimental results for temperatures $\gtrsim$ 0.7 K.

In conclusion, these data provide strong evidence for Rayleigh-like scattering of phonons by ^{3}He quasiparticles in dilute mixtures as suggested by the Baym and Ebner theory. The second sound velocity data, however, suggest the necessity for more detailed theoretical work on the elementary excitation spectrum, particularly in the region near the ^{4}He roton minimum.

References

(1) V. Narayanamurti, R. C. Dynes, and K. Andres, Phys. Rev. B 11, 2500 (1975).

(2) V. Narayanamurti and R. C. Dynes, Phys. Rev. B 12, 1731 (1975).

(3) S. J. Rogers, Phys. Rev. B 3, 1440 (1971).

(4) G. Baym and C. Ebner, Phys. Rev. 164, 235 (1967).

(5) As pointed out in Ref. 6, the phonon absorption rate proposed by Baym and Ebner is expected to be negligible in the temperature and concentration range considered here.

(6) R. L. Rosenbaum, J. Landau, and Y. Eckstein, J. Low Temp. Phys. 16, 131 (1974).

(7) R. B. Kummer, V. Narayanamurti, and R. C. Dynes, submitted to Phys. Rev. B.

(8) C. M. Varma, Phys. Lett. A 45, 301 (1973); M. J. Stephen and L. Mittag, Phys. Rev. Lett. 31, 923 (1973).

THE EFFECTIVE INTERACTION BETWEEN ^{3}He QUASI-PARTICLES ON THE SURFACE OF LIQUID ^{4}He*

D. O. Edwards, J. D. Feder and V. S. Nayak

The Ohio State University
Columbus, Ohio 43210

Measurements of surface tension and of the velocity of surface sound[1] have shown that less than a monolayer of ^{3}He adsorbed on the surface of superfluid ^{4}He behaves like a two-dimensional gas of ^{3}He 'quasiparticles' with an effective mass M about 1.3 times the real mass m_3. The interaction potential between the surface quasiparticles has been found to be quite weak. When interpreted in terms of an 'effective interaction' $V^s(q)$, the analog of the BBP[2] interaction V(q) between ^{3}He dissolved in bulk ^{4}He, the data give[1]

$$V^s(q=0) = +(0.5\pm2.0) \times 10^{-31} \text{erg cm}^2. \qquad (1)$$

The sign and magnitude of $V^s(q)$ are important because they determine whether the adsorbed ^{3}He will become a two-dimensional BCS superfluid at a sufficiently low temperature. In this paper we use some naive theoretical arguments to augment our information about $V^s(q)$ and the possibility of superfluidity.

CALCULATION OF $V^s(0)$

For bulk liquid helium, Baym,[3] using first order perturbation theory, showed that exchanging one or two ^{3}He atoms (of opposite spin) for ^{4}He at constant total number density gives an energy linear in the number of ^{3}He: $E(N_3) = NE_4 + N_3\delta E_1$. On expanding

*Supported by NSF Grant Number DMR75-19546-A01.

the liquid to restore the pressure to the original ^{4}He ground-state value, he obtained a term in the energy proportional to N_3^2 from which he deduced

$$V(0) = -\alpha^2 m_4 s^2/n_4 \,, \tag{2}$$

where s is the velocity of sound, n_4 the number density in the ^{4}He ground state and $(1+\alpha)$ the ratio between the ^{3}He and ^{4}He partial volumes.

If we consider the analogous problem for the surface where the ^{3}He atoms, which are exchanged for ^{4}He, are localized and bound to the (flat) surface of the liquid, the pressure and chemical potential μ_4 remain constant during the exchange and there is no term in $V^s(0)$ analogous to $-\alpha^2 m_4 s^2/n_4$. However, in the initial perturbation step of the calculation, the energy change on substituting ^{3}He is not linear in N_3 (in the same order of approximation as that used by Baym).

To demonstrate this we use the Feynman-Lekner variational method.[4] The wave-function of N ^{4}He atoms, in which either atom 1 or atoms 1 and 2 are replaced by ^{3}He is chosen to be $f_1\phi$ or $f_{12}\phi$, where $\phi = \phi(\vec{r}_1 \ldots \vec{r}_N)$ is the ground state of pure ^{4}He and $f_1 = f(\vec{r}_1)$ and $f_{12} = f(\vec{r}_1, \vec{r}_2)$ are functions to be varied. In both cases the condition that the expectation value of the energy be a minimum can be written in the form of a Schrodinger equation, with the one- or two-body 'wavefunctions'

$$\psi_1 \equiv \sqrt{\rho_1} f_1; \qquad \psi_{12} \equiv \sqrt{\rho_{12}} f_{12} \tag{3}$$

where ρ_1 and ρ_{12} are the one- or two-particle densities in the ^{4}He ground state:

$$\rho_1 \equiv N \int \phi^2 d\vec{r}_2 \ldots d\vec{r}_N; \qquad \rho_{12} \equiv N(N-1) \int \phi^2 d\vec{r}_3 \ldots d\vec{r}_N \tag{4}$$

The 'wavefunctions' in (3) have the property that the probability density for atom 1 or for atoms 1 and 2 is propotional to $|\psi_1|^2$ or $|\psi_{12}|^2$. The Hamiltonian in the one-particle Schrodinger equation contains the potential energy function $\hbar^2 U(\vec{r}_1)/2m_3$ where

$$U(\vec{r}_1) = \nu_1 + \frac{1}{\sqrt{\rho_1}} \nabla_1^2 \sqrt{\rho_1} \tag{5}$$

and $\nu_1 \equiv \nu(\vec{r}_1)$ is proportional to the kinetic energy in the ^{4}He ground state:

$$\rho_1\nu_1 \equiv -1/4N\int\phi\nabla_1^2\phi d\vec{r}_2\ldots d\vec{r}_N \tag{6}$$

It has been shown in several calculations[5] that this potential, which has a minimum near the surface, has a bound state which is in good agreement with the binding energy for surface ^{3}He. In the two-body 'Hamiltonian' the potential is

$$U_{12} = \sum_{i=1}^{2} [\nu_{12}^i + \frac{1}{\sqrt{\rho_{12}}} \nabla_i^2 \sqrt{\rho_{12}}] \tag{7}$$

where

$$\rho_{12}\nu_{12}^i \equiv -1/4N(N-1)\int\phi\nabla_i^2\phi d\vec{r}_3\ldots d\vec{r}_N . \tag{8}$$

When $\vec{r}_1$ and $\vec{r}_2$ are far apart, $U_{12} \to U_1 + U_2$ so that $(U_{12}-U_1-U_2)$ can be thought of as the interaction potential between two ^{3}He quasiparticles. At extremely small distances it is the same as the bare interatomic potential.

To get Baym's result in 3D or $V^s(0)$ in 2D we calculate the expectation value of the energy using the two-body 'Hamiltonian' and the approximate wavefunction:

$$f_{12} = f_1f_2; \qquad \psi_{12} = \sqrt{\rho_{12}}f_1f_2 = (\rho_{12}/\rho_1\rho_2)^{\frac{1}{2}}\psi_1\psi_2 \tag{9}$$

where $\psi = f\sqrt{\rho}$ is the solution to the one-body Schrodinger equation. In bulk helium (no surface), $U_1 = \nu_1 = \text{constant} - (2m_3/\hbar^2)(\frac{1}{4}T_4)$ where T_4 is the KE per atom in the ^{4}He ground state. The one-body function is $f_1 = \text{constant}$, and the approximate two-body wavefunction (9) gives

$$E(N_3) = NE_4 + \tfrac{1}{4}N_3T_4 \tag{10}$$

which is Baym's result. For ^{3}He localized on the surface we find

$$V^s(0) = A[E(2) - 2E(1)] = A_o h^2/2m_3 \tag{11}$$

where A is the area of the surface and the dimensionless quantity A_o is given by[6]

$$A_o = A^{-1}\int d\vec{r}_1 d\vec{r}_2 \psi_1^2\psi_2^2 g_{12} \sum_{i=1}^{2}[\nu_{12}^i - \nu_i - (\frac{\nabla_i g_{12}}{g_{12}})(\frac{\nabla_i f_i}{f_i})] \tag{12}$$

and where we have introduced $g_{12} = \rho_{12}/\rho_1\rho_2$ and normalized ψ:

$$\int\psi^2 dz = A^{-1}\int\psi^2 d\vec{r} = 1. \qquad (13)$$

Without complicated numerical calculations, we cannot deduce even the sign of A_o from equation (12), which in contrast to (2), the equation for bulk helium, allows the possibility of a repulsive interaction. This is in agreement with the tentative result from experiment (1), which can be expressed as

$$A_o = +(0.45 \pm 1.8) \ . \qquad (14)$$

BACKFLOW AND SURFACE TENSION EFFECTS

To the result for $V^s(0)$ in the previous section must be added the effect of the ^{4}He backflow caused by the motion of the ^{3}He quasiparticles. For ^{3}He in the bulk of the liquid this has been done[2,7] quite successfully by treating the quasiparticles as tiny spheres of volume ω whose dipolar velocity fields give rise to a hydrodynamic enhancement of the mass, $m-m_3 = \frac{1}{2}\omega n_4 m_4$, and a hydrodynamic dipole-dipole interaction which depends on the velocities of the interacting quasiparticles. In the surface problem we can estimate the dipole interaction by assuming the ^{3}He displace hemispheres[8] of superfluid of volume ω_s such that

$$M - m_3 = \tfrac{1}{2}\omega^s n_4 m_4 \qquad (15)$$

The empirical value of M gives $2\omega_s \simeq n_4^{-1} = v_4$, the volume of a ^{4}He. The corresponding dipolar interaction has a matrix element between two-quasiparticle states with momenta $\vec{p}_1 - \vec{q}$, $\vec{p}_2 + \vec{q}$ and $\vec{p}_1$, $\vec{p}_2$ (See Ref. 7 for a fuller explanation of the method and notation) which is

$$A^s_{\uparrow\downarrow}(\vec{p}_1,\vec{p}_2,\vec{q}) \equiv V^s_d(\vec{p}_1,\vec{p}_2,\vec{q})/A = -(A^s_d/An_4m_4)(\vec{p}_1\cdot\hat{q})(\vec{p}_2\cdot\hat{q})$$

$$\times\ (\tfrac{1}{2}q/\hbar)\,\zeta(bq/\hbar) \qquad (16)$$

The dipolar contribution to $V^s(\vec{q})$, which we have called V^s_d in (16), depends on the momenta of the quasiparticles $\vec{p}_1$ and $\vec{p}_2$ as well as on the momentum transfer $\vec{q}$. The number A^s_d is given by

$$A^s_d = \tfrac{1}{2}(3n_4m_4\omega^s/M)^2 \sim 1.1, \qquad (17)$$

The length b is related to the width of the surface ^{3}He ground-state wave-function ψ, and the function $\zeta(bq/\pi)=1$ for $bq/\hbar \to 0$ and decreases slowly and monotonically to zero as $bq/\hbar$ becomes large.

The ^{4}He backflow also produces a distortion of the liquid surface which gives a surface-energy contribution both to the energy of a single quasiparticle and to the effective interaction. These effects are proportional to high powers of quasiparticle momenta and we have neglected them. To the same approximation we have treated the surface as a rigid boundary in estimating the dipolar part of the interaction.

DISCUSSION - SUPERFLUIDITY

If the two interacting surface ^{3}He have momenta on the edge of a degenerate 'Fermi circle,' then $p_1 = p_2 = p_F$. Substituting $\vec{p}_1\cdot\vec{q} = -\vec{p}_2\cdot\vec{q} = \frac{1}{2}q$ and adding $V^s(0)$ and V^s_d we obtain the 'local' interaction on the Fermi edge:

$$V^s(q) = (\hbar^2/2m_3)\ [A^s_o + 1/8A^s_d(m_3/n_4m_4)(q/\hbar)^3\zeta(qb/\hbar)] \qquad (18)$$

where $q \sim 2p_F\sin\frac{1}{2}\theta$, and θ is the angle between $\vec{\rho}_1$ and $\vec{\rho}_2$. To discuss the possibility of superfluidity we follow the method used by BBP[2] and others[7] in 3 dimensions and assume a 'cut-off' energy in BCS theory proportional to the Fermi energy, say $ck_BT_F(= c\hbar^2N^s/M$ where N^s is the surface ^{3}He number density). According to Patton and Zaringhalam[9] $c \sim 0.05$ in 3D. Expanding (18) in $\cos\ell\theta$, the eigenfunctions of the pair angular momentum $\hbar\ell$ in 2 dimensions:

$$V^s(q) = \tfrac{1}{2}V_o + \sum_{=1}^{\infty} V_\ell\cos(\ell\theta), \qquad (19)$$

we find that V_ℓ is <u>negative</u> for $\ell=1$. In the BCS approximation in 2D the transition temperature for $\ell=1$ pairing is

$$T_c = cT_F\exp(-4\pi\hbar^2/M|V_1|). \qquad (20)$$

It is unlikely that BCS theory is correct in 2D because of the effect of fluctuations which will smear out the transition, but it is plausible that (20) gives the temperature range in which superfluidity may be observed and that a state resembling the A-phase of bulk ^{3}He with the pair angular momentum perpendicular to the surface will occur at very low temperatures.

To estimate T_c we have taken $c = 0.05$ and $|\psi|^2$ to be Gaussian with a standard deviation $b = 1.1\text{Å}$. This is a fit to the wave function given by Saam.[5] If $M = 1.33m_3$, then from (15) and (17), $A_d^s \simeq 1.1$ and for 3/4 of a layer, $N_s = 4.8 \times 10^{14}\text{cm}^{-2}$, we have $T_F = 1.8\text{K}$ and $T_c \sim 0.03\text{mK}$. This value is, of course, very sensitive to the input parameters. For instance, if $M = 1.39m_3$ then $A_d \simeq 1.4$ and $T_c \sim 0.3\text{mK}$ for 3/4 of a layer, or 2mK for a full layer.

Although the numerical estimates of T_c should not be taken seriously, the important point is that we have a specific model for the effective interaction which can be compared to experiments, and that the model is favorable for $\ell=1$ pairing.

We would like to acknowledge the assistance given by Professor W. F. Saam at the beginning of this work, and some useful discussions with Professor C. Ebner and Dr. W. Gully.

References

(1) D. O. Edwards, S. Y. Shen, J. R. Eckardt, P. P. Fatouros and F. M. Gasparini, Phys. Rev. B 12, 892 (1975).

(2) J. Bardeen, G. Baym and D. Pines, Phys. Rev. 156, 207 (1967).

(3) G. Baym, Phys. Rev. Lett. 17, 952 (1966).

(4) J. Lekner, Phil. Mag. 22, 669 (1970).

(5) e.g. W. F. Saam, Phys. Rev. A 4, 1278 (1971); C. C. Chang and M. Cohen, Phys. Rev. A 8, 3131 (1973).

(6) This formula has also been obtained independently by W. F. Saam (private communication).

(7) C. Ebner and D. O. Edwards, Physics Reports 2 C, 77 (1971).

(8) The assumption that the ^{3}He displace hemispheres is not strictly necessary, as we shall show in a more extended discussion to be submitted to Physical Review.

(9) B. R. Patton and A. Zaringhalam, Physics Letters A 55, 95 (1975).

OBSERVATION OF BALLISTIC ROTONS IN LIQUID ^{4}He*

D. O. Edwards, G. G. Ihas and C. P. Tam

Physics Department, The Ohio State University

Columbus, Ohio 43210

The excitations produced when a pulsed ^{4}He atomic beam strikes the surface of liquid ^{4}He held near T $\sim$ 0K have been measured as a function of direction in the liquid and time elapsed since the beam was generated. Comparison of the data with a theory which assumes that a single atom is converted into a single excitation indicates that the observed signals are due to ballistic phonons and rotons, although the experimental velocity and angular distributions of the excitations are displaced from the theoretical ones. The direction of the displacements suggests that multiple ripplon production occurs as an atom approaches the liquid surface, as predicted by Echenique and Pendry.

I. INTRODUCTION

The scattering of a beam of ^{4}He atoms from a liquid ^{4}He surface has recently been studied.[1] For an incident beam with a nearly Maxwellian velocity distribution corresponding to a temperature of $\sim$ 0.6K and a liquid surface at $\sim$ 0K it was found that
(a) more than 95% of the atoms condense (are not scattered),
(b) the remaining atoms are elastically scattered, with a probability $R(k,\theta)$ which depends only on the vertical momentum of the atom $\hbar k\cos\theta$, and (c) there is no discontinuity in $R(k,\theta)$ at the 'roton threshold' - the minimum atomic kinetic energy which, added

* Supported by a grant from the National Science Foundation, Number DMR 75-19546.

to the latent heat, is sufficient to produce a roton - indicating that multiple production of excitations by a single atom must occur. In a later experiment,[2] the reflection coefficient for ^{3}He atoms was measured and found to have a simple relationship to that for ^{4}He. This is in agreement with the idea that the van der Waals potential outside the liquid is the dominating factor in determining the reflection probability. A calculation by Echnique and Pendry[3] has indeed shown that the van der Waals interaction with the liquid allows the incoming atom to lose energy by the production of low energy ripplons before it reaches the surface.

The present experiment[4] was performed to discover what becomes of the atom and the rest of its energy after being captured by the liquid. It indicates that a large amount of this energy is converted into single phonons or rotons with transverse momentum being nearly conserved, and that these phonons and rotons propagate ballistically through the liquid. Although a propagating collective 'roton pulse' has been observed by Dynes et al.[5] and Balibar,[6] this experiment and later unpublished work by Balibar[7] are the first observations of the ballistic propagation of single rotons through liquid ^{4}He.

II. APPARATUS

The scattering apparatus used has been described elsewhere.[8] Briefly, it consists of two flat carbon transducers (8 mm x 8 mm), each mounted on a 40 mm long arm which can be rotated about a common pivot point using its own superconducting stepping motor. This system is contained in a can which is filled to the pivot with liquid ^{4}He and cooled by a dilution refrigerator to $\sim$ 39 mK.

In the present experiment, one transducer is used to generate the atomic beam in the vacuum above the liquid by being pulsed electrically. Each pulse of 0.95 ergs evaporates a small fraction of the ^{4}He film which covers the transducer. Measurements have shown that the resulting beam is nearly Maxwellian ($T_M \approx 0.8$ K) and has an angular distribution of approximately $(\text{cosine})^{5/2}$. The other transducer is used to detect the excitations in the liquid by monitoring the power required to maintain it at a constant temperature ($\sim$ 80 mK). The generator is pulsed several times a second and the detected signal averaged over many thousands of pulses. Both the atomic beam and the resultant beam of excitations are collimated by a window (11 mm wide by 15 mm high and centered at the pivot) in a screen which separates the generator from the detector.

III. CALCULATION

The expected signal was calculated by assuming that each atom directly converts into a single phonon or roton (if possible), conserving both energy and transverse momentum, that the excitation passes to the detector without scattering or decomposing, and that the detector is a perfect absorber. This was done for various combinations of generator and detector angles and flight times using the known atomic beam composition, liquid excitation spectrum, and apparatus geometry. The results for the generator at 42° to the liquid normal and a total flight time of 0.8 msec are shown in the figure as a function of detector position.

The dashed curve assumes all the atoms are converted into phonons; the full curve assumes conversion into rotons on the high-momentum side of the minimum in the excitation curve (where the velocity is parallel to the momentum); the dotted curve corresponds to 'anomalous' rotons, on the left of the minimum where the momentum is in the opposite direction to the group velocity. The calculations of the roton curves ignore those atoms which have insufficient energy to produce rotons. The low-angle parts of the dashed and solid theoretical curves are due to 'leakage' of excitations around the bottom of the screen. In principle the theory should have taken into account the decay of energetic phonons due to the 3-phonon process, caused by the upward curvature of the dispersion curve. The resultant broadening of the phonon beam is hard to estimate with presently available data.

IV. RESULTS

Due to multiple scattering of excitations from the walls and screens of the apparatus caused by a large Kapitza resistance, the data contain a diffusely scattered signal which begins at 0.6 msec and reaches a broad plateau after 3 msec, corresponding to an increase in bath temperature of 0.8 mK. The delay of 0.6 msec agrees with the combined time-of-flight of the fastest atoms in the beam plus that of phonons with one wall reflection. The comparison with theory shown in Fig. 1 was done at a total flight time $t=0.8$ msec and generator angle $\Theta=42^o$, minimizing the effect of the diffuse scattering while maintaining a good signal-to-noise ratio. The following features are also revealed by other values of Θ and t: (1) The data clearly show the broad phonon and narrower roton distributions expected from single atom-excitation conversion. (2) The signal is $\sim 10^3$ smaller than the calculated power, due to the Kapitza resistance of the receiver, which means that only $\sim 10^{-3}$ of the incident excitations are absorbed. (3) Both angular distributions are shifted to a slightly larger angle than predicted by theory. This is exactly what is expected

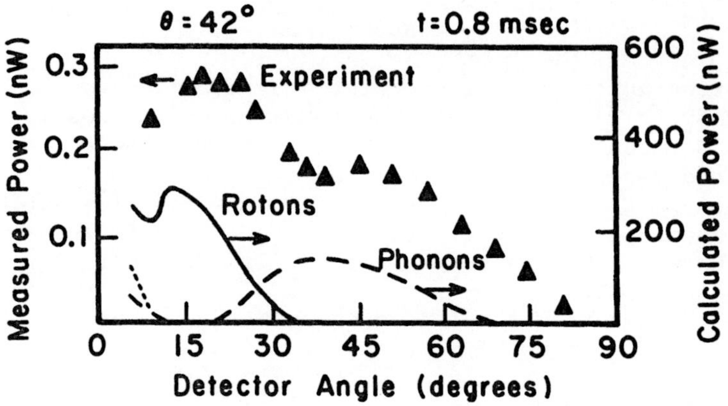

Figure 1. Detector power (nanowatts) for a total time-of-flight of 0.8 msec and the beam generator at 42°, plotted against the detector angle. Solid circles (left scale) are experimental points. The calculated curves assume single atom-excitation conversion, conserving energy and transverse momentum, for (dashed curve) phonons and (solid curve) rotons on the high momentum side of the roton minimum. The dotted line is for conversion into 'anomalous' rotons, whose momenta are below the roton minimum and antiparallel to their group velocity.

if the incoming atom loses substantial amounts of energy but not transverse momentum through the production of ripplons.

The data analysis has not been carried further, both because of the theoretical difficulties and the adverse effects of the diffuse background in the experiment. However, it is already clear that atoms of sufficient energy do produce rotons in the liquid, despite the loss of energy by ripplon production during their approach to the surface, and that these rotons do traverse large distances in the liquid.

References

(1) D. O. Edwards, P. Fatouros, G. G. Ihas, P. Mrozinski, S. Y. Shen, F. M. Gasparini and C. P. Tam, Phys. Rev. Lett. 34, 1153 (1975).

(2) D. O. Edwards, P. P. Fatouros and G. G. Ihas, Phys. Lett. 59A, 131 (1976).

(3) P. M. Echenique and J. B. Pendry, Phys. Rev. Lett. 37, 561 (1976).

(4) A more complete description of our work has been submitted to the Physical Review.

(5) R. C. Dynes, V. Narayanamurti and K. Andres, Phys. Rev. Lett. 30, 1129 (1973).

(6) S. Balibar, Phys. Lett. 51A, 455 (1975).

(7) S. Balibar, Doctoral Dissertation, University of Paris (Laboratoire de Physique de l'Ecole Normale Supérieure) 1976.

(8) J. R. Eckardt, D. O. Edwards, F. M. Gasparini and S. Y. Shen, Proc. of 13th Intern. Conf. on Low Temp. Physics, ed. Timmerhaus, O'Sullivan and Hammel, (Plenum Press, New York, 1974), p. 518.

SECOND SOUND ATTENUATION IN ROTATING ^{4}He

K. N. Zinov'eva

Institute for Physical Problems, Academy of Sciences of the USSR, Moscow

S. Y. Tsakadze

Institute of Physics, Georgian Academy of Sciences, Tbilisi

Abstract: Experiments on second sound in ^{4}He in an axial mode resonator at T = 1.53 K were performed in the presence of steady supracritical heat flow ($\sim 10^{-3}$ w/cm^2). It is shown that there is an additional attenuation of second sound on bended vortex lines. After rotation is stopped this attenuation grows because of random distribution of vortices in heat flow.

It is known that second sound propagating in superfluid He in the presence of heat flow shows an extra attenuation added to the usual one caused by the viscosity and heat conductivity of the normal components.[1,2,3,4,5]

This additional attenuation is a threshold phenomenon and it arises starting from some critical heat flow density which is very small ($\sim 10^{-3}$ w/cm^2). At small heat flow densities this attenuation is linear with heat power.[2,3]

An additional attenuation of second sound is also seen in rotating He. It was shown in the works of Hall and Vinen[4] that second sound propagating transversally to the rotation axis (radial mode resonator) has an additional dissipation due to heat excitations scattering on normal cores of the vortex lines. This attenuation grows linearly with rotation speed or with the density of vortices. On the contrary, the propagation of second sound in direction of vortex lines is not accompanied by any extra dissipation.

The additional attenuation due to heat flow and vortex lines are of the same order of magnitude as the comparison shows. If one also takes into account that the character of attenuation dependence on heat flow density and density of vortices is similar, then one comes by the idea that the physical reason for both attenuations is the same.

To check this suggestion we studied the propagation of second sound in a rotating axial mode resonator at presence of a heat flow. A stainless steel resonator 90 mm long, 15 mm inner diameter, with 0.3 mm wall thickness, 1.5 mm thick bottom and cap were machined from stainless steel. The transducer and receiver, flat bifillar spirals, were glued to the bottom and cap of the resonator. The transducer was made of 50 μ constantan wire and had resistance of 300 Ω. The receiver (thermometer) was made of 40 μ phosphor bronze wire with resistance of 40 Ω. The cap and bottom were hard soldered to the body of the resonator. Leads were led into the resonator by platinum-glass tightening. The resonator was filled with helium through a 4 cm long, 0.3 mm inner diameter capillary, soldered into cap. A wire was pulled tightly into the capillary to make weaker contact with the outer volume. The Q-factor of the resonator was 200.

Second sound was generated and detected by usual thermal method.[1,2,3] The experiments were run at a second sound frequency 224 Hz, temperature of 1.52 K and rotation speeds up to 1.6 $sec.^{-1}$ The support with resonator was rotated by a motor through a stuffing-box.

First experiments showed that the resonant amplitude of the second sound noticeably grows at rotation. The effect was strongly pronounced at a bigger power irradiated in transducer and thermometer. Figs. 1 and 2 show the variation with time of the second sound resonant amplitude in immovable and rotating resonator at different heat densities of transducers and thermometer. We supposed that this effect could be explained by the quantum turbulence of heat flow. In order to check this we studied the dependence of the second sound attenuation in rotating He on magnitude of steady heat flow density.

The experiment was performed in the following way (Fig. 3a,b). At the beginning a power of 10^{-4} w/cm^2 was put into the receiver and transducer to generate and detect second sound (Fig. 3 a). At moment t_1 rotation of resonator at speed 1 sec^{-1} was started. No change in the second sound amplitude was seen. The rotation was stopped at t_2. Then at t_3 a steady heat flow with density 3.6 x 10^{-3} w/cm^2 was switched on. This led to a considerable fall of the second sound amplitude in the immovable resonator. The rotation of the resonator which was started at t_4 was followed by an increase of the second sound amplitude. When the rotation was stopped again

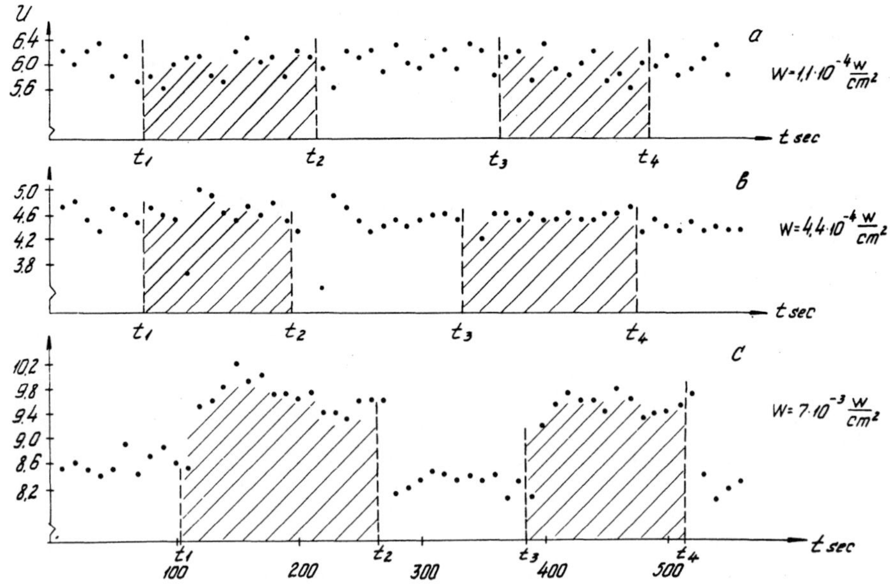

Figure 1. The variation with time of the resonant amplitude of second sound in immovable and rotating resonator (shaded intervals) for 3 heat flow densities. Heat flow is generated by transducer (heat flow density due to thermometer is 1.3×10^{-5} w/cm^2). Rotating speed $\omega = 1.05$ sec^{-1}.

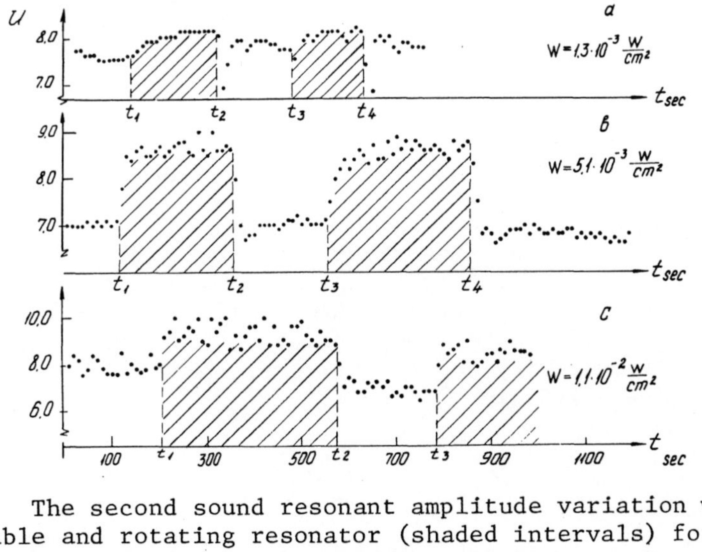

Figure 2. The second sound resonant amplitude variation with time in immovable and rotating resonator (shaded intervals) for 3 densities of heat flow generated by power dissipation in thermometer (heat flow density due to transducer is 4.4×10^{-7} w/cm^2). Rotating speed $\omega = 1.05$ sec^{-1}.

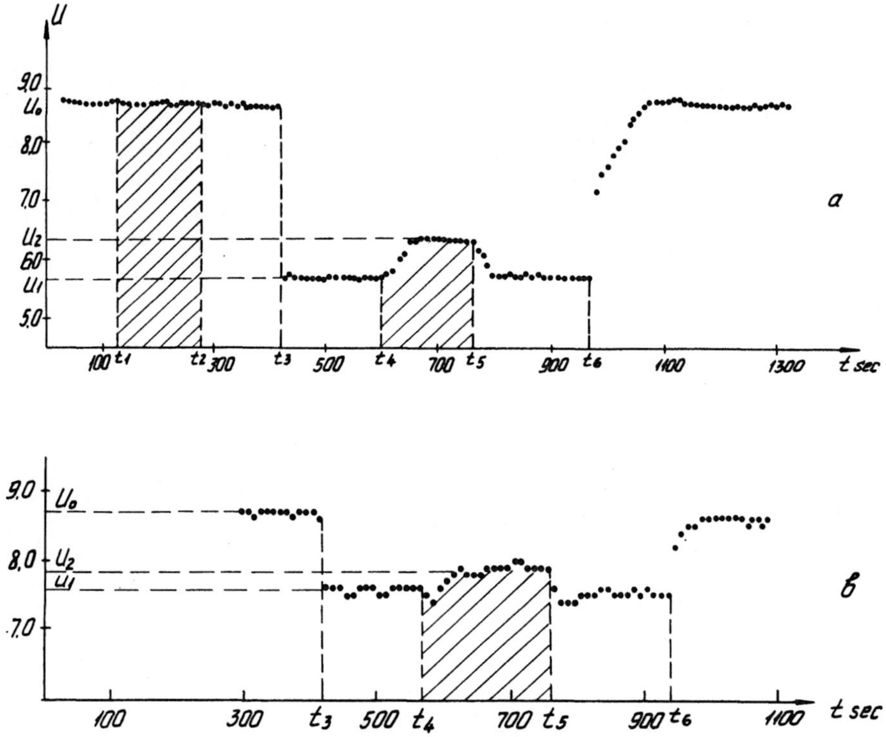

Figure 3 a,b. The second sound resonator amplitude variation with time in immovable and rotating resonator (shaded intervals), when additional axial steady heat flux is on (a) 3.6×10^{-3} w/cm^2, (b) 0.9×10^{-3} w/cm^2. Heat flow density due to transducer 2×10^{-4} w/cm^2. Heat flow density due to thermometer 1.3×10^{-5} w/cm^2. Rotating speed $\omega = 1.05$ sec^{-1}.

(t_5) and heat flow switched off (t_6), the second sound amplitude returned to the starting value. In Fig. 3b the amplitude of the time dependence for heat flow density of 0.9 x 10^{-3} w/cm^2 is shown.

The similar dependences were seen at other temperatures and frequencies, but at the low frequency the effect was stronger.

One should pay attention to the following: when the power was put on (at t_3) the second sound amplitude fell down rather quickly (< 10 sec), but after the power was switched off (at t_6) one could see comparably long relaxation ($\sim$ 100 sec). The beginning and end of rotation are followed by relaxation processes of approximately equal duration ($\sim$ 50 sec).

The study of the effect dependence on supplied power showed that the second sound "sees" rotation starting from heat flow density of approximately 10^{-3} w/cm^2 (0.6 + 0.8 x 10^{-3} w/cm^2) independently of where this heat flow was generated - on the side of transducer or receiver. When the heat flow grows, the relative change of the second sound amplitude $\Delta U/U$ (where $\Delta U = U_2 - U_1$; $U = U_1$) grows too.

For example, for $W = 10^{-2}$ w/cm^2 $\Delta U/U$ = 20% at ω = 1.05 sec^{-1}. Fig. 4 shows relative second sound resonant amplitude growth as a function of heat flow density. However, total attenuation in

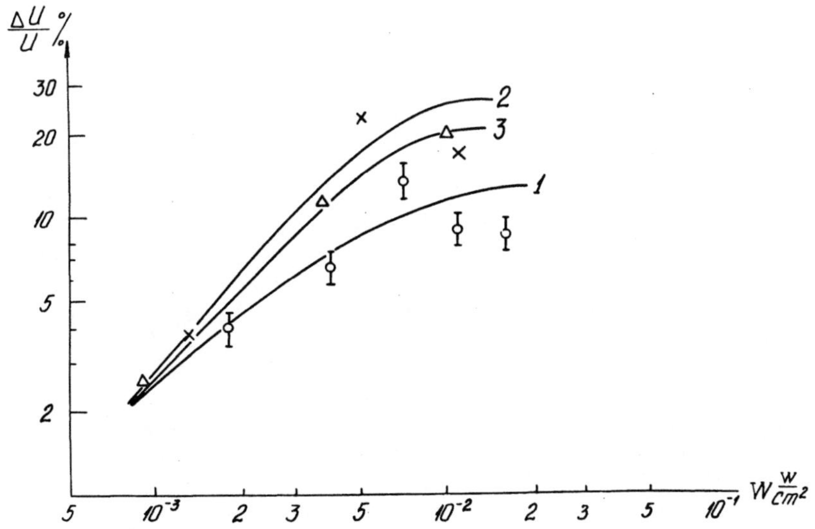

Figure 4. Relative second sound resonant amplitude growth as a function of heat flow density, caused by: transducer (curve 1, Φ), thermometer (curve 2-X) and putting on additional steady heat flow (curve 3-Δ). Rotation speed = 1.05 sec^{-1}.

steady liquid $U_o - U_1/U_o$ grows simultaneously with respect to subcritical regime (U_o).

In Fig. 5 the dependence of the relative growth of the second sound amplitude on rotation speed at heat flow density of 7×10^{-3} w/cm^2 is shown. Thus with the presence of heat flow in an axial mode resonator, the second sound attenuation grows both in immovable and in rotating resonator, but the attenuation in rotating resonator is always less than in the immovable one. The second sound velocity stays unchanged to the accuracy of 1%.

To explain attenuation in an immovable liquid we shall compare our results with the previous ones of 3. It is clear that in both cases we deal with setting of critical regime in a counter flow of normal and superfluid components when vortices are generated. In our previous work[3] we got for the critical velocity:

$$v_{sc} = 5 \times 10^{-3} \text{ cm/sec } (d = 5.3 \text{ mm } T = 1.27 \text{ K})$$

in these experiments

$$v_{sc} = 3 \times 10^{-3} \text{ cm/sec } (d = 15 \text{ mm } T = 1.52 \text{ K})$$

(d - resonator diameter).

These results are in satisfactory agreement with the results of other experiments and with a theoretical curve of Feynman.

Thus we come to the conclusion that the heat flow more intensive than 10^{-3} w/cm^2 turbulizes immovable liquid helium II by generating vortices. Randomly oriented vortices make the medium turbid. This explains the additional attenuation in supercritical regime. Rotation decreases the second sound attenuation by partial involving of turbulent vortices in regular motion, the medium becomes "transparent".

It is worth noting that in uniformly rotating He II, as was shown in Ref. 6, a heat flow of big enough value can bend and even destroy a regular vortex lattice. The instability follows from the fact that any disturbance of a vortex in the presence of the supracritical axial heat flow infinitely grows.

In that work,[6] a theoretical treatment of the problem was done. Estimation of critical heat flow density gives a value 10^{-3} w/cm^2. It also gives that at this heat flow at temperature of ~ 1 K the deformation of a vortex line is about 1 cm, which inevitably should

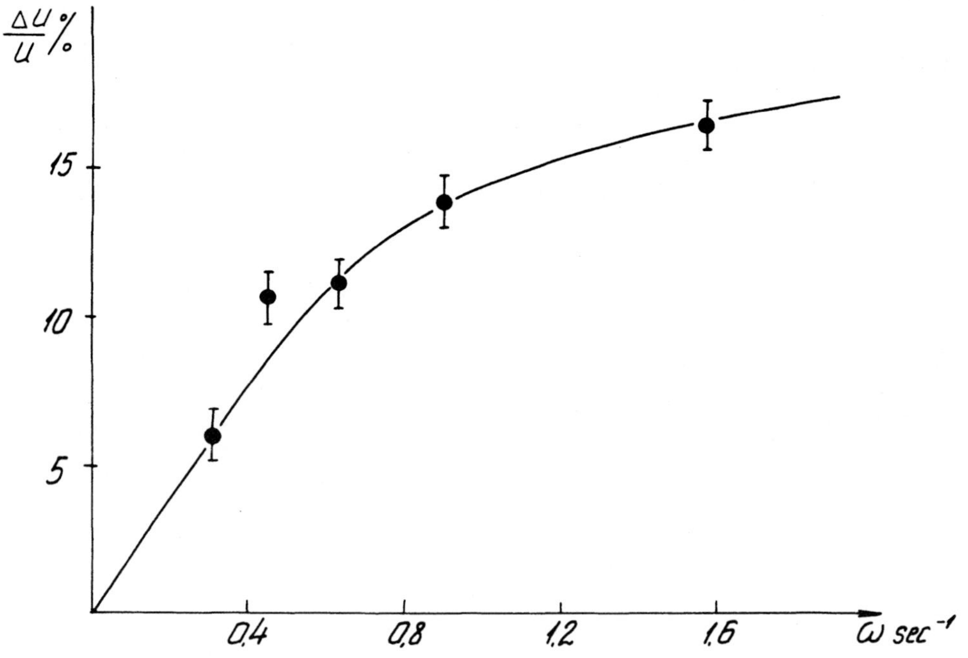

Figure 5. Relative second sound resonant amplitude growth as a function of rotating speed at heat flow density from transducer 7×10^{-3} w/cm^2. Heat flow density from thermometer 1.3×10^{-5} w/cm^2.

cause second sound attenuation.

Thus the second sound attenuation in supracritical regime when second sound propagates along the rotation axis, could be explained by bending and even destruction of the vortices. After the rotation is stopped the vortex lines do not disappear but gradually lose regularity, randomly distribute in volume and become completely enmeshed.

We are greatly indebted to Profs. V. P. Peshkov and E. L. Andronikashvili for fruitful discussions and to Yu. G. Mamaladze and D. S. Tsakadze for helpful advice. In addition, we are greatly indebted to Prof. W. Glaberson for a very interesting discussion.

References

(1) V. P. Peshkov, JETP, 16, 1000 (1946).

(2) K. N. Zinov'eva, JETP, 25, 235 (1953).

(3) K. N. Zinov'eva, JETP, 31, 31 (1956).

(4) H. E. Hall and W. F. Vinen, Proc. Roy. Soc. A238, 204 (1956).

(5) V. P. Peshkov, V. B. Stryukov, JETP, 41, 1443 (1961).

(6) William I. Glaberson, Warren W. Johnson and Richard M. Ostermeier. Phys. Rev. Lett. 33, 1197 (1974).

PHASE SEPARATION IN ROTATING ^3He-^4He SOLUTIONS

W. I. Glaberson and R. M. Ostermeier

Department of Physics, Rutgers University
New Brunswick, N. J. 08903

It is by now well accepted that ions interact with vortex lines through a hydrodynamic potential of the sort first suggested by Donnelly.[1] This potential is essentially the hydrodynamic kinetic energy deficit in the system associated with the presence of the ion in the vicinity of the vortex line. Donnelly[1] and Donnelly and Roberts[2] treated the capture of an ion by a vortex line by considering the ion as a Brownian particle immersed in a gas of quasiparticles and involved in a process of sedimentation into the vortex potential wells. These ideas lead to reasonable agreement with the experimental capture cross-section data obtained by Douglass,[3] Tanner,[4] and Springett, et al.[5]--but only over a very narrow temperature region. At the higher temperatures, above ∿1.6K, thermal activation out of the vortex well gives rise to an apparent cross-section much smaller than predicted (the temperature corresponding to the appearance of this phenomenon is termed the "lifetime edge"). At lower temperatures, below ∿1.4K, the cross-section is also much smaller than predicted for reasons which, when we started our investigation, were not at all clear.[6] We became interested in this anomaly and began a series of experiments to probe this problem as well as other details of the ion-vortex interaction.

Using a rotating ^3He refrigerator, we first extended capture cross-section measurements to temperatures below 1 K. We found that the cross-section decreased very rapidly as the temperature was lowered and became too small to be observed below ∿0.75K. We were able to account quantitatively for our data by considering the ballistic motion of the ions in the vortex potential. The stochastic theory originally used for calculating cross-sections is valid at higher temperatures, where the ion diffusion length is small compared to the vortex well width, but breaks down completely at temperatures below ∿1.3K.[7] Arguing that the addition of ^3He impurities to the system would decrease the ion diffusion length and therefore restore the validity of the original stochastic theory down to our lowest temperature, we repeated the cross-section measurements in ^3He-^4He solutions and found this indeed to be the case.[8]

The experiments discussed so far involved the relatively large negative ions. The smaller positive ions interact much more weakly with vortices and thermal activation out of vortices becomes important at much lower temperatures (i.e., the lifetime edge for positive ions is lower). We were not at all surprised by our inability to observe bare positive ion capture in pure ^4He since, at temperatures low enough to be below the lifetime edge, ballistic effects would prevent any measurable capture. That we could not observe bare positive ion capture in ^3He-^4He solutions at temperatures below what we thought was the lifetime edge, however, was a mystery to us.

The lifetime edge for positive ions was obtained from experiments on the escape of ions from vortex rings in large electric fields.[9] In order to be certain that we were using the correct lifetime edge in interpreting our data, we performed a direct measurement of escape from vortex lines in weak fields.[10] Because high temperature negative ion escape is, in some ways, qualitatively different from low temperature positive ion escape (the ratio of the ion diffusion length to the vortex well "width" is much smaller than one in the former case and much larger than one in the latter), the anomalous electric field dependence of the escape that we observed led to an interesting and new physical picture of the escape process. The temperature of the lifetime edge, however, was approximately where we expected it, and we retained the mystery of our inability to see positive ion capture in ^3He-^4He solutions. It should be pointed out that escape experiments were carried out in pure ^4He with ions that were trapped by vortices after first making charged vortex rings, which interact very strongly with vortex lines.[11]

At this time, we harbored suspicions that the explanation of the anomalously low positive ion capture in ^3He-^4He solutions was that ^3He atoms condensed onto the cores of the vortices, increased the core radius, and thereby decreased the binding energy of the ion to the line. This suspicion was reinforced by our observation of an anomaly[12] in the extrinsic lifetime discussed at length by Douglass[13] and DeConde, Williams, and Packard.[14] The idea of ^3He condensation into the vortex cores was further reinforced by observations by Williams and Packard[15] of a decreasing lifetime for positive ions with increasing ^3He concentration in dilute $^3He-^4He$ mixtures. Furthermore, using thermodynamic arguments, Rent and Fisher[16] and Ohmi, Tsuneto, and Usui[17] predicted that ^3He condensation and indeed phase separation should occur in the vicinity of vortex cores.

In order to directly test this interesting idea we set up to measure the mobility of ions trapped on vortices along the vortices in $^3He-^4He$ solutions.[18] If ^3He atoms were to condense onto the vortex cores, they would manifest themselves in an increased drag experienced by the ion motion along the vortex line.

Our measurements for 0.27% concentration are shown in Fig. 1. The drag experienced by free ions--those not trapped on vortex lines--decreases rapidly as the temperature is lowered (and the thermal excitations freeze out) until, at the lowest temperatures, the relatively temperature independent drag is dominated by the scattering of ^3He quasiparticles. The solid line in the figure is an extrapolation from low concentration measurements.[19] The trapped ion drag also decreases rapidly with decreasing temperatures at high temperatures but, at some "critical" temperature, the drag turns around and then increases as the temperature is lowered.

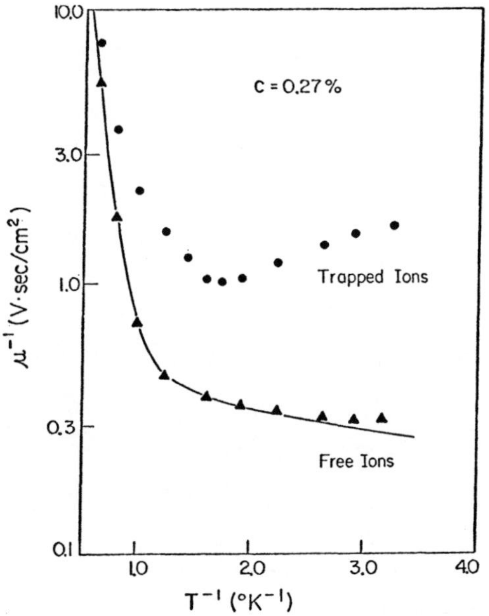

Figure 1. Temperature dependence of the inverse mobility for negative free ions (triangles) and trapped ions (circles). The curve is an extrapolation from low-concentration free-ion data.

We extract the vortex contribution to the ion drag by subtracting the inverse mobility of the free ions from that of the trapped ions. The data for negative ions in several different ^{3}He concentrations are shown in Fig. 2. Similar results for the positive ion were obtained at the lowest concentrations and are shown in Fig. 3. At higher temperatures, most of the data, independent of concentration, lie essentially on the same curve. The rather good agreement between these universal curves for the two ion species and the vortex wave drag calculations of Fetter and Iguchi,[20] shown as the dashed lines in Figs. 2 and 3, suggests that the drag in this temperature range is due in large part to the scattering of thermally activated vortex waves.

The "critical" temperature, where deviation from universal behavior becomes apparent, is smaller for smaller concentrations and, most important, is the same for the two ions species. This last observation proves that the "critical" temperature is associated with the vortex and not with changes in the ion structure. In Fig. 4 we have plotted the ambient ^{3}He concentration as a function of the critical temperature at which the vortex drag deviates from its universal high-temperature behavior. Also shown in Fig. 4 is a calculation which predicts the onset of the condensation of ^{3}He atoms onto the vortex cores. This calculation, which contains no adjustable parameters, is based on the following

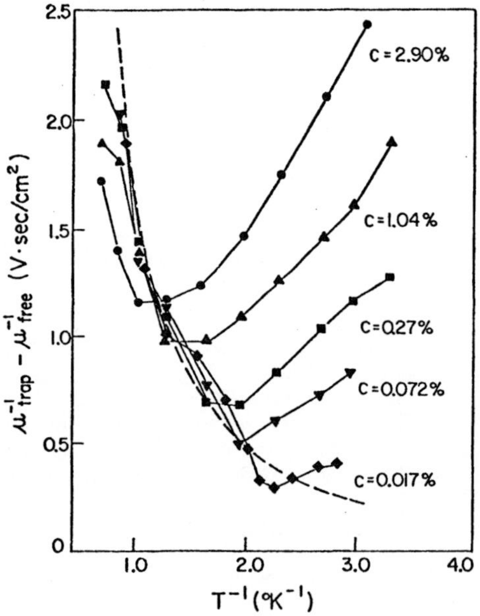

Figure 2. Temperature dependence of the vortex contribution to the inverse mobility of the negative ion. The dashed curve is the calculated vortex-wave contribution. The error in the data is estimated at ±6%.

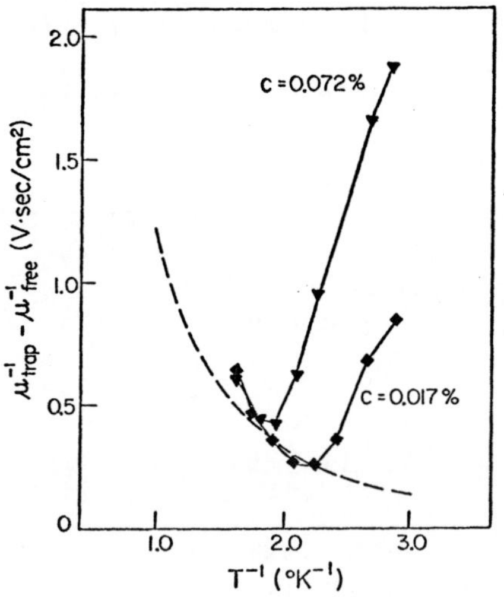

Figure 3. Temperature dependence of the vortex contribution to the inverse mobility of the positive ion. The dashed curve is the calculated vortex-wave contribution.

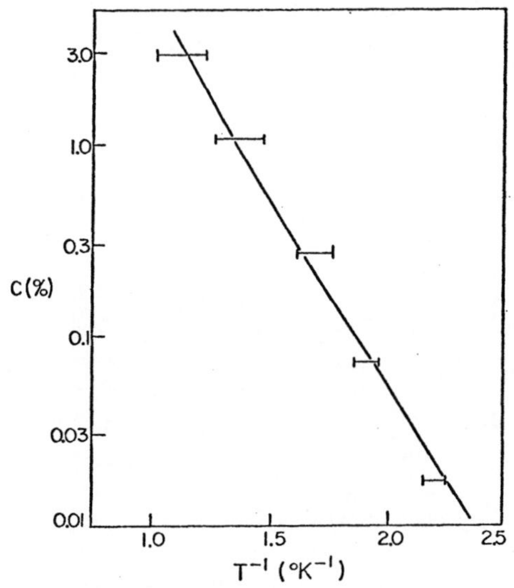

Figure 4. Ambient concentration versus critical temperature for ^3He condensation onto vortex cores. The curve is discussed in the text.

argument. In the presence of a vortex line, we assume that the radial dependence of the ^3He number density is given by

$$n(r,T) = n_\infty \exp\left[-U(r,T)/k_B T\right] \tag{1}$$

where n_∞ is the ambient ^3He number density and $U(r,T)$ is an effective hydrodynamic potential related to the kinetic energy of the superfluid displaced by a ^3He atom. We argue that condensation will occur when the maximum ^3He density given by Eq. (1), $n(0,T)$, becomes equal to the critical density $n_{ps}(T)$ measured at phase separation in bulk solutions. The curve in Fig. 4 is thus given by

$$n_\infty = n_{ps}(T_c)\exp[+U(0,T_c)/k_B T_c] \ . \tag{2}$$

We have used the formalism of Parks and Donnelly[21] for $U(r,T)$, the binding energy of a spherical impurity to the vortex line. The effective hydrodynamic radius of a ^3He atom in solution a_3 is determined from the known effective mass m_3^* by

$$m_3^* = m_3 + \frac{1}{2}\frac{4}{3}\pi a_3^3 \rho \ . \tag{3}$$

Using these results, we find that the ^3He atom-vortex binding energy varies smoothly from about -2.7 to -3.1 K over the temperature range indicated in figure 4. The validity of Eq. (3) and the Parks-Donnelly formalism, both based on continuum fluid mechanics, is quite uncertain when applied to a ^3He atom which, in solution, can displace only one or two ^4He atoms. However, it is interesting and somewhat reassuring to note that Ohmi and Usui[22] have investigated the ^3He binding energy in the Hartree-Fork approximation and have found the value -3.5 K, consistent with our results.

In this calculation we have neglected any pressure dependence of $n_{ps}(T)$ and the possibility of an interphase surface tension. The measured pressure variation of $n_{ps}(T)$ at positive pressures[23] suggests, however, that these effects would tend to cancel. Other more serious objections to this semiphenomenological argument can certainly be raised. When one consideres that the thermal de Broglie wavelength of a ^3He atom is on the order of 15-20 Å, the use of continuum mechanics and bulk properties in describing a phenomenon which is probably localized over only a few angstroms is rather questionable. It is difficult, however, to attribute the sharp increase in drag to anything other than ^3He condensation, and it seems rather unlikely that the good agreement between our data and this simple and physically appealing, though admittedly crude, calculation is entirely fortuitous.

We now proceed to a brief discussion of ion mobility at temperatures below the "critical" temperature. In Fig. 5 we plot the contribution to the ion drag from condensed ^{3}He atoms as a function of reduced inverse temperature $\varepsilon = (T_c/T-1)$. Note first that μ_3^{-1} for the positive ion is 2-3 times larger than for the negative ion. This is unique among known ion drag mechanisms in liquid helium.

The increase in the drag with decreasing temperature below T_c seems to indicate a growing ^{3}He-rich core. Using ideas similar to those in the previous discussion, we argue that phase separation will occur at a distance r_c from the center of the vortex at a temperature T when

$$n_{ps}(T) = n_\infty \exp[-U(r_c,T)/k_BT] \tag{4}$$

Eliminating n_∞ by means of Eq. (2) and inserting the Parks-Donnelly result for U, we can then solve numerically for r_c in terms of T and T_c. The number of ^{3}He atoms per unit length of line can be determined from

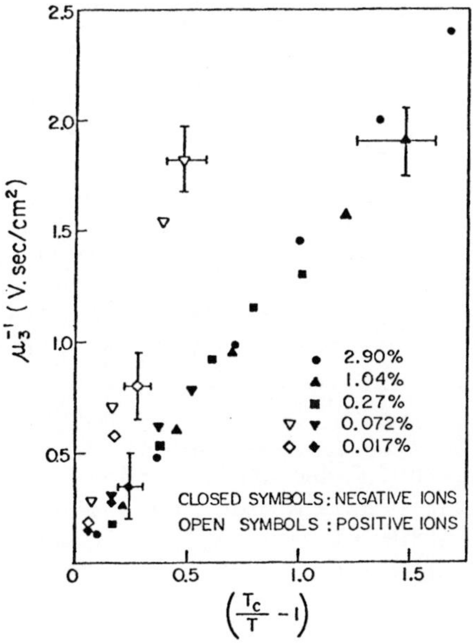

Figure 5. The condensed ^{3}He contribution to the vortex drag for the positive and negative ion versus the reduced inverse temperature.

$$n_L = \pi r_c^2(T,T_c)n_{3R}(T) \tag{5}$$

where we take $n_{3R}(T)$ to be the ^{3}He number density in the bulk ^{3}He-rich phase. These arguments obviously suffer the same flaws as the condensation theory given earlier, but the ideas are probably qualitatively correct and n_L, though not exhibiting a universal behavior, is consistent with the data of Fig. 5, indicating a core growth with decreasing temperature.

Before developing these ideas further, it would be of interest to estimate μ_3^{-1} simply assuming specular reflection and using our results for n_L. As the ion moves along the line with velocity v_D, $n_L v_D$ ^{3}He atoms will scatter off the ion per unit time. If, in addition, we assume that the condensed ^{3}He can be treated in the effective mass approximation and that the atoms obey Maxwell-Boltzmann statistics, then each ^{3}He atom will exchange momentum $2(m^* k_B T)^{1/2}$ with the ion. The total rate of change of the ion's momentum is thus

$$dP/dT \cong 2(m^* k_B T)^{1/2} n_L v_D \; . \tag{6}$$

But this is equal to the external force, so that

$$\mu_3^{-1} \cong 2(m^* k_B T)^{1/2} n_L/e \; . \tag{7}$$

If our approximations are correct, then m^* in Eq. (7) will probably be of the same order as that in dilute solutions, $2.34m_3$. Using this result, we find that Eq. (7) predicts a μ_3^{-1} lying somewhere between the data for the positive and negative ions.

The difference between the positive and negative ion mobilities can be understood in terms of the ion-^{3}He interaction. The smaller positive ion experiences a large repulsive interaction and the negative ion an attractive interaction. If a ^{3}He atom is to make its way from in front of the moving ion to the other side, it must mover further from the vortex core in the case of the positive than for the negative ion. This leads to a larger effective reflection coefficient for a ^{3}He atom incident upon a positive ion and therefore a larger positive ion inverse mobility. A detailed theoretical consideration of the mobility based on these simple notions[24] yields results consistent with our data.

References

(1) R. J. Donnelly, Phys. Rev. Lett. 14, 39 (1965).

(2) R. J. Donnelly and P. H. Roberts, Proc. Roy. Soc. A312, 519 (1969).

(3) R. C. Douglass, Phys. Rev. Lett. 13, 791 (1964).

(4) D. J. Tanner, Phys. Rev. 152, 121 (1966).

(5) B. E. Springett, D. J. Tanner, and R. J. Donnelly, Phys. Rev. Lett. 14, 585 (1965).

(6) For a more complete discussion, see Experimental Superfluidity by R. J. Donnelly, W. I. Glaberson, and P. E. Parks (Univ. of Chicago Press, Chicago, Illinois, 1967).

(7) R. M. Ostermeier and W. I. Glaberson, Phys. Letters 49A, 223 (1974).

(8) R. M. Ostermeier and W. I. Glaberson, J. Low Temp. Phys. 20, 159 (1975).

(9) A. G. Cade, Phys. Rev. Lett. 15, 238 (1965); W. I. Glaberson and W. W. Johnson, J. Low Temp. Phys. 20, 313 (1975).

(10) R. M. Ostermeier and W. I. Glaberson, Phys. Lett. 51A, 348 (1975). Similar results were obtaind independently by G. A. Williams, K. DeConde, and R. E. Packard, Phys. Rev. Lett. 34, 924 (1975).

(11) K. W. Schwarz and R. J. Donnelly, Phys. Rev. Lett. 17, 1088 (1966).

(12) R. M. Ostermeier and W. I. Glaberson, Phys. Lett. 51A, 403 (1975).

(13) R. L. Douglass, Phys. Lett. 28A, 560 (1969).

(14) K. DeConde, G. A. Williams, and R. E. Packard, Phys. Rev. Lett. 33, 683 (1974).

(15) G. A. Williams and R. E. Packard, Phys. Rev. Lett. 35, 237 (1975).

(16) L. S. Rent and I. Z. Fisher, Soviet Physics-JETP 28, 375 (1969).

(17) T. Ohmi, T. Tsuneto, and T. Usui, Prog. Theor. Phys. 41, 1395 (1969).

(18) For a more detailed description of the experiment and results, see R. M. Ostermeier, E. J. Yarmchuk and W. I. Glaberson, Phys. Rev. Lett. 35, 957 (1975); and R. M. Ostermeier and W. I. Glaberson, J. Low Temp. Phys. 25, 317 (1976). Results above 1 K were reported by W. I. Glaberson, J. Low Temp. Phys. 1, 289 (1969).

(19) K. W. Schwarz, Phys. Rev. A6, 1947 (1972).

(20) A. L. Fetter and I. Iguchi, Phys. Rev. A2, 2067 (1970); and I. Iguchi, Phys. Rev. A4, 2410 (1971).

(21) P. E. Parks and R. J. Donnelly, Phys. Rev. Lett. 16, 45 (1966).

(22) T. Ohmi and T. Usui, Prog. Theor. Phys. 41, 1401 (1969).

(23) K. N. Zinov´eva, Soviet Physics-JETP 17, 1235 (1963).

(24) See reference 18; and W. Huang and A. J. Dahm, Phys. Rev. Lett. 36, 1466 (1976).

RESULTS FROM A GAUGE THEORY OF SUPERFLUIDITY IN ^{4}He

J. Chela Flores

Centro de Física, I.V.I.C.
Apartado 1827, Caracas

Abstract: We complete our previous gauge theory of superfluidity of ^{4}He and use it to evaluate the liquid structure factor for small values of the momentum transfer.

In a previous work[1] we suggested to approach the question of Bose-Einstein condensation by constructing a two-fluid model in a Lagrangian formalism supplemented with gauge fields. A second fluid, besides the underlying condensate Hartree-liquid model is to be added by gauging the theory, the gauge field being interpreted as a hydrodynamic velocity of the second fluid (depletion); following this line of thought we find the following set of field eqns.

$$\partial_t \sigma + \partial_{\underset{\sim}{x}} \cdot \left[(\sigma-\rho)\, \underset{\sim}{u}\right] = 0 \tag{1}$$

$$\partial_t \rho + \partial_{\underset{\sim}{x}} \cdot \left[\rho \underset{\sim}{v} + \rho \underset{\sim}{u}\right] = 0 \tag{2}$$

where (σ, u), (ρ, v) are the density and velocity of the depletion and condensate, respectively.

In ref. (1) a term was omitted from the equation of motion for the macrowave function when real and imaginary parts were taken; the correct form of the equations is as shown in Eqs. (1) and (2). We further observe that the gauge field $\underset{\sim}{u}$ introduced in ref. (1) is still ambiguous up to a sign in the Lagrangian. However, in the present work we have chosen $-\underset{\sim}{u}$ instead of $+\underset{\sim}{u}$ so that a physical solution to eqns. (11) and (12) below is possible.

For the bulk fluid the continuity equation is preserved

$$\partial_t(\rho + \sigma) + \partial_x \cdot \left[\rho \underset{\sim}{v} + \sigma \underset{\sim}{u} \right] = 0 \tag{3}$$

This indeed was how the form of Eq. (1) was obtained previously.

A further equation was discussed in ref. (1), namely a Bernoulli type of equation (analogous to the one in Gross' paper for zero depletion[2])

$$\partial_t \phi = \frac{1}{2} \underset{\sim}{v}^2 + F + \underset{\sim}{u} \cdot \underset{\sim}{v} , \tag{4}$$

where

$$F = \frac{1}{m} (Uf^2 - E_v) - \partial_x^2 \sqrt{\rho}/2m^2 \sqrt{\rho} + \frac{1}{4} \underset{\sim}{u}^2 , \tag{5}$$

ϕ being related to the phase S of the macrowave function ψ by

$$\phi = -m^{-1}(E_v t + S)$$

and E_v is an integration constant. It is interesting to go back to first principles and find an analytic solution for the pair correlation function g(r). Let us therefore consider Eq. (4) in the limit of low fluid velocities

$$\underset{\sim}{v}^2 << 1, \ \underset{\sim}{u} \cdot \underset{\sim}{v} << 1, \ \partial_{\underset{\sim}{x}} \cdot u << 1 \tag{6}$$

and also suppose time independence.

We find that a solution to the Bernoulli-like Eq. (4) is given by,[3]

$$\rho(r) \equiv F^2 F^2_{\infty} \tanh^2 \left[\sqrt{\rho(\infty) mU} \ (r_o - r) \right] . \tag{7}$$

In the same approximation we find that Eq. (1) yields,

$$\underset{\sim}{u} \cdot (\partial_{\underset{\sim}{x}} \rho - \partial_{\underset{\sim}{x}} \sigma) = 0 \tag{8}$$

From Eq. (8) we infer by integration (we assume, for the liquid, spherical symmetry),

$$\rho(r) - \sigma(r) = \text{constant.} \tag{9}$$

As r tends to approximately 2Å, g(r) tends to zero and, by the same token,

$$\rho_T(r) \equiv g(r)\ \rho_T(\infty) \tag{10}$$

also tends to zero; here we consider macroscopically ^{4}He (from the point of view of the total density ρ_T) as a monatomic classical fluid characterized by ρ_T and temperature T. But considering Eqs. (9) and (10) we infer that

$$\rho(r) - \sigma(r) = 0 \tag{11}$$

since the densities of the two fluids should also become zero in the region of "hard spheres" of the ^{4}He atoms ($r_o \lesssim 2$Å).

Combining Eqs. (1) and (11) with the known solution for $\rho(r)$ in Eq. (7) we may infer[3]

$$g(r) = \tanh^2\left[\Lambda(r_o - r)\right] \tag{12}$$

where we use the convenient notation $\Lambda = \sqrt{\rho(\infty)mU}$. Since g(r) must vanish at $r \cong 2$Å, we find that the constant of integration $r_o \cong 2$Å.

Our solution (12) for g(r) has the correct qualitative shape since it vanishes at $r \cong 2$Å and does not tend to 1 as $r \to \infty$, the range being given by the parameters m, $\rho(\infty)$, U of the theory. Physically we may understand the similarity of (12) with a g(r) of a gas since we are describing correctly a liquid which shows B.E. condensation just like a gas of bosons and both physical systems do sometimes show similar macroscopic behavior.

Let us consider the expression for the structure factor[4,5]

$$S(Q) = 1 + (4\Pi\rho_T/Q) \int_0^\infty \operatorname{sech}^2[\Lambda(r_o - r)]\ r\mathrm{Sin}\ (Qr)\ dr\ . \tag{13}$$

We Fourier-transform our solution in order to find directly S(Q) and find that, for an effective interaction potential for condensate particles given by $(\Lambda r_o)^{-1} \sim 0.193$, we have,[3]

$$S(Q = 0) \equiv S_o = 0.05 \qquad (14)$$

We further find that the functional dependence in Q is quadratic

$$S(Q) \equiv S_o + S_2 = S_o + \alpha Q^2 \qquad (15)$$

where $\alpha \sim 0.33(\mathring{A}^2)$, in good agreement with X-ray data[6-8] and neutron diffraction data.[11] In Table I we compare the results of the calculations with the experiment of Gordon et al.[6] We conclude by saying that the gauge theory supplemented with a U-potential is able to explain the X-ray data very well.[6] More recent data, also with X-rays.[7,8] lie somewhat above the Gordon et al. experiment, but these new experiments are not in agreement amongst themselves, particularly around $Q \sim 0.7\mathring{A}^{-1}$. Also the neutron diffraction experiment of Mozer et al.[9] lies somewhat above our quadratic curve, particularly for very low $Q(0.2\mathring{A}^{-1})$; however, this may not be too significant in view of the well known difficulties with small angle neutron diffraction implied at these values of Q.

$Q(\mathring{A}^{-1})$	S(Q) THEORY	S(Q) AT T = 1.4K[6]
0.10	0.053	0.058
0.20	0.063	0.070
0.30	0.080	0.087
0.40	0.106	0.104
0.50	0.133	0.125

Table I. Functional variation of S(Q) in the very low-Q region $S \equiv 0.05 + (0.3A^2)\ Q^2$ is also denoted as $S \equiv S_o + S_2$; beyond $Q \sim 0.75\ A^{-1}$ the S(Q) theory breaks down.

References

(1) J. Chela-Flores, Journal of Low Temp. Phys. 21, 305 (1975).

(2) E. P. Gross, J. Math. Phys. 4, 195 (1963).

(3) J. Chela-Flores (to be published).

(4) N. S. Gingrich, Rev. Mod. Phys. 15, 90 (1943).

(5) G. E. Bacon, Neutron Diffraction, Third Edition (Clarendon Press, Oxford, 1975), chapter 16, p. 544.

(6) W. L. Gordon, C. H. Shaw and J. G. Daunt, J. Phys. Chem. Solids 5, 117 (1958).

(7) E. K. Achter and L. Meyer, Phys. Rev. 188, 291 (1969).

(8) R. B. Hallock, Phys. Rev. A 5, 320 (1972).

(9) B. Mozer, L. A. de Graaf and B. Le Neindre, Phys. Rev. A 1, 448 (1974).

THE FREE SURFACE OF ROTATING ^{4}He THICK FILMS*

P. L. Marston† and W. M. Fairbank

Department of Physics, Stanford University
Stanford, California

Abstract: Surface profiles of He II with an initial thickness of 6-50 μm were measured for a rotating and stationary cylindrical container. Equilibrium porfiles are consistent with $\langle v_{s\theta} \rangle = v_{n\theta} = \omega r$, $v_{nr} > 0$ and $\langle v_{sr} \rangle < 0$. Metastable depressions created after sudden spin-up appear to be a superfluid vortex with a circulation of several hundred h/m.

Measurements of surface profiles of He II contained in rotating and stationary cylinders were used to infer equilibrium and metastable superfluid rotational states.[1,2] This paper summarizes the experimental method and the results of those measurements. Prior to rotation, the equilibrium liquid thickness at the center of the cylinder was typically 20 μm. In comparison with previous measurements of profiles of He II in rotating containers, the film thickness in this experiment is much greater than the 300 Å film studied by Vittoratos and Meinke[3], but it is significantly less than for the bulk measurements of Osborne[4] and Meservey.[5] The significance of initial thickness in this experiment was that it permitted the creation of a metastable flow configuration: the center region consisted of a thin film surrounded by a large-quantum-number superfluid vortex of bulk He II. The metastable surface depressions are apparently a manifestation of the Bernoulli pressure of a superfluid vortex.

---

*Work supported by NSF Grant DMR75-15628
†Present address: Department of Engineering and Applied Science, Yale University, New Haven, Connecticut.

The principal method of measuring the surface profiles was Fizeau optical interferometry.[1,2] This is perhaps the first time optical interferometry has been used to measure the surface contour of any rotating liquid. He II was contained in a 2.15 cm diameter copper cylinder and was illuminated vertically by the expanded beam from a 6 mw laser having a wavelength $\lambda = 0.6328$ μm. Light reflected from the free surface interferes with light reflected from the cylinder bottom and produces fringes of constant liquid thickness ζ. Adjacent fringes normally indicate a change in thickness of $\lambda/2n = 0.3072$ μm where $n = 1.030$ is the refractive index of He II. To reduce thermal gradients, the bucket bottom consisted of a single piece of crystaline quartz. It was anti-reflection coated on the upper side since its reflection coefficient had to be similar to the free He II surface if there was to be adequate fringe contrast. The local height of the free surface Z is the sum of the liquid thickness ζ and the substrate height ψ. Optical measurement techniques[2] made it possible to align the optical beam direction, the axis of rotation, the gravitational acceleration, and the normal to the center of the substrate, to be parallel to within 10^{-4} radians. The profile ψ of the quartz substrate was determined with Twyman-Green interferometry[2] to have a parabolic term $\alpha_s r^2$ where r = the radial coordinate and $\alpha_s = -4.7$ f/cm^2 (1 f = 0.3072 μm = 1 Fizeau fringe). It was important to know that the substrate was slightly curved to interpret correctly the equilibrium fringe patterns.[6]

The sealed cylinder was immersed in a helium bath whose temperature T was controlled by a diaphragm manostat. For $T < T_\lambda$, the drift in the liquid thickness for near equilibrium conditions was less than 1 f/2 minutes. Temperature gradients were sufficiently great when $T > T_\lambda$ to cause significant drift in the liquid thickness. The rotation period was measured by independent photoelectric and electromechanical methods with an accuracy of 0.5% and the angular velocity ω could be traced with a chart recorder.

After the cylinder was set gradually into rotation, the fringe pattern usually stabilized in 30 seconds or less. Photographs of fringes taken when T = 2.16 K ($\rho_s/\rho = 0.09$) indicated that both the stationary and rotating equilibrium profiles were indistinguishable from those calculated for a classical liquid of the same surface tension and density. The equilibrium surface of a classical liquid which both wets the cylinder walls and is sufficiently thick that the Van der Waals attraction of the container is insignificant is given by:[2,7]

$$Z(r) = (1 - 2^{\frac{1}{2}}\omega^2 b/g)h_0[I_0(r/a) - 1] + \omega^2 r^2/2g \qquad (1)$$

where: ω = the angular velocity of rotation, b = the cylinder radius, g = 980.0 cm/sec^2, the capillary constant $a=[\sigma/(\rho-\rho_v)g]^{\frac{1}{2}}$ =

0.4657 mm, I_0 is a modified Bessel function, and $h_0 = a[4(2\ b/a)^{\frac{1}{2}}\tan(\pi/8)\exp(-b/a - 2 + 2^{\frac{1}{2}})]$. For $\omega = 0$, the measured profile was within 1 fringe of (1) in the observed region of $r < 8.5$ mm. The surface height above that at $r = 0$ was less than 1 fringe for $r < 6.5$ mm. With T = 2.16 K photographs with the cylinder rotating gave a similar agreement with (1). Fig. 1 is a profile measured along a diameter when $\omega = 1.616$ radians sec^{-1} and the liquid thickness at $r = 0$ given by $\zeta_0(\omega) = 10$ µm. The deviation from parabolicity caused by the wetting of the cylinder walls is found to be in good agreement with the nonparabolic term in (1). Points with $R < 0$ were located at $r = |R|$ in the opposite direction from the center but along the same diameter as the points with $R > 0$.

Equilibrium profiles were also measured for several $T < 1.8$ K ($\rho_s/\rho > 0.7$) with $\omega = 0$ and with $\omega \geq 0.60$ sec^{-1}. Fig. 2 is a profile measured along a diameter when $\omega = 1.221$ sec^{-1} and $\zeta_0(\omega) = 6$ µm. In the region $r < 5$ mm the measured profile was accurately fitted to $Z = \alpha r^2$ with $\alpha = 0.69\alpha_c$, where $\alpha_c = \omega^2 r^2/2g$ is the classical parabolic coefficient. Fig. 3 shows that for all measurements in the low temperature region, the curvature of the He II near the center of the cylinder was less than the classical prediction by approximately 4×10^{-4} cm^{-1} for all ω. On the other hand, the curvature with T = 2.16 K is shown to be classical. The dashed line in Fig. 3 corresponds to $\alpha = (\rho_n/\rho)\alpha_c$ at T = 1.7 K

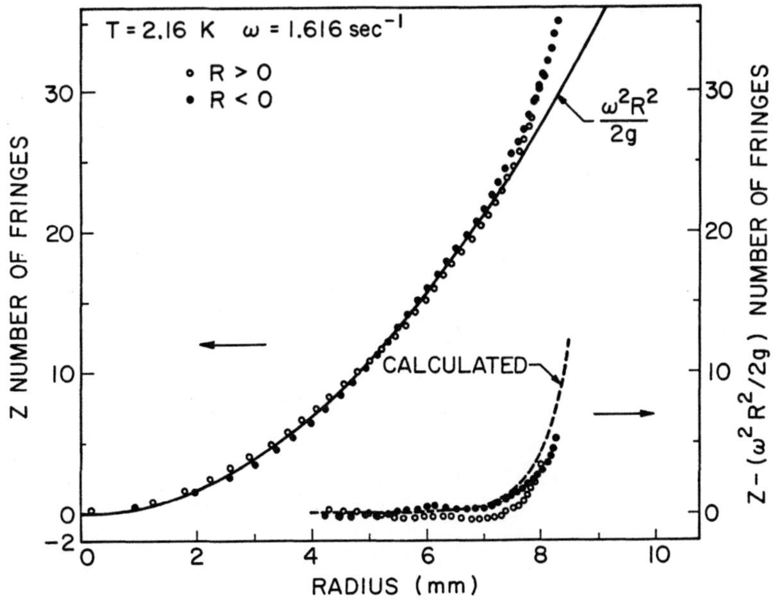

Figure 1. Equilibrium profile with T = 2.16 K measured relative to the liquid height on the axis of rotation. The dashed curve was calculated from the first term of eq. (1).

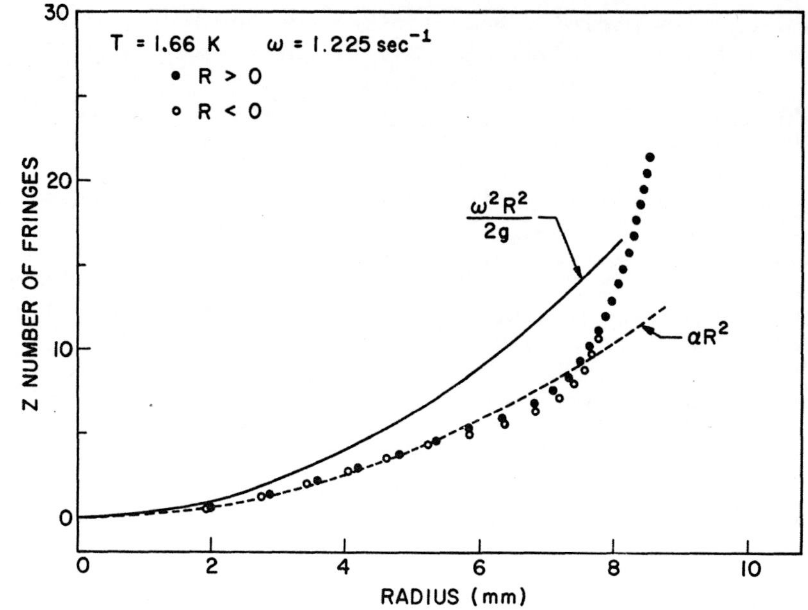

Figure 2. Equilibrium profile with T = 1.66 K measured relative to the liquid height on the axis of rotation. The dashed curve is least-square fit to the measured points having $|R| < 5$ mm.

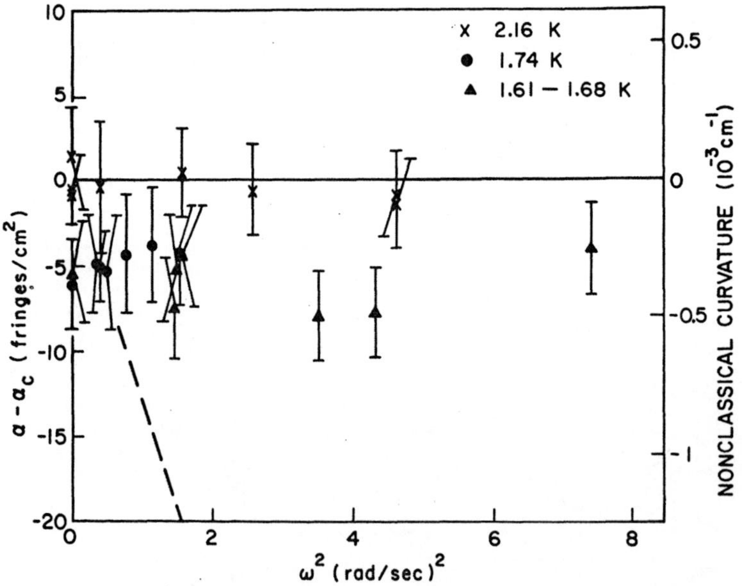

Figure 3. Difference between the fitted α and the calculated classical α. Each α was fitted to the $r < 5$ mm points of a measured profile. One standard deviation error bars arise from the least-square fit and the measurements of ω and ψ.

which would describe the profile in the energetically unfavorable case in which $v_s = 0$. For an equilibrium uniform distribution of singly quantized vortex lines (SQVL): $\langle v_{s\theta} \rangle = \omega r$ and we expect $\alpha = \alpha_c$ at all temperatures.[5] The nearly constant difference between α_c and the measured α observed when $\rho_s/\rho > 0.7$ can be attributed to radial internal convection ($v_{nr} > 0$, $\langle v_{sr} \rangle < 0$) induced by the absorbtion of background thermal radiation[2] by the substrate. The equilibrium measurements are consistent with the density of SQVL being the expected value of $2\omega\kappa^{-1}$ where $\kappa = h/m$ is the circulation quanta.

Fig. 4 is the fringe pattern from which Fig. 2 was obtained. The fringes are seen to be of high contrast. A calibration grid with a line spacing of 0.5 mm is also shown. It was placed in a plane optically equivalent to that of the substrate. Though Fizeau interferometry revealed the average change in the surface caused by rotation, it was not sufficiently sensitive to observe the array of SQVL. The depression in the free surface at each singly quantized vortex line is believed to be only 50 Å deep when T = 1.7 K.[8]

The equilibrium film thickness at the center of the cylinder $\zeta_0(\omega)$ was measured by measuring the lowest ω for which $\zeta_0(\omega) \lesssim 1$ f. With the temperature and volume of liquid in the cylinder held

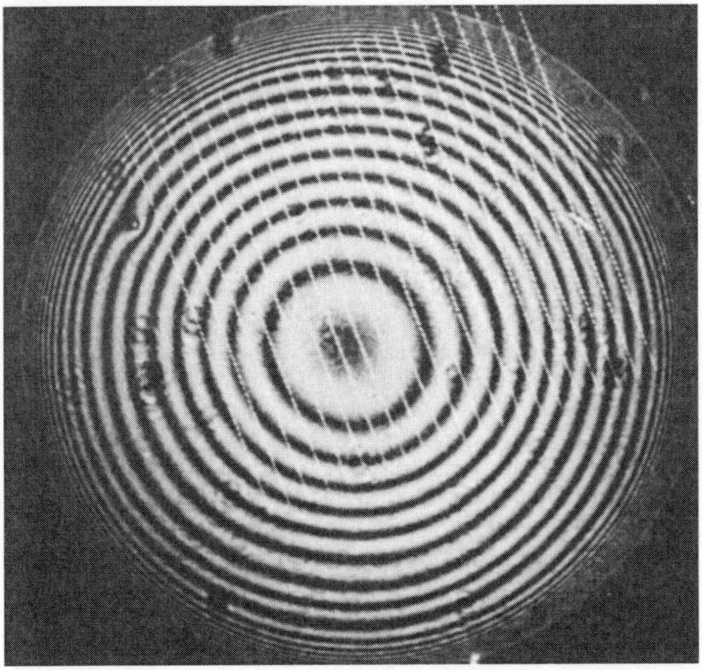

Figure 4. Photograph of fringe pattern from which Fig. 2 was taken. Cylinder wall is out of view.

constant, ω was gradually increased. When $\zeta_0(\omega) \simeq 0.4\ \mu m$, the lowest part of the free surface was no longer parabolic since the Van der Waals attraction of the substrate was greater than the earth's force of gravity. Using the constant volume condition and the empirical result that at fixed T, $\alpha - \alpha_c$ is approximately independent of ω (Fig. 3) a volume integration gives for $\omega < \omega_0$:

$$\zeta_0(\omega) = (\omega_0^2 - \omega^2)b^2/4g \tag{2}$$

where $\zeta_0(\omega_0) \equiv 0$. The derivation of (2) also assumes that except for the wetting of the cylinder walls, the surface is parabolic. It was verified that ω_0 could be identified with the lowest ω at which fringes no longer originated at $r = 0$ as ω was increased. This measurement of the equilibrium thickness is in effect a measurement of the liquid volume. It was used in the study of metastable vortex profiles described in the remainder of this paper.

Fig. 5 shows an example of a metastable surface depression taken 175 sec after ω was held constant at $\omega_2 = 3.22\ sec^{-1}$ with $T = 1.80$ K. In contrast to Fig. 4, the fringe spacing near the central region of bulk liquid is less than the fringe spacing at larger radii. Thus, in the inner region: $d^2A/dr^2 < 0$ (with $r=0$ as the center of the depression). The normal fluid is locked to rotate with the substrate within a relaxation time[2,5] $\rho_n\zeta^2/\eta \simeq 10^{-2}$ sec and hence $v_{n\theta} = \omega_2 r$. For a simplified model in which a superfluid vortex is centered in the cylinder, $v_{s\theta} = \Gamma/2\pi r$ where Γ is the circulation. A profile calculation based on an extension of the Bernoulli equation to account for the non-irrotational flow of the normal fluid gives[2]:

$$Z(r) = (\rho_n/\rho)\omega_2^2 r^2/2g - (\rho_s/\rho)\Gamma^2/8\pi^2 g r^2 \tag{3}$$

where surface tension, the wetting of the walls, and the Van der Waals attraction of the substrate has been omitted. For sufficiently large Γ, (3) gives $d^2Z/dr^2 < 0$. Unfortunately (3) would probably not fit the observed depressions since they were not centered in the cylinder. Nevertheless it was possible to estimate Γ as follows: It is energetically favorable that for $r >$ some r_1, there is an array of SQVL with density $2\omega_2\kappa^{-1}$ and there $\langle v_{s\theta}\rangle = \omega_2 r$. In that region, $Z(r) = \alpha r^2 +$ constant, so for some r near r_1, $d^2Z/dr^2 = 0$. If $\langle v_{s\theta}\rangle$ is to be continuous at r_1 then $\Gamma = 2\pi\omega_2 r_1^2$. In Fig. 5, the inflection radius is approximately 2.2 mm giving $\Gamma \simeq 980\ \kappa$ and $d^2Z/dr^2 < 0$ in (3) for $r < r_1$.

The center-most region of the metastable depressions consisted of only a thin film of He II with a thickness $\zeta \ll 1$ f. That region

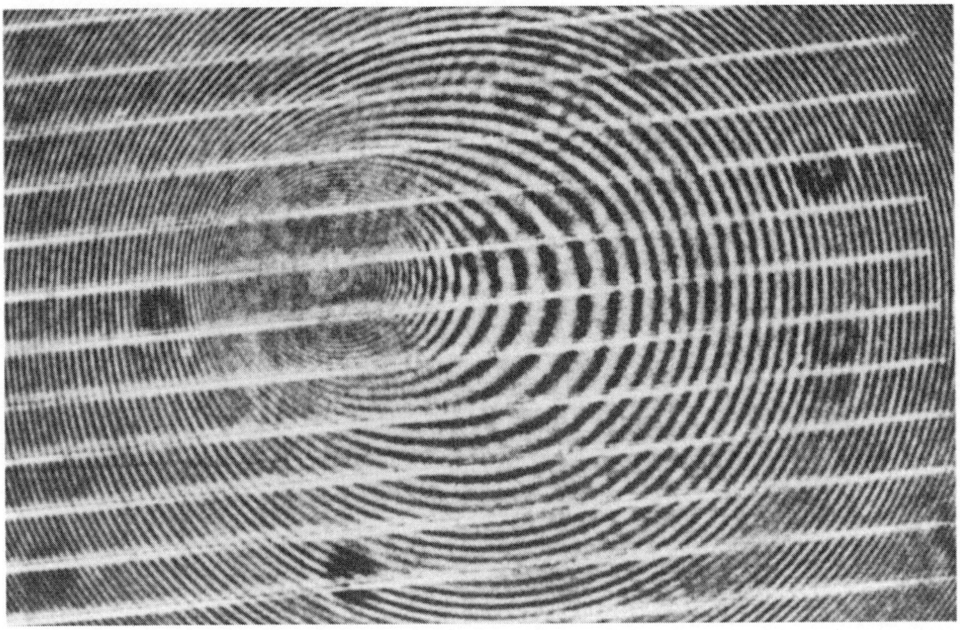

Figure 5. Photograph of a vortex depression taken at c in Fig. 6A. The distance between the parallel lines is 0.5 mm.

moved with the rotating substrate. The trapping of circulation in the bulk liquid is probably associated with the pinning of a high density of SQVL in the thin film.

The angular velocity record of the depression shown in Fig. 5 was typical of the history of many such depressions observed when $T \leq 1.80$ K. It is the short spin-up cycle in Fig. 6A and it was preceded by the long spin-up cycle in Fig. 6B. In each cycle, ω was held constant at ω_i during the interval $t_i \leq t \leq t_i'$, $i = 1,2$, and $t_i' < t_{i+1}$. During the $i = 1$ interval the bulk liquid was seen to be spun to the wall of the container as expected since $\omega_1 >> \omega_0$ of eq. (2); $t_1' - t_1 \simeq 15$ sec for all cycles. For example, in Fig. 6, $\omega_1 = 14.4$ sec^{-1} and $\omega_0 = 4.01$ sec^{-1} where ω_0 was measured during the $i = 0$ interval of the long spin-up cycle (the short cycle has no $i = 0$ interval).

Letting the time be masured relative to the initial instant of motion in each cycle, the short spin-up cycles had $t_1 < 2$ sec while the long spin-up cycles had $t_1 > (t_1 - t_0') \gtrsim 20$ sec and $d\omega/dt$ was nearly constant during the t_0', t_1 interval of spin-up. The depressions observed in the short spin-up cycles were

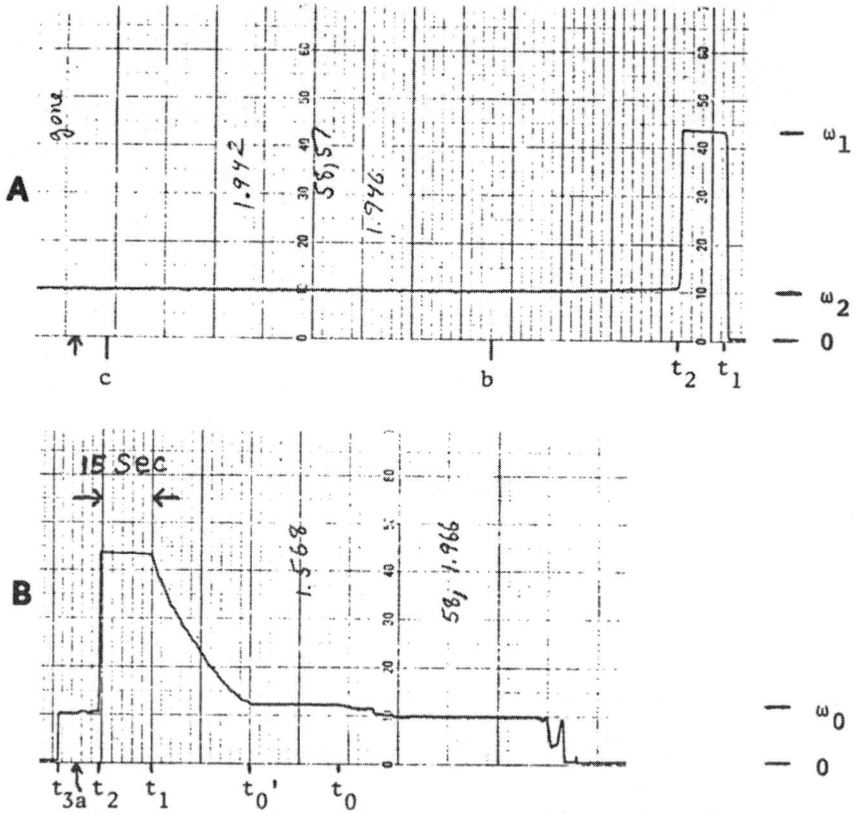

Figure 6. Chart record of ω for (A) short spin-up cycle and (B) long spin-up cycle.. Time increases from right to left.

metastable; ω_2 was selected to be less than ω_0 so that the stable equilibrium condition was for bulk He II to entirely cover the substrate with a central thickness of $\zeta_0(\omega_2)$ and a parabolic profile. For Figs. 5 and 6, the initial ($\omega = 0$) He II thickness was $\zeta_0(0) = 47$ μm and if the corresponding volume were in equilibrium with $\omega = \omega_2$, the minimum thickness should be $\zeta_0(\omega_2) = 17$ μm from (2). Contrast this to the depression in Fig. 5 which has $\zeta \leq 0.3$ μm at its center. Fig. 5 was photographed at $t_2 + 175$ sec of Fig. 6A (point c); it is similar to one at $t_2 + 57$ sec (point b).

Approximately 20 observations were made with T = 1.65 - 1.80 K, $t_1 < 2$ sec, $\omega_2 \simeq 12 \pm 2$ sec^{-1}, and a range of $\zeta_0(\omega_2) < \zeta_0(0) < 50$ μm. In each cycle a depression was still present at $t_2 + 120$ sec and sometimes bulk He II had not yet covered the substrate at $t_2 + 9$ min. In the same temperature interval, with $t_1 > 20$ sec the formation of metastable depressions was suppressed. For example, a photograph taken at $t_2 + 7$ sec during the slow spin-up cycle of Fig. 6B (point a) was similar in appearance to Fig. 5; bulk liquid had covered the entire substrate with a nearly parabolic profile.

There is additional evidence that the depressions observed in the 1.6 - 1.8 K temperature interval were caused by a large superfluid vortex. At T = 2.16 K, 10 cycles were observed with $t_1 < 2$ sec and ω_0, ω_1, and ω_2 similar to their low temperature values. No depressions were seen which indicates that they are not produced when ρ_s/ρ is small. Furthermore, the calculated profiles of a single superfluid vortex in a stationary cylinder ($v_n = 0$) provided a good fit to depressions which persisted when the cylinder rotation was stopped after t_2' of short spin-up cycles at T = 1.68 K.[1,2] The only free parameter in the fits were the circulation . With $\Gamma \simeq 400\ \kappa$, the maximum error in the fits was only about 1 f over the entire thickness of the films (6 μm). These Γ are similar in magnitude to that estimated for Fig. 5. Finally, the observations that metastable depressions are created when $t_1 < 2$ sec but usually not when $t_1 > 20$ sec might be explained as a critical velocity phenomena. With $t = t_1 < 2$ sec, near the center of the cylinder: $\omega_1 r > v_c(\zeta)$ where $v_c(\ )$ is the local superfluid critical velocity for film flow relative to the substrate. The bulk He II was seen to not yet be spun to the walls so that $\zeta \simeq 10$ μm and[9] $v_c(\zeta) \simeq (\zeta\ \mathrm{cm})^{-1/4} \simeq 4$ cm/sec. For the cases with $t_1 > 20$ sec, the film was observed to thin during the t_0', t_1 spin-up interval (Fig. 6B) so that when $t = t_1$, $\zeta << 1$ f and $\omega_1 r < v_c(\zeta) \simeq 25$ cm/sec. This suggestion is consistent with the reported observation that a film with $\zeta = 300$ Å rotates only where $\omega r > v_c$.

References

(1) P. L. Marston, in "Proc. of the 14th Internat. Conf. Low Temp. Phys.," edited by M. Krusius and M. Vuorio (North-Holland, Amsterdam, 1975), p. 268.

(2) P. L. Marston, Ph.D. Dissertation, Stanford University (1976).

(3) E. Vittoratos and P. P. M. Meincke, Phys. Rev. Lett. 34, 796 (1975).

(4) D. V. Osborne, Proc. Roy. Soc. (London) A63, 909 (1950).

(5) R. Meservey, Phys. Rev. 133A, 1471 (1964).

(6) Since this correction was not known when ref. 1 was written, $\alpha_s r^2$ should be added to the profiles reported there.

(7) R. R. Turkington and D. V. Osborne, Proc. Roy. Soc., 82, 614 (1963).

(8) K. C. Harvey and A. L. Fetter, J. Low Temp. Phys. 11, 473 (1973).

(9) W. M. Van Alphen et. al, Phys. Lett. 20, 474 (1966).

THE DECAY OF PERSISTENT CURRENTS IN UNSATURATED SUPERFLUID ^{4}He FILMS*

K. K. Telschow,† D. T. Ekholm, and R. B. Hallock

Department of Physics and Astronomy
University of Massachusetts
Amherst, Mass. 01003

Abstract: Persistent currents of unsaturated ^{4}He films are observed to decay at a rate which depends strongly on film thickness. Preliminary results on the temperature dependence of the decay rate are reported. Possible models which might explain the observed features of the decays are discussed.

Recently the first observations of the decay of unsaturated ^{4}He film persistent currents in an open geometry as a function of film thickness were reported.[1] Those measurements were conducted primarily at T = 1.45K. We wish to report here preliminary measurements of the temperature dependence of the decay of these currents at a thickness of 8.7 atomic layers.

The apparatus is the same as has been described previously[1] and the reader is referred to Ref. 1 for further details concerning the apparatus. A schematic representation of the glass substrate flow path[1] is shown in Fig. 1. For this work, persistent film currents were made to flow around the ring by applying power, $\dot{Q}$, to the wire-wound heater Q. This creates a persistent current since application of sufficient heat to Q causes the flow velocity v_{s1} along the shorter flow path to become dissipative while v_{s2} continues to increase, thus increasing the circulation around the ring. When the heater Q is shut off, circulation remains and the

*Supported by the National Science Foundation through Grant DMR76-08260.
†Present address, Department of Physics and Astronomy, Southern Illinois University, Carbondale, Illinois.

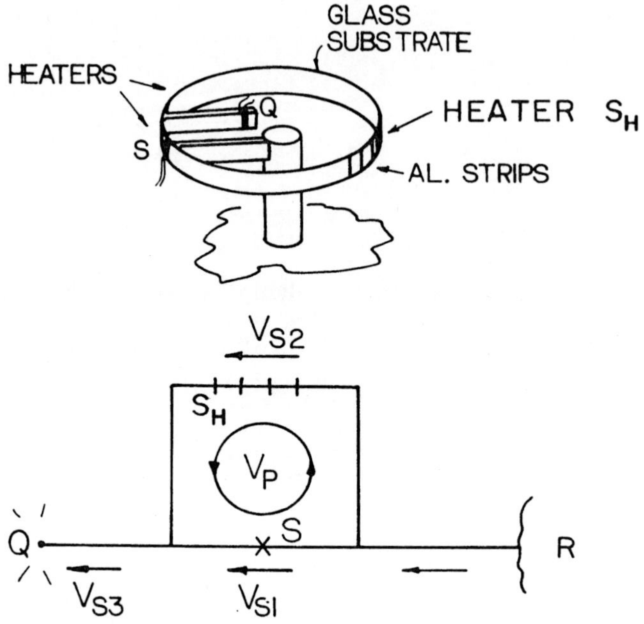

Figure 1. Representation of the pyrex flow path used for these measurements.[1] The lower portion of the figure shows a more schematic equivalent picture.

trapped velocity v_P can be studied by the techniques of Doppler-shifted third sound.[2] The presence of the third sound does not appear to enhance the degradation of the persistent currents.[3]

An example of decays of persistent currents of various starting velocities is shown in Fig. 2. There is no apparent systematic dependence of the observed fractional decay per decade on the initial ($v = v_o$ for $t = t_o = 1$ min) velocity of the persistent current. Thus, with

$$v_s = v_o - B \log t/t_o \ , \tag{1}$$

this means

$$v_s = v_o(1 - \xi \log t/t_o), \tag{2}$$

where the fractional decay per decade, ξ, is independent of v_o. We can conclude then that

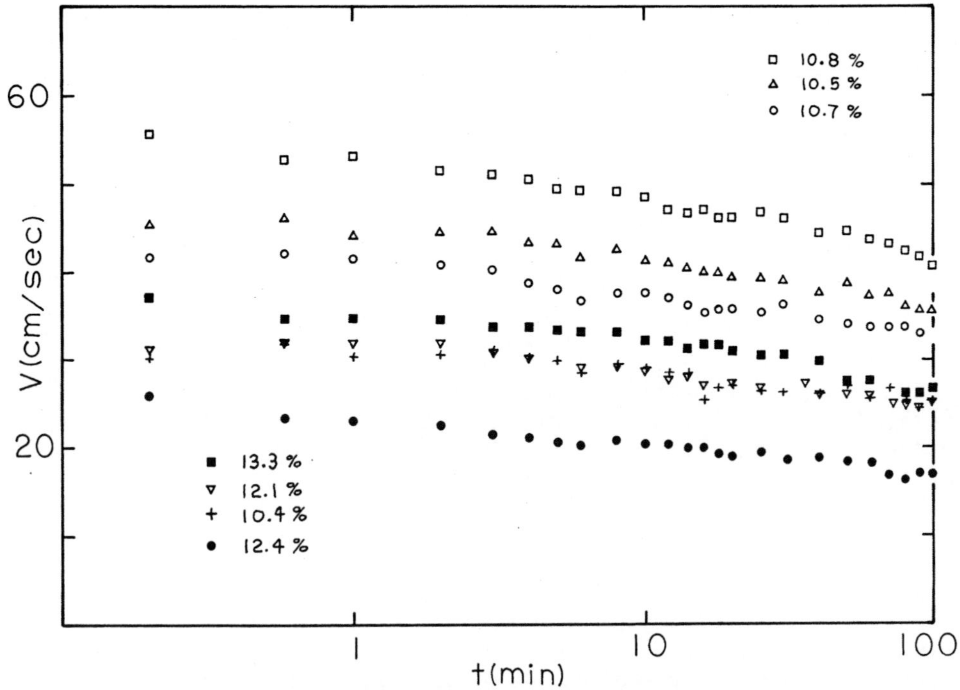

Figure 2. Observed superfluid film flow velocity around the ring as a function of time. Various different starting velocities are used. The numbers beside the symbol key represent the observed fractional decay per decade, ξ, for the particular measurement. These measurements were made at a film thickness d = 8.7 atomic layers and a temperature T = 1.52K.

$$\frac{dv_s}{dt} = -\frac{\xi v_o}{t_o} \exp\left[-\frac{2.303}{\xi}\left[1-(v_s/v_o)\right]\right] . \qquad (3)$$

where ξ has a strong dependence on film thickness.[1]

If one assumes eq. (3) to be consistent with the general fluctuation[4] picture where

$$\frac{dv_s}{dt} = -K A \nu_o e^{-E_a/kT} , \qquad (4)$$

where K is the quantum of circulation (h/m), A the film cross sectional area and ν_o an attempt frequency, then the activation energy, E_a, is observed to be linear in velocity. Since the behavior characterized by Eq. (2) is observed over changes in v_s/v_o as large as 50% in some cases, this represents a situation somewhat different from the usual one where only small changes in v_s away from a critical value are observed (and hence one can no longer sensibly expand[4] E_a about $v_s = v_c$).

One may speculate as to possible mechanisms which might give rise to superfluid deceleration consistent with Eq. (3). One such mechanism is analogous to the phenomena known as flux creep in superconductors first described by Anerson.[5] One imagines vortex lines in the film oriented perpendicular to and pinned to the substrate. In such a model the deceleration of the superfluid would presumably result from the motion of these vortex lines. If this is in fact a realistic description it might be possible to observe a flow regime at higher values of v_s than used to date which would be equivalent to the phenomena of flux flow in superconductivity. A mechanism which involves vortices such as this is attractive since eq. (3) shows that at any time dv_s/dt depends on v_o and hence the helium currents have a memory.

We have recently begun to measure the temperature dependence of the fractional decay per decade, ξ, at fixed thickness. Our preliminary results for this temperature dependence at a film thickness of d = 8.7 atomic layers are shown in Fig. 3. Each point shown is obtained from at least eight separate measurements of at the particular temperature. The best reproducibility was evidenced at T = 1.52K.

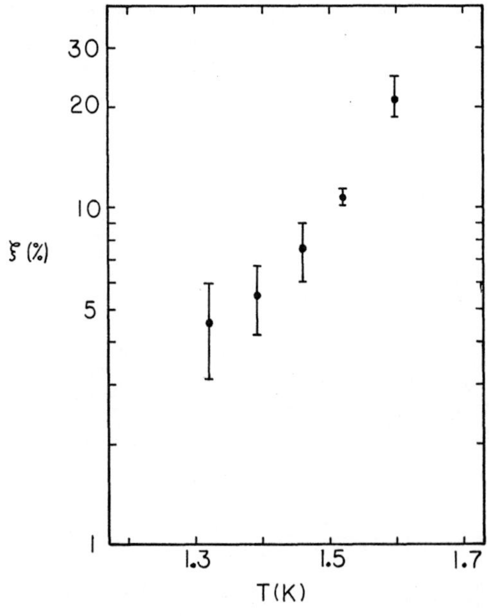

Figure 3. Temperature dependence of the fractional decay per decade, ξ, at d = 8.7 layers.

We know of no obvious explanation for the strong dependence of ξ on both film thickness and temperature. The previously observed[6] dependence of $\langle\rho_s\rangle/\rho$ on film thickness and $\langle\rho_s\rangle/\rho T$ on temperature are not adequate to fully describe the behavior of ξ over the range of thicknesses and temperatures studied to date.

References

(1) K. L. Telschow and R. B. Hallock, Phys. Rev. Lett. 37, 1484 (1976).

(2) K. R. Atkins and I. Rudnick, Progress in Low Temperature Physics, ed. C. J. Gorter (North-Holland, Amsterdam, 1970), Vol. 6, Chap. 2. See also, K. L. Telschow, I. Rudnick and T. G. Wang, Phys. Rev. Lett. 32, 1292 (1974).

(3) The observed (R, K. Galkiewicz, K. L. Telschow and R. B. Hallock, J. Low Temp. Phys. 26, 147 (1977)) fact that the presence of third sound does not destroy the stability of stable persistent currents is perhaps understandable. That such should also be the case for persistent currents which decay is perhaps somewhat of a surprise.

(4) S. V. Iordanskii, Zh. Eksp. Theor. Fiz 48, 708 (1965) [Sov. Phys. JETP 21, 467 (1965)]; and J. S. Langer and M. E. Fisher, Phys. Rev. Lett. 19, 560 (1967). See also, J. S. Langer and J. D. Reppy, Progress in Low Temperature Physics, ed. C. J. Gorter (North-Holland, Amsterdam, 1970), Vol. 6, Chap. 1.

(5) P. W. Anderson, Phys. Rev. Lett. 9, 309 (1962).

(6) J. H. Scholtz, E. O. McLean and I. Rudnick, Phys. Rev. Lett. 32, 147 (1974).

ELEMENTARY EXCITATIONS IN VERY THIN SUPERFLUID ^{4}He FILMS*

J. E. Rutledge, W. L. McMillan, and J. M. Mochel

Department of Physics and Materials Research Laboratory
University of Illinois
Urbana, Illinois 61801

We present high precision measurements of third sound velocity in ^{4}He films between 0.1 K and 1.5 K for superfluid surface densities between .16 atomic layers and 5.25 atomic layers. By reformulating Landau's quantum hydrodynamics using only the area density we have been able to use these measurements to describe an elementary excitation spectrum and to add a surface roton branch.

For thin films with a superfluid surface density of order one atomic layer the Ginzburg-Pitaevskii[1] healing length concept is not applicable at any temperature. Even the film thickness D (used here with units of atomic layers) is a fuzzy concept and is definable only with an accuracy of a monolayer. For monolayer films the film thickness is not a macroscopically definable quantity and it cannot be used in the macroscopic hydrodynamic theory. By allowing the helium film to close upon itself on the inner surface of a quartz capsule, we have been able to establish a third sound resonance.[2] As a result we have been able to measure the third sound velocity to 1 part in 10^5. The observed T^3 dependence[3] of the changes in the square of the third sound velocity at low temperatures suggested a two-dimensional spectrum of Landau elementary excitations (surface phonons). The thermally excited surface phonons form a normal fraction which is pinned to the substrate;

---

*Work supported by the National Science Foundation under Grant DMR-76-01058 and DMR-72-03291.

the resulting reduction in the superfluid surface density explains the T^3 dependence of $c_3^2(T)$. This point of view is common to bulk ^4He where, below 1 K, Landau's elementary excitations[4] describe the normal fraction. This interpretation, for restricted geometries, is also argued for by Padmore and Reppy.[5] Chester[6] has pointed out that the order of magnitude and temperature dependence of the time-of-flight third-sound velocity measurements of Scholtz et al.[7] above 0.8 K can be explained by a surface roton contribution to the normal surface density. Thus the surface phonon and surface roton have been discussed previously.

QUANTUM HYDRODYNAMICS

We now reformulate quantum hydrodynamics using surface quantities. We define a complex order parameter $\psi(\vec{x})$ which is proportional to the wave function of the condensate so that the surface density at T = 0K of superfluid atoms is

$$\sigma(\vec{x}) = |\psi(\vec{x})|^2. \tag{1}$$

The usual quantum mechanical current density is then

$$\vec{j}_s(\vec{x}) = \mathrm{Re}[\frac{\hbar}{im} \psi \nabla\psi] \equiv \sigma(\vec{x})\vec{v}_s(\vec{x}) \tag{2}$$

where m is the helium atom mass. Note that there is no conceptual difficulty in defining a condensate wave function for a two-dimensional system at T = 0K.

There are several terms in the energy of this quantum state. The kinetic energy of the moving film is

$$H_1 = \int d^2 x \frac{\hbar^2}{2m} |\vec{\nabla}\psi|^2 \tag{3}$$

where the integral is taken over the physical surface area. The Van der Waals binding of the film to the substrate can be represented by the following expression

$$H_2 = \int d^2 x \frac{A}{2(a + \sigma)^2} \tag{4}$$

where A and a are constants. The chemical potential term is

$$H_3 = - \int d^2 x \, \mu\sigma \quad . \tag{5}$$

In addition to these three "obvious" terms we include a fourth term of the form

$$H_4 = \int d^2 x \frac{1}{2} B(\sigma) \; (\vec{\nabla}\sigma)^2 \tag{6}$$

where B may be a function of surface density. For thick films we can write $\sigma(\vec{x}) = \rho_o d(\vec{x})$ where ρ_o is the bulk particle density and d is the film thickness. Then we have

$$H_4 = \frac{1}{2}B\;(\infty)\;\rho_o^2 \int d^2 x \; (\vec{\nabla}d)^2 \tag{7}$$

which is just the surface energy with $B(\infty)\,\rho_o^2 = \beta_o$, the surface tension. For bulk helium $\beta_o = .378$ ergs cm^{-2}. Thus, for thick films, H has a simple physical interpretation as a surface energy. With these energy terms we can obtain the equation of motion and compute the excited states obtaining

$$(h\omega_k)^2 = \frac{3A\;_o\hbar^2 k^2}{m(a+\sigma_o)^4} + \left[\frac{\hbar^2 k^2}{2m}\right]\left[1 + \frac{4Bm\sigma_o}{\hbar^2}\right] \tag{8}$$

where $\sigma_o = |\psi_o|^2$ is the average superfluid surface density of the film. The frequencies given by Eq. (8) are the frequencies of the collective modes of the condensate which are the elementary excitations of the thin film. These elementary excitations are surface density waves with a linear dispersion and a sound velocity

$$c^2 = [3A\sigma_o/m(a+\sigma_o)^4] \tag{9}$$

at long wavelengths and an upward dispersion for shorter wavelengths. We call these elementary excitations surface phonons. In the thin film limit the elementary excitation spectrum approaches the free particle energy for large k.

The theory is based on macroscopic quantum hydrodynamics and the elementary excitation spectrum. Eq. (8) is expected to be valid only for wave numbers less than 1/(interparticle spacing) and less than 1/(film thickness). For short wavelengths and thicker films we expect our surface wave excitation to go over smoothly to the classical ripplon excitation.[8] We can modify our dispersion relation easily so that it goes over to this limit correctly.

$$(\hbar\omega_k)^2 = \left(\frac{3A\sigma_o\hbar^2 k^2}{m(a+\sigma_o)^4} + \left(\frac{\hbar^2 k^2}{2m}\right)^2 \left[1 + \frac{4Bm\sigma_o}{\hbar^2}\right]\right)\tanh\left[\frac{k\sigma_o}{\rho_o}\right] / \left[\frac{k\sigma_o}{\rho_o}\right] \tag{10}$$

This correction is not very important for the range of film thicknesses and temperatures where the surface phonons are the dominant excitations.

We observe in our experiment an excitation with an energy gap which we will call a surface roton. This presumably is related closely to the two-dimensional roton studied theoretically by Padmore using the Feynman-Cohen method.[9] A simple, intuitive picture of this excitation is that of a bound pair of vortices of opposite circulation, with the surface roton being the smallest or most tightly bound pair permitted by quantum mechanics. We will assume a phenomenological dispersion relation in the surface roton region

$$\hbar\omega_k = \Delta + \frac{\hbar^2(k-k_o)^2}{2m^*} \tag{11}$$

in close analogy with the bulk roton dispersion relation. The two branches of the dispersion curve cross at a wavenumber k_c. The thermodynamics are insensitive to the details of this crossover region; we retain the surface phonon branch for $k < k_c$ and the surface roton branch for $k > k_c$.

THIRD SOUND VELOCITY

At finite temperature the third sound wave is a long wavelength surface density wave accompanied by a temperature wave. The thermally excited elementary excitations behave as a "normal fluid" which is pinned to the substrate. The superfluid surface density $\sigma_s(T)$ can be calculated using an argument due to Landau.[4] The thermodynamic energy $E(\sigma_o,T)$ is a well-defined function of surface density σ_o and temperature. The work done to increase the surface density by an amount $d\sigma_o$ is

$$dW = \left.\frac{dE}{d\sigma_o}\right]_s d\sigma_o \ . \tag{12}$$

Now using the standard macroscopic hydrodynamic argument the third sound velocity under adiabatic conditions is

$$c_3^2(T) = K(T)\sigma_s(T)/m \tag{13}$$

where the adiabatic elastic constant is

$$K(T) \equiv \left.\frac{d^2E}{d\sigma_o^2}\right] \quad . \tag{14}$$

This assumes that the thermal boundary resistance between the substrate and the elementary excitations in the film is large enough that the thermal equilibration time is much longer than the period of the third sound wave; this condition is satisfied at low temperatures in the present experiment. One can define an isothermal third sound velocity, but we will not need this quantity. We find that the third sound velocity approaches the surface phonon velocity at low temperature with corrections proportional to T^3.

$$c_3^2(T) \simeq c_3^2 \left[1 - \frac{1.202T^3}{2\pi m\hbar^2 c^4 \sigma_o} \left(\frac{3}{2} - \frac{\sigma_o^2}{c} \frac{\partial^2 c}{\partial \sigma_o^2} \right) \right] \tag{15}$$

At higher temperature there are deviations from the T^3 law due to dispersion and the integrals involved in the derivation of $c_3(T)$ must be performed numerically. It is convenient to write

$$c_3^2(T) \simeq c^2(1-\alpha(T)T^3) \tag{16}$$

and to discuss the quantity $\alpha(T)$. At low temperature only the linear portion of the excitation spectrum is involved and $\alpha(T)$ is a constant. At intermediate temperatures the upward dispersion of the excitation spectrum is important in reducing the number of excitations and $\alpha(T)$ decreases. Finally at high temperature (0.6°K) the surface roton branch is excited and $\alpha(T)$ increases exponentially.

ANALYSIS

For small k, E(k) is given by Eq. (10) and requires three parameters; σ_o, a, and B. For large k, E(k) is given by Eq. (11) and also requires three parameters; Δ, k_o, and m*. With E(k) given in these two domains $\sigma_s(T)$ and K(T) are generated numerically and $c_3(T)$ is calculated via Eq. (13) and compared with the experimental results as shown in Fig. 1 for D = 1.77 atomic layers.

Because of the dominant T^3 behavior, comparison is also made with the function

$$\alpha(T) = \frac{(1 - c_3^3(T)/c^2)}{T^3} \tag{17}$$

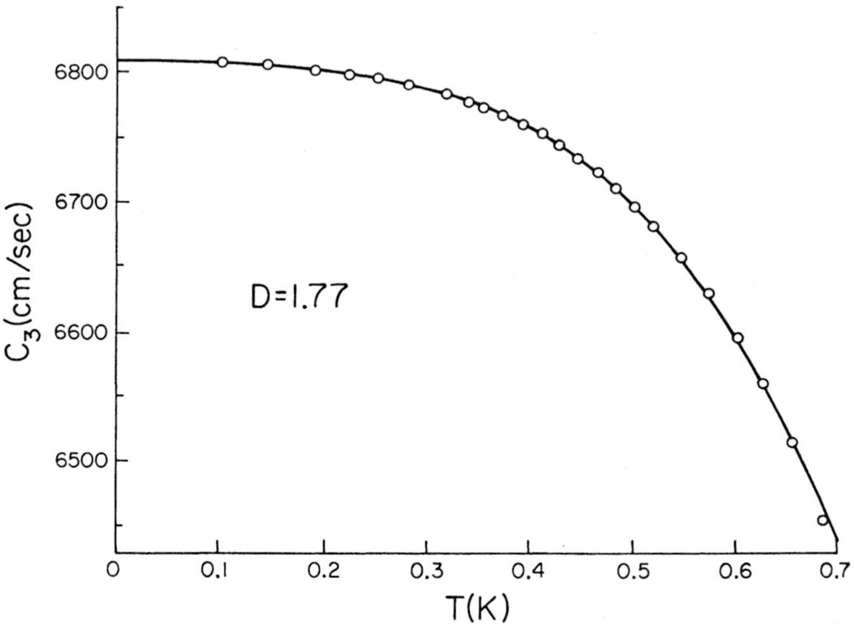

Figure 1. The open circles are the measurements of third sound velocity *versus* temperature. At low temperatures the resolution is 2 parts in 10^5; note the expanded scale on the ordinate. The solid line is our model of two dimensional superfluidity discussed here. D is the coverage of ^{4}He in atomic layers.

as shown in Fig. 2. In fact, the experimental curve $\alpha(T)$ motivated the attempt to make detailed calculations of E(k) for thin films. The remarkable precision of our measurement alows us to measure the deviations of E(k) from a linear dispersion. Had the scatter in the measurements of the third sound velocity below 0.6 K been as large as 1 part in 10^3, the experimental $\alpha(T)$ would have been scattered about a horizontal line and $E = \hbar ck$ would have been consistent with our measurements.

Since the net coverage, D, is determined from the vapor pressure measurements there are only two independent parameters needed for small k since $D = a + \sigma_o$. However, we can also determine a directly from the 3rd sound data. Since measured values of $c_3(T)$ are within 1% of c we can, for this analysis, let c equal the coldest measured value of the third sound velocity. At T = 0, we can use Eq. (9) to give

$$c^2 D^4 = \left[\frac{3A}{m}\right] (D - a) \tag{18}$$

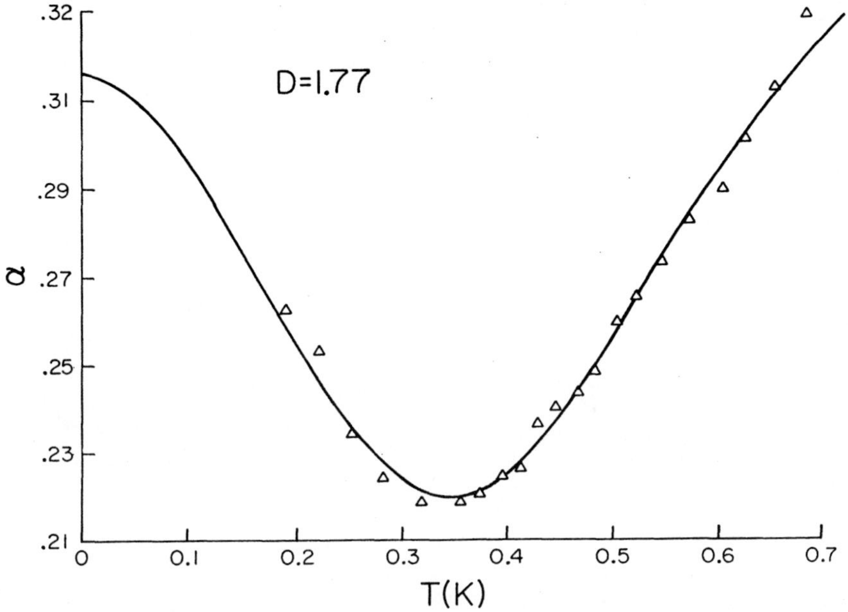

Figure 2. The open triangles are from experiment and the solid line from the theory discussed here. α is defined by Eq. (17) and would be independent of temperature if dispersionless two-dimensional phonons were the only excitations. At low temperatures the downward trend is due to surface tension; at higher temperatures the upward trend is due to surface roton excitations.

In Fig. 3 the intercept gives a value for a of $1.25 \pm .05$ atomic layers. a is considered as that portion of the ^{4}He coverage which does not participate in third sound wave motion. No more can be said about a. We cannot tell whether a is the coverage of frozen helium or of some temperature independent normal fluid, or a combination of both. From the slope of the line drawn in Fig. 3, A is 13.7 K which is within 6% of the calculated value for an argon substrate of 14.5 K. In practice then, there is only one adjustable parameter for T less than about .4 K, the surface energy, B. However, c becomes our second working parameter for low k because D and a are known to only a few percent and c must be adjusted to a part in 10^5 for the analysis of the surface tension. Of the three adjustable parameters for high k, the effective mass m* and the momentum of the roton minimum k_o can be combined into a single parameter ($m^*k_o^6$). Since we cannot separate m* from k_o in this analysis we have arbitrarily fixed m* to 0.2 m, the value of the roton effective mass for bulk helium. A partial justification for this choice is the resulting behavior of k_o as a function of coverage. To leave k_o fixed and vary m* during the fitting procedure would have produced unphysically small values for m* of 10^{-3}m.

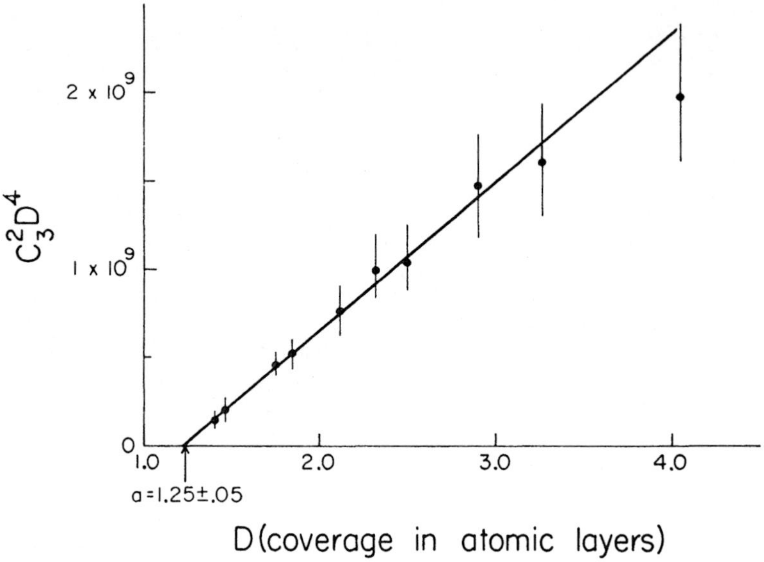

Figure 3. This plot tests the validity of Eq. (18) near T = 0 and the intercept gives a = 1.25 $\pm$ .05 atomic layers. a is the coverage of ^{4}He which does not participate in third sound wave motion. The vertical error bars result from an uncertainty of the coverage of $\pm$ 5%.

CONCLUSIONS

Table I summarizes the results of fitting theory to experiment. The errors represent the range over which a good fit can be made while optimizing the other parameters. In Fig. 4 the resulting E(k) is plotted for three thicknesses. The intermediate range of E(k) is not shown since we have no information or model to reveal this region.

Perhaps the most satisfying result is in the thick film limit where surface tension, energy gap, and roton minimum approach the bulk values as indicated by the arrows in Fig. 5 through Fig. 7 (the thickest film is not shown in these Figures but appears in Table I). There was no *a priori* reason, within the theory, for this to happen.

The straight line in Fig. 4 has a slope of c for each film thickness. The rise of the surface phonon branch above this line illustrates the effect of the free particle-like excitations for very thin films and, in addition, ripplons for thicker films. As

Table I. The results of fitting theory to experiment. c is the surface phonon velocity for k = 0 and the third sound velocity for T = 0. B is the surface energy in the thick film limit. Δ is the energy of the roton minimum and k_o is the momentum of the roton minimum.

Run	D(0) (cm) ±5%	c (cm/sec) ±.1	B (erg/cm^2) ±.025	(K) ±.2	k_o ($\AA^{-1}$) ±.1
I	1.41	6340.0	.05	2.40	0.8 ± .5
II	1.45	6281.3	.02	3.15	1.35 ± .1
III	1.77	6807.2	.12	3.35	0.9
IV	1.85	6622.1	.27	4.90	1.6
V	2.11	6267.2	.23	5.26	1.9
VI	2.35	5936.4	.30	5.22	2.0
VII	2.49	5111.2	.34	5.40	1.8
VIII	2.89	4612.8	.32	5.32	1.8
IX	3.31	3657.1	.33	8.4 ± .5	1.85
X	4.05	2722.5	.37	8.0 ± .5	1.8
XI	6.58	1455.8	.37	8.0 ± 1	1.9 ± .2

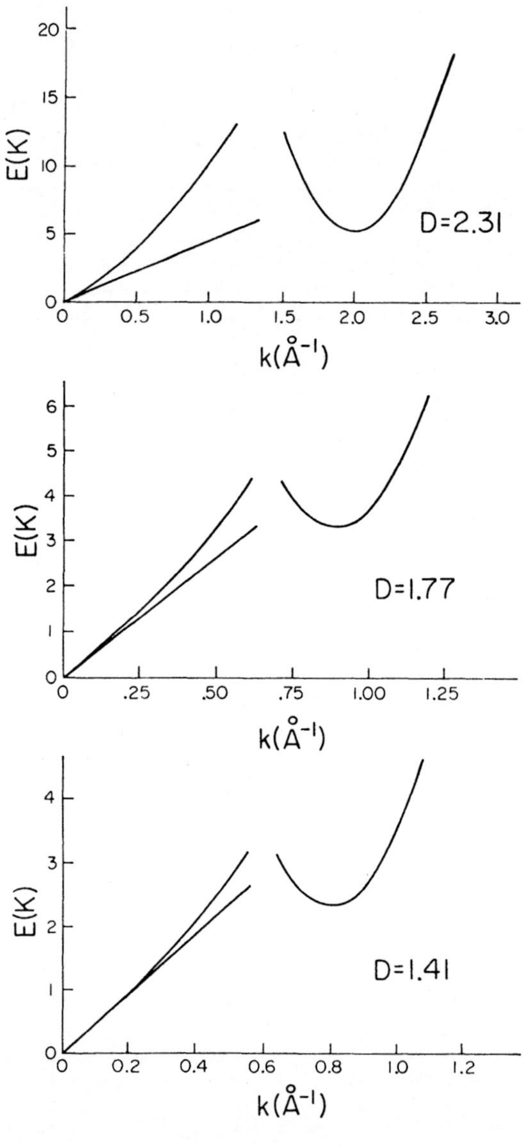

Figure 4. E(k) for three coverages of ^{4}He based on a fit of theory to experiment. There is no component of k perpendicular to the film. Near k = 0 the slope approaches the T = 0 third sound velocity indicated by the striaght line. For large k the surface energy causes an upward bending which is larger for thicker films. For large k, a surface roton contribution is required to describe the experimental results.

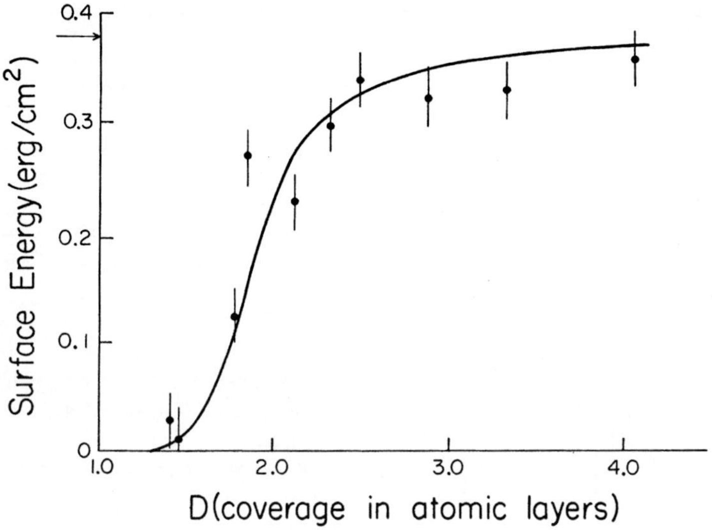

Figure 5. Surface energy, the coefficient of the square of the gradient of the coverage, required to fit experiment. The arrow at the top of the ordinate indicates the surface energy of bulk ^{4}He. The surface energy is near zero while the film is still superfluid indicating a 2-D gas-like state.

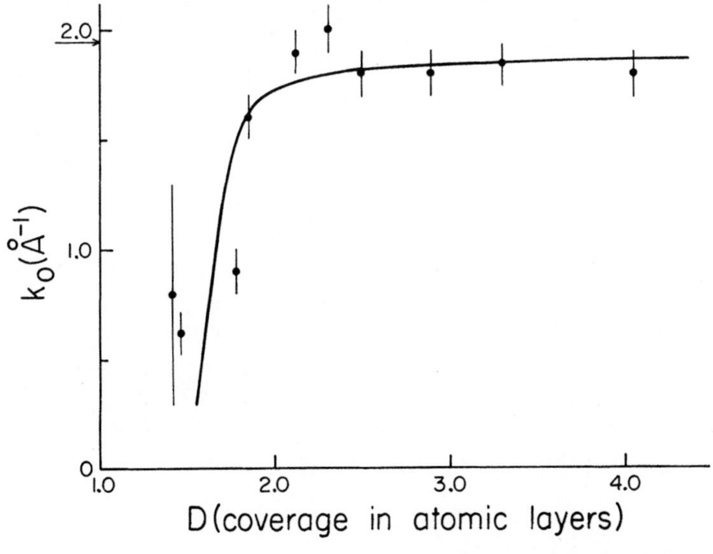

Figure 6. The momentum of the roton minimum, k_o, required to fit experiment. The arrow at the top of the ordinate indicates the value of k_o for bulk ^{4}He. The fall of k_o with smaller D indicates, as does surface tension, that the superfluid is becoming more gas-like as D approaches a.

shown in Fig. 5 when D approaches a, the superfluid surface energy falls to zero.

Both the roton gap and roton minimum fall toward zero along with the surface tension. Again, the both turn downward in the vicinity of a as shown in Fig. 6 and Fig. 7. Both the reducton in surface tension and the reduction in k_o indicate a decrease in the surface density of the mobile portion of the film. The energy gap behaves in a rather peculiar fashion, having a constant value of 5.3 $\pm$ 1 K between a D of two and three atomic layers. Finally, for larger D the gap abruptly jumps to near its bulk value.

This theory permits a quantitative interpretation of our data in terms of an elementary excitation spectrum. The fact that the experimental $\alpha(T)$ curves go through a minimum shows that the real dispersion curve must bend upward for small k and then bend down again for larger k. We have forced our theoretical dispersion to bend properly by replacing the surface phonon branch by a surface roton branch for $k > k_c$ and rather arbitrarily assumed that the roton effective mass was the same as that for bulk rotons. The details of our dispersion curve may not be right in that the

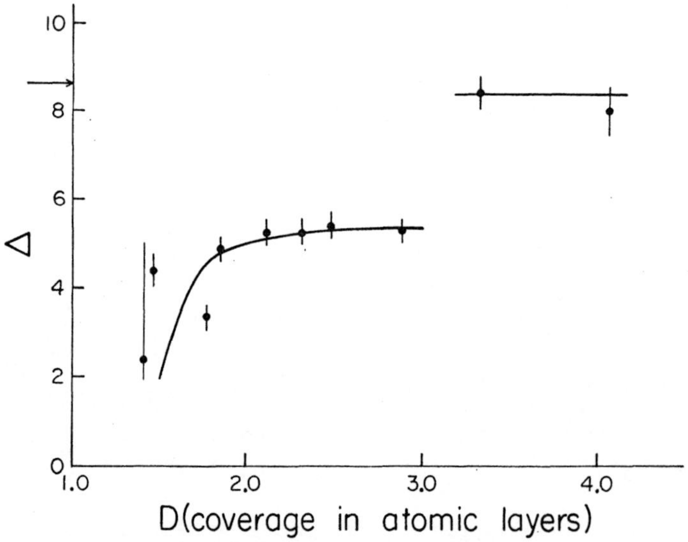

Figure 7. The energy, Δ, of the roton minimum required to fit experiment. The arrow at the top of the ordinate indicates the value of Δ for bulk ^{4}He. The average value of Δ between 2 and 3 atomic layers if 5.3 $\pm$ .1 K. The sharp jump to the bulk value of Δ is not understood.

region around k_c may not be as sharply peaked and the roton minimum may be shallower than the dispersion curves shown in this paper, especially in the thinner films. The surface tension, roton gap and roton momentum from the fitting all go to their proper bulk helium values as the film thickness. As the surface density falls toward a, the surface energy and the roton minimum move toward zero. In this respect, the films begin to resemble a two-dimensional superfluid gas. Thus we have shown that the third sound measurements provide a detailed, quantitative probe of the elementary excitation spectrum in thin superfluid helium films.

References

(1) V. L. Ginzburg and L. P. Pitaevskii, Soviet Phys. - JETP 7, 858 (1958).

(2) B. Ratnam and J. M. Mochel, Phys. Rev. Lett. 25, 711 (1970).

(3) T. E. Washburn, J. E. Rutledge, and J. M. Mochel, Phys. Rev. Lett. 34, 183 (1975).

(4) L. Landau, J. Phys. (U.S.S.R.) 5, 71 (1941).

(5) T. C. Padmore and J. D. Reppy, Phys. Rev. Lett. 33, 1410 (1974).

(6) M. Chester and L. Eytel, Phys. Rev. B 13, 1069 (1976).

(7) J. H. Scholtz, E. O. McLean, and I. Rudnick, Phys. Rev. Lett. 32, 147 (1974).

(8) C. G. Kuper, Physica 22, 1291 (1956); 24, 1009 (1956).

(9) T. C. Padmore, Phys. Rev. Lett. 32, 826 (1974).

NMR ON ^{3}He ADSORBED IN GRAFOIL* AT NEAR MONOLAYER COMPLETION: THE SOLID PHASE

B. P. Cowan,† J. R. Owers-Bradley, A. L. Thomson and M. G. Richards

Sussex University
Brighton, England

The effect of mobility of adsorbed atoms on thermal measurements such as specific heat and vapor pressure isotherms is not very direct. NMR relaxation times, however, do provide temporal information of a quite direct kind. Existing relaxation time data come principally from three studies. Rollefson,[1] using graphitized carbon black as substrate, carried out cw NMR measurements at 20.5 MHz. He concluded that for fractional coverages x above 0.7 of a completed monolayer the NMR line which is narrowed by motion at 4 K, broadens as the sample is cooled, reaching the rigid lattice value for the case of x = 0.9 by about 2.5 K. Grimmer and Luszcynski,[2] using pulsed NMR methods on ^{3}He adsorbed in grafoil, found that T_2 at 1 K and 4 K for a monolayer was about half as long at 20 MHz as at 10 MHz, which rules out dipole-dipole coupling as the main relaxation mechanism. Hedge, Lerner, and Daunt[3] working at 5.5 MHz in grafoil found a similar change to Rollefson in the NMR line width as a function of temperature for a sample near monolayer completion (x = 0.96), but the low temperature line width was still a factor 4 lower than the rigid lattice line width, suggesting some quantum tunnelling was occurring.

---

*Grafoil is a trademark of a product marketed by Union Carbide, Carbon Products Division, 270 Park Avenue, New York.

†Now at Centre d'Etudes Nucleaires de Saclay, BP No. 2, F-91190, Gif-sur-Yvette, France.

The data reported here relate to samples of ^3He adsorbed in grafoil. The form of the sample and other experimental details are described in a companion paper. The data were taken at 0.3, 1.0 and 2.0 MHz and with the angle β between the DC field H_0 and the normal to the grafoil sheets set at π/2 unless otherwise stated.

The phase diagrams can be explored either by varying the temperature or the fractional coverage x (monolayer completion is determined by the point B criterion applied to a 4.2 K isotherm). Fig. 1 shows the variation of the longitudinal and transverse relaxation times, T_1 and T_2 respectively, as x is increased from below 0.7 to above 1.1, the temperature remaining constant at 1.0 K. Data points above 10^{-3} sec were obtained from spin echoes, the relaxation being exponential; T_2 values below 10^{-3} sec were obtained from the line width of the cw adsorption signals, the line shape being assumed to be Lorentzian.

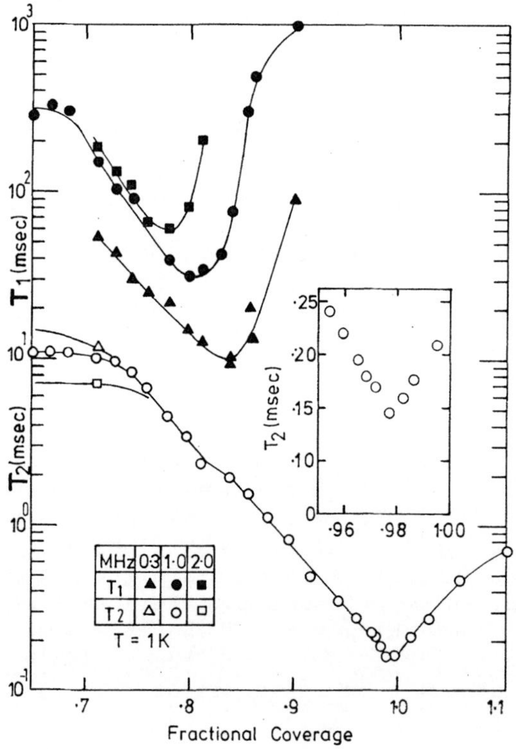

Figure 1. Longitudinal (T_1) and transverse (T_2) NMR relaxation times measured at 1 K as a function of fractional coverage x. x=1 corresponds to a completed monolayer using the point B criterion applied to a 4.2K isotherm. The inset figure gives more detail of the minimum in T_2 at $x \approx 1$ for the same conditions as the main figure except that the angle between the DC field H_o and the normal to the grafoil sheets is π/4 rather than π/2.

The shape of the data in Fig. 1 strongly recalls the behavior of a spin system whose correlation time increases from a value below ω^{-1} to times well above this, where ω is the Larmor frequency at which the measurements are made. Solid ^{3}He displays[4] very similar behavior as either the temperature or pressure is varied. In that system at about 1 K, the chief relaxation process is quantum tunnelling of neighboring spins which modulate the diplar field. The exchange frequency J_{3d}, which results from overlapping of single particle localized atomic wavefunctions, falls from about 10^8 sec^{-1} on the melting curve to 10^5 sec^{-1} at a pressure of about 300 atmospheres. A similar theory[5] applied to the present system yields the pair tunnelling frequency J_{2d} through

$$1/T_2 = 0.26\ M_2/J_{2d}$$

M_2 being the second moment of the NMR line. The non-adiabatic contribution to $1/T_2$ is related to $1/T_1$ and its presence can just be

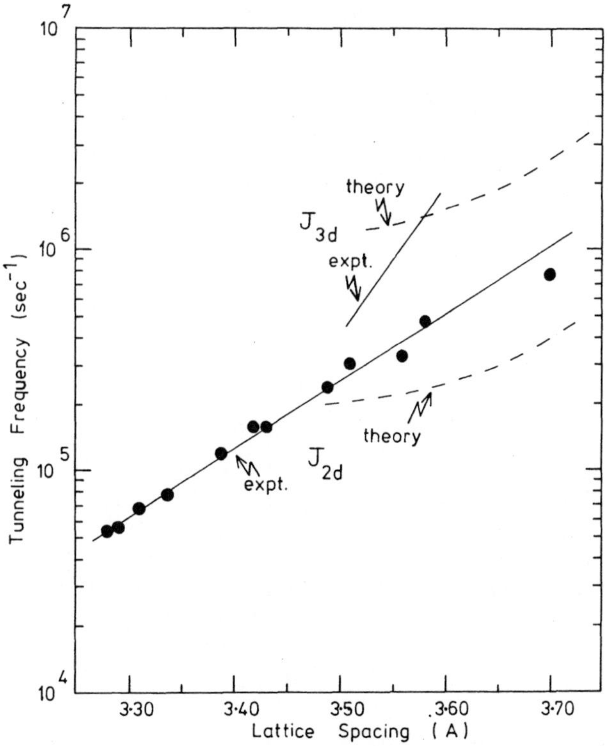

Figure 2. Values of the quantum tunnelling frequency for ^{3}He atoms in the hcp solid (J_{3d}) and adsorbed on grafoil (J_{2d}) as a function of nearest neightbor spacing. References are given in Ref. 6.

detected in the slight sideways shift in the T_2 points at about $x = 0.8$.

Fig. 2 shows[6] the resulting J_{2d} assuming a triangular lattice of ^3He spins coupled through their dipolar field. Also shown are J_{3d} and theoretical results based on models developed by Mullin and colleagues.

Observations of T_1 minima are particularly useful because they provide unambiguous values of correlation times τ_c at the corresponding values of x ($\tau_c \approx 1/\omega$ at the minimum). Also, through the approximate relation $(T_1)_{min} = \omega/M_2$, which yields $M_2 \approx 2 \times 10^{-8}$ sec^{-2} for $x \approx 0.8$, we get support for the assumption that dipolar coupling is the principal relaxation mechanism.

While the T_1 minimum and the T_2 variation for $0.72 < x < 0.98$ are well understood, outside this range there are some features that are difficult to interpret. On the low coverage side, specific heat data suggest[7] that we should see a solid-liquid transition at $x \approx 0.7$. Such a transition should be accompanied by a large change in mobility and it is surprising to see only a change in the slope of the T_1 and T_2 values as x is reduced. The process responsible for relaxation in the fluid phase is probably[8] diffusive motion of spins in the spatially varying demagnetizing fields caused by the misaligned crystallites of graphite whose susceptibility is highly anisotropic. The problem of melting is discussed further in connection with Fig. 4.

The sharp minimum in T_2 at $x = 0.98 \pm 0.01$ arises because the quantum mobility falls as the areal density increases, but when it becomes energetically favorable for additional gas to create a second layer of atoms, these atoms will be highly mobile and, probably through first-second layer exchange, they shorten the correlation time of atoms in the first layer and hence increase T_2. T_2 is proportional to the rate at which spins leave the first layer in such a process and this would be proportional to the number of atoms in the second layer, i.e. to $x - 0.98$. Fig. 3 shows this relation to be obeyed, with different slopes, for three values of the angle β.

It is interesting to note that T_2 never fell below 1 m sec before the grafoil was cleaned by heating it in vacuo for 24 hours at 1000°C. During this treatment gas equivalent to about a third of a monolayer was pumped away. Presumably the impurity atoms prevented the adsorbed helium from forming a homogeneous high density monolayer: the condition necessary for the short T_2 observed after cleaning. The data in Fig. 1 contrast sharply with that for ^3He adsorbed on Vycor glass[9] where T_2 varies with coverage in just the opposite way, probably because low coverage films consist of atoms

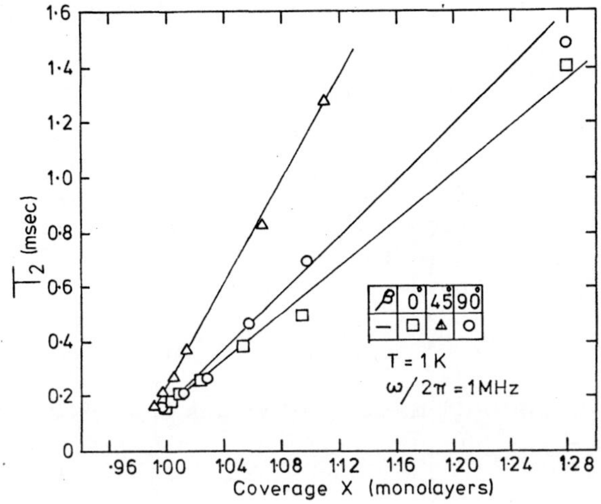

Figure 3. T_2 at coverages above a monolayer for various angles β between the DC field direction H_o and the normal to the grafoil sheets.

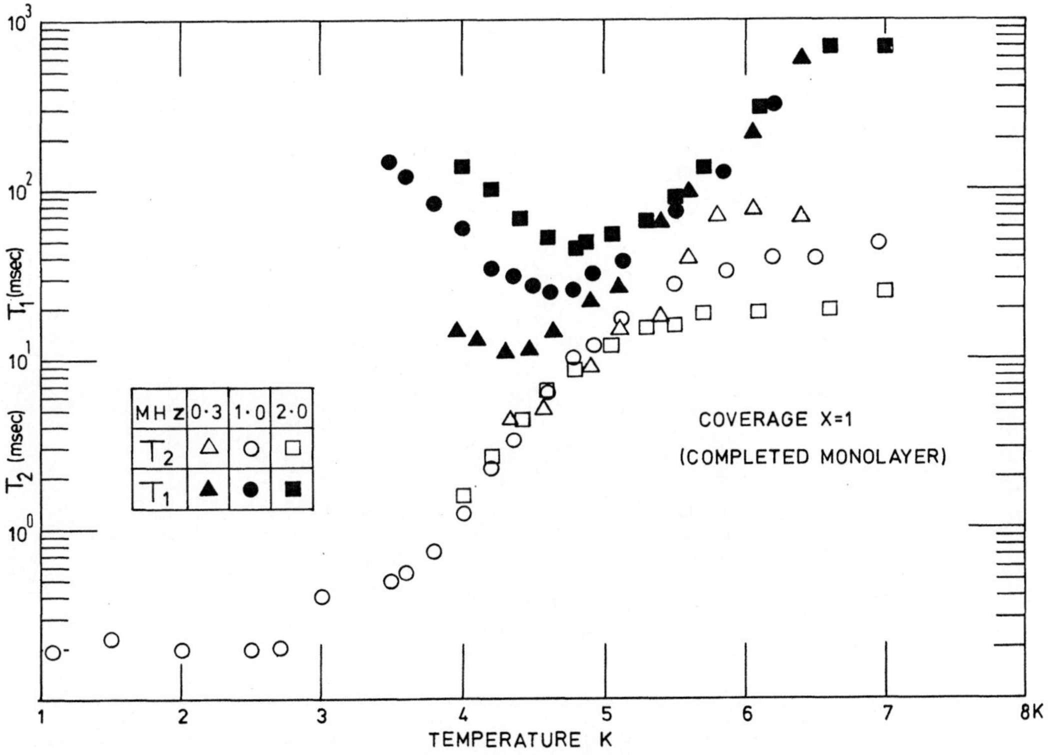

Figure 4. T_1 and T_2 as a function of temperature for a completed monolayer ($x \approx 0.98$).

situated in low energy sites where they are immobile. On approaching a monolayer mobility increases, but monolayer completion is not signified by any change in T_2.

The variation of T_1 and T_2 with temperature for a completed monolayer (x = 0.98) is shown in Fig. 4. The similarity to Fig. 1 is very striking with 1/T replacing x as abscissa. The temperature independence of T_2 from 1 to 3 K confirms the quantum mechanical

(a) Promotion of atoms into a second layer where they would be highly mobile. The effect on the first layer is either (i) to create vacancies or (ii) to increase the lattice spacing.

(b) The creation and motion of vacancies in the first layer.

(c) Thermally activated tunnelling in the first layer.

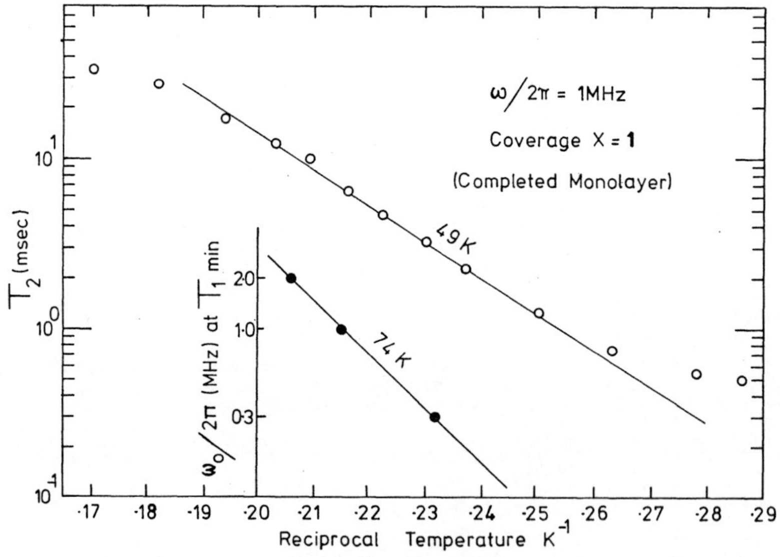

Figure 5. T_2 as a function of reciprocal temperature for a completed monolayer ($x \approx 0.98$). In the inset figure, the frequency for which a T_1 minimum occurs is plotted against the reciprocal temperature at which it is observed. The data points come from Fig. 4.

nature of the tunnelling at 1 K and suggests that there is little promotion of atoms to the second layer in this region. The steep increase in T_2 between 3.5 and 5.5 K is plotted as Fig. 5 to show that the main part of the increase can be attributed to a thermally activated process with an activation temperature of 49 $\pm$ 5 K. However, the data at the high temperature end are affected by a new process discussed below and the data may not fit an Arrhenius equation with a single activation energy. This possibility is supported by the plot of $\ell n 1/\tau_c$ against $1/T$ where τ_c is obtained from the T_1 minima by assuming $\omega\tau_c$ = constant $\approx$ 1 at the minimum. Since the T_1 data are apparently less affected by the new process occurring above 5 K, the activation temperature of 74 $\pm$ 6 K obtained may be a better estimate. The thermally activated process could be due to

Process (a) (ii) would not be described by a single activation energy since a significant percentage (say 20%) of promoted atoms would be required and the activation energy would change from 0 for the first promoted atom ($x = 0.98$) to a value appropriate for $x = 0.8$. By studying the variation of activation energy W with coverage from $0.7 < x < 1.0$ one could distinguish between process (a) where W would increase as x was reduced and processes (b) and (c) where W would have the opposite type of x dependence.

The behavior of T_2 above 5 K in Fig. 4 is very similar to that below $x = 0.72$ in Fig. 1. However, while melting is expected[7] in the latter case at about this coverage, a completed monolayer is not expected[10] to melt until about 7 K. In both cases the problem is that a separation of T_2 values measured at different frequencies occurs very close to where T_1 minima are observed. Elsewhere, it is shown that $\partial 1/T_2/\partial\omega^2$ is inversely proportional to the diffusion coefficient D and both at 6 K ($x = 0.98$) and $x = 0.7$ ($T = 1$ K) the values obtained for the coefficient $\partial 1/T_2/\partial\omega^2$ are comparable with that for the sample filled with bulk liquid ^{3}He for which D is known[11] to be $\sim 10^{-4}$ cm^2 sec^{-1}. This implies that the time taken to travel an interparticle distance ($\sim$3 Å) is 10^{-11} sec. However, at 4.8 K ($x = 0.98$) or at $x = 0.78$ ($T = 1$ K) where a T_1 minimum is observed at 2 MHz, we have τ_c (the correlation time for dipolar coupling) $\approx 1/\omega \approx 10^{-7}$ sec. This model, therefore, implies a change by 10^4 in the mobility for a small change in T or x. An alternative model can be constructed by assuming a range of graphite-produced local fields categorized according to their correlation lengths. There are likely to be small field changes at the crystallite boundaries, i.e. every[12] $\sim$200 A, and much larger changes every[8] 10^{-4} cm associated with misaligned platelets. Such a distribution can effectively be represented by a power spectrum such that $J(k)dk$ is the mean square z field for wave numbers between k and $k + dk$, z being the H_o field direction. For a given value of D, a range of K values between k_o and ∞ will contribute to transverse relaxation, where k_o is determined by setting the time taken to travel a distance $2\pi/k_o$, i.e. π^2/Dk_o^2, equal to

$$\left[\gamma^2 \int_{k_0}^{\infty} J(k)\,dk \right]^{-\frac{1}{2}} ,$$

this being the motional narrowing (more correctly here, "motional broadening") criterion. Thus as D increases, a greater fraction of the power spectrum contributes to the line width and this offsets the fact that the motion being faster might be expected to narrow the line.

It was hoped that the ^{3}He-grafoil system might offer an opportunity of studying the interesting anisotropies that arise[5] from the way that motion parallel and perpendicular to the DC field H_0 have different effects on NMR relaxation processes. Also there is an anisotropy of a factor of 4 in M_2. These anisotropies will be reduced by the spread of c axis directions of the graphite crystallites (estimated[12] to be 30°) and by the fraction of randomly oriented crystallites (estimated to be 44%). Data was collected under most of the conditions described in this paper for the three values of β, 0, $\pi/4$ and $\pi/2$. The most significant features of the data are recorded in Table I.

Further work is planned using UCAR oriented graphite grade XYZ where the range of crystallite directions is reported[13] to be 3°. This will allow effective study of the anisotropies and in addition will lead to greatly reduced graphite local fields.

Orientation β	0°	45°	90°
$d\ \ln T_2/dX$	15·5 $\pm$·5	17·9 $\pm$·5	17·9 $\pm$·5
T_2 (min)	·145 ms $\pm$·01	·175 ms $\pm$·01	·145 ms $\pm$·01
dT_2/dX, $X>1$	4·3 ms $\pm$·3	9·4 ms $\pm$·3	5·2 ms $\pm$·3
T_1 (min) as f(X)	24 ms +1 - 5	28 ms +1 - 5	31 ms $\pm$1
T_1 (min) as f(T)	18 ms $\pm$1	19 ms $\pm$1	25 ms $\pm$1
T_2 at X = 0.7	5·8 ms	9·5 ms	10·0 ms

Table 1

Stimulating communications from Professor J. G. Dash and R. A. Guyer are warmly acknowledged. The theoretical treatment of J_{2d} is due to Professor W. J. Mullin who has been a frequent source of guidance and encouragement. Experimental work has been made possible by grants from the Science Research Council who have also provided research studentships for B.P.C. and J.R.O.-B.

References

(1) R. J. Rollefson, Phys. Rev. Lett. 29, 410 (1972).

(2) D. P. Grimmer and K. Luszczynski, in "Low Temperature Physics, LT-13", (Proc. 13th Int. Conf. Low Temp. Phys., Boulder, Col., 1972) (Plenum Press, New York, 1973).

(3) S. G. Hegde, E. Lerner and J. G. Daunt, Phys. Lett. 49A, 437 (1974).

(4) M. G. Richards, Adv. Mag. Res. 5 305 (1971).

(5) W. J. Mullin, D. J. Creswell and B. P. Cowan, J. Low Temp. Phys. 25, 247 (1976).

(6) B. P. Cowan, M. G. Richards, A. L. Thomson and W. J. Mullin, to be published in Phys. Rev. Lett.

(7) M. Bretz, J. G. Dash, D. C. Hickernell, E. O. McLean and O. E. Vilches, Phys. Rev. A8, 1589 (1973); for the melting curve of ^{3}He films in grafoil, see the Ph.D. thesis of S. Hering (University of Washington, 1975, unpublished).

(8) D. L. Husa, D. C. Hickernell and J. E. Piott in "Monolayer and Submonolayer Helium Films," edited by J. G. Daunt and E. Lerner, (Plenum Press, New York, 1973) p. 133.

(9) D. F. Brewer, D. J. Creswell, Y. Goto, M. G. Richards, J. Rolt and A. L. Thomson, Ref. 8, p. 101.

(10) M. Bretz, G. B. Huff and H. G. Dash, Phys. Rev. Lett. 28, 729 (1972).

(11) J. R. Gaines, K. Luszczynski and R. E. Norberg, Phys. Rev. 131, 901 (1963).

(12) J. K. Kjems, L. Passell, H. Taub, J. G. Dash and A. D. Novaco, Phys. Rev. B 13, 1446 (1976).

(13) M. Bretz, private communication.

RECENT ADVANCES IN DILUTION REFRIGERATION

A.Th.A.M. de Waele, A. B. Reekers and H. M. Gijsman

Eindhoven University of Technology
Eindhoven, The Netherlands

1. INTRODUCTION

Since the experiments of Das *et al.*[1], the technique of reaching low temperatures by diluting ^3He in ^4He has been improved with regard to cooling power,[2] temperature range,[3,4] and reliability. Machines have been developed circulating ^3He, ^4He[5] or both.[6]

In this paper we will describe recent efforts to reach lower temperatures. In 1976 the successful operation was reported of two ^3He-circulating refrigerators maintaining temperatures below 4 mK continuously. In the machines the ^3He flowing into the (final) mixing chamber was precooled in two different ways.

In the first method, used by Frossati and Thoulouze,[3] continuous heat exchangers, with a large surface area and a free passage for the liquid, were employed.[7] The lowest reported temperature is about 3.5 mK continuously. We will discuss this method only briefly.

In the second method, developed in our laboratory, the refrigerator is equipped with more than on mixing chamber. The ^3He flowing into the final mixing chamber is precooled by the other mixing chambers. This method will be discussed in more detail.

2. CONTINUOUS HEAT EXCHANGERS

A heat exchanger can be considered as continuous when:

I *The heat conducted in the liquid in the direction of the flow (or opposite to it) can be neglected compared to the heat exchanged.*

II *The heat conducted by the heat-exchanger body, parallel to the flows, can be neglected.*

An exchanger satisfying these conditions has a long and slender geometry. Viscous heating might become a problem. Therefore, the third condition is:

III *Viscous heating must be minimized.*

In order to have a good efficiency, there are some additional requirements:

IV *There must be sufficient surface area in the exchanger.*

When the surface area is provided by sintered powder:

V *The thermal resistance of the sinter sponge and of the helium in it must be sufficiently small, but not too small (see condition II).*

These five conditions are discussed by several authors.[2,8,9,10] The requirements I and III also apply to interconnecting tubes. The dimensions of the flow channels can be estimated with an order of magnitude calculation assuming that the heat conduction in the direction of the flow is independent of the flow rate $\dot{n}$, and that the temperature in the channel changes with a factor of about 2 over a length l_c. Condition I then gives for concentrated ^{3}He and temperatures below 40 mK

$$\frac{A_c}{l_c} \kappa_{oc} << 12 \dot{n} T^2 , \tag{1}$$

where T is a typical temperature of the liquid, A_c is the cross-sectional area of the liquid (also including the liquid in the sinter sponge), and $\kappa_{oc} = 3.6 \times 10^{-4}$ in SI-units. Condition III gives

$$Z_c \frac{\eta_{oc}}{T^2} \dot{n}^2 V_c^2 << 12 \dot{n} T^2 ; \tag{2}$$

where Z_c is the flow impedance, $\eta_{oc} = 2 \times 10^{-7}$ in SI-units, and $V_c = 37 \times 10^{-6}$ m^3 is the molar volume of pure liquid ^{3}He. For a tube with diameter D_c, the area $A_c = \pi D_c^2/4$, and the impedance $Z_c = 128\, l_c/(\pi D_c^4)$. Eqs. (1) and (2) can then be combined to

$$\frac{\pi D_c^2 \kappa_{oc}}{48\, \dot{n}\, T^2} << l_c << \frac{12\, \pi\, T^4 D_c^4}{128\, \eta_{oc}\, \dot{n}\, V_c^2} \quad .$$

From this inequality the following numerical results can be derived:

$$2.4 \times 10^4 \left(\frac{D_c}{\text{mm}}\right)^2 \frac{\mu\text{mol/s}}{\dot{n}} \left(\frac{\text{mK}}{T}\right)^2 << \frac{l_c}{\text{mm}} << 1.0 \times \left(\frac{T}{\text{mK}} \frac{D_c}{\text{mm}}\right)^4 \frac{\mu\text{mol/s}}{\dot{n}} \quad . \tag{3}$$

A value for l_c satisfying (3) can only be found when

$$\frac{D_c}{\text{mm}} \left(\frac{T}{\text{mK}}\right)^3 >> 155. \tag{4}$$

Taking e.g.

$$\frac{D_c}{\text{mm}} \approx 300 \left(\frac{\text{mK}}{T}\right)^3 \tag{5}$$

as the minimum value for D_c, we can choose:

$$\frac{l_c}{\text{mm}} \approx 5.10^9 \frac{\mu\text{mol/s}}{\dot{n}} \left(\frac{\text{mK}}{T}\right)^8 \quad . \tag{6}$$

For the dilute stream one can similarly derive for temperatures below 20 mK

$$4.4 \times 10^3 \left(\frac{D_d}{\text{mm}}\right)^2 \frac{\mu\text{mol/s}}{\dot{n}} \left(\frac{\text{mK}}{T}\right)^2 << \frac{l_d}{\text{mm}} << 0.15 \left(\frac{T}{\text{mK}} \frac{D_d}{\text{mm}}\right)^4 \frac{\mu\text{mol/s}}{\dot{n}} \quad , \tag{7}$$

$$\frac{D_d}{\text{mm}} \left(\frac{T}{\text{mK}}\right)^3 >> 170 \quad , \tag{8}$$

$$\frac{D_d}{\mathrm{mm}} \approx 300 \left(\frac{\mathrm{mK}}{T}\right)^3 \text{ or } \frac{T}{\mathrm{mK}} \approx 6.7 \left(\frac{\mathrm{mm}}{D_d}\right)^{1/3} , \tag{9}$$

and

$$\frac{l_d}{\mathrm{mm}} \approx 0.8 \times 10^9 \frac{\mu\mathrm{mol/s}}{\dot{n}} \left(\frac{\mathrm{mK}}{T}\right)^8 \approx 198 \frac{\mu\mathrm{mol/s}}{\dot{n}} \left(\frac{D_d}{\mathrm{mm}}\right)^{8/3} . \tag{10}$$

Although the constants in (5) and (9), (6) and (10) are of the same order, the dimensions of the tubes of the dilute phase will be larger, because the temperatures are lower.

The conditions II, IV, and V can be treated similarly. However, they depend on the materials and on the geometry in consideration. They do not lend themselves to a short general discussion.

In 1976 the type of heat exchanger discussed here was further developed by the group in Grenoble, to be used in a dilution refrigerator with the purpose of reaching 3 to 4 mK continuously. Frossati and Thoulouze[3] satisfied condition II by using an epoxy housing for the heat exchanger. The thermal contact between the dilute and the concentrated flows was provided by a thin silver foil and sintered sub-micron silver powder. A magnetic temperature of 3.2 mK in the continuous mode was reported. It is their impression that by increasing the exchange surface or by improving the exchange mechanism a temperature in the 2 mK region may be reached.

3. THE CASCADE MIXING CHAMBERS

The cooling power of a mixing chamber can be used to cool a stream of concentrated ^{3}He. This is achieved most efficiently by leading the ^{3}He directly through a mixing chamger (MC-1), thus avoiding the Kapitza resistance. When this ^{3}He is subsequently diluted in a second chamber (MC-2), a temperature (T_2) can be maintained in MC-2, lower than the temperature T_1 of MC-1. Still lower temperatures can be reached when the process of partly diluting the ^{3}He is repeated in MC-2 by adding more mixing chambers to the system. In principle, the number is unlimited.

In our laboratory 5.5 mK was reached continuously with two mixing chambers using a dilution unit that reaches 13 mK with one mixing chamber.[4,12] Schumacher reached 3.2 mK in a double mixing chamber using only one continuous heat exchanger.[11]

4. INTERNAL CIRCULATION

In principle, it would be possible to lead the dilute flows of the mixing chambers separately to different stills, but it is preferable to combine them just after the mixing chambers. The dilute phases of the mixing chambers are then connected by a continuous path of superfluid ^4He. In a stationary situation the chemical potential of the ^4He must be constant along the path. This means that in the cascade mixing chambers the difference between the osmotic pressures Π and the pressures p (at equal heights) is the same in the dilute phases of all mixing chambers.

The phase boundaries will, in general, be at unequal heights. The osmotic-pressure difference also introduces an "internal circulation" in the mixing chambers.

In Fig. 1 two vessels, denoted MC-1 and MC-2 , with temperatures T_1 and T_2 (with $T_1 > T_2$), are depicted. They both contain dilute and concentrated mixtures. The concentrated phases are connected by a tube with negligible flow impedance and the dilute phases by a tube with flow impedance Z_1. As the result of the temperature difference there will be an osmotic-pressure difference $\Delta\Pi$ given by

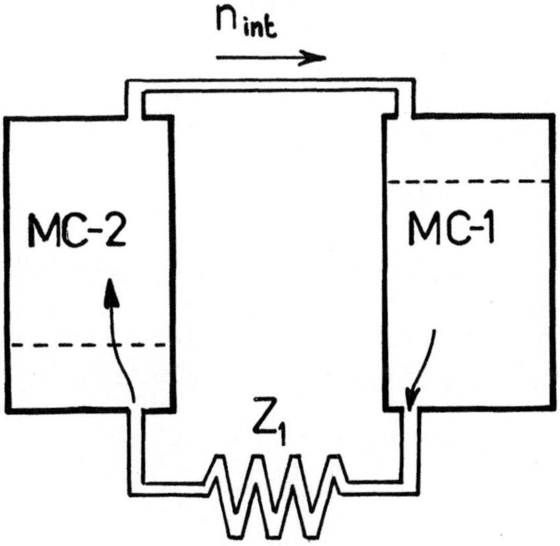

Figure 1. Internal circulation in a double mixing chamber without externally applied ^3He flow.

$$\Delta\Pi = \Pi_o(T_1^2 - T_2^2). \tag{11}$$

In this expression Π_o is a constant, equal to 10^5 in SI-units. In a stationary state $\Delta\Pi$ is balanced by a hydrostatic pressure difference Δp:

$$\Delta\Pi = \Delta p = g(\rho_d - \rho_c)\ \Delta h. \tag{12}$$

Here Δh is the difference in height of the two phase boundaries, g is the gravitational acceleration and ρ_d and ρ_c are the densities of the dilute and concentrated phases. The ^3He flows through Z_1 from MC-1 to MC-2 driven by the pressure difference Δp. The internal circulation rate $\dot{n}_{int}$ is given by

$$\Delta p = \Pi_o(T_1^2 - T_2^2) = \frac{\eta_{od}}{T_1^2} Z_1\ \dot{n}_{int}\ V_d\ . \tag{13}$$

In (13) $\eta_{od} = 5 \times 10^{-8}$ in SI-units, and $V_d = 430 \times 10^{-6}\ m^3$ is the volume of one mol ^3He in the dilute solution.

In MC-1 ^3He is diluted and produces cooling; in MC-2 the reverse process takes place. The internal circulation tends to equilize the two temperatures. In order to keep the temperatures constant, heat has to be supplied to MC-1 and extracted from MC-2.

In a double mixing chamber an external circulation is applied to this system. The net flow pattern can be regarded in a simplified mode as the sum of the external circulation (completely diluted in MC-2) and the internal circulation described above.

5. THE DOUBLE MIXING CHAMBER

A schematic drawing of a double mixing chamber is given in Fig. 2. The equations governing the system are the following:

$$q_1 + 12\ \dot{n}_t\ T_i^2 + 96\ \dot{n}_1\ T_1^2 + 12\ \dot{n}_2\ T_1^2 \tag{14}$$

$$q_2 + 12\ \dot{n}_2\ T_1^2 = 96\ \dot{n}_2\ T_2^2 \tag{15}$$

$$\dot{n}_1 + \dot{n}_2 = \dot{n}_t \tag{16}$$

$$\Delta\Pi = R_1\,\dot{n}_1\,V_d - R_2\,\dot{n}_2\,V_d \tag{17}$$

$$\Delta\Pi = \Pi_o\,(T_1{}^2 - T_2{}^2) \tag{18}$$

$$\Delta\Pi = g\,(\rho_d - \rho_c)\,\Delta h + R_c\,\dot{n}_2\,V_c\ . \tag{19}$$

The meaning of the symbols is as follows: The lower indices refer to MC-1 or MC-2; q_α is the total amount of heating power in MC-α including the heat originating from viscous heating and heat conduction; T_i is the temperature of the ^{3}He leaving the last heat exchanger; $\dot{n}_\alpha$ is the number of moles per second diluted in MC-α; R_α is the flow resistance of the dilute outlet tube; R_c is the flow resistance of the concentrated outlet tube of MC-1; $\dot{n}_t$ is the total molar flow rate.

Eqs. (14-19) constitute a system of 6 equations and 6 variables ($\dot{n}_1$, $\dot{n}_2$, T_1, T_2, $\Delta\Pi$, Δh). In general, q_1, q_2, T_i, R_1, R_2 and R_c are complicated functions of the variables. They also depend on the flow rate, dimensions of the tubes and on the characteristics of the heat exchangers. In order to get some insight into the behavior of the system, we will assume that viscous heating and heat conduction can be neglected. Furthermore, it will be assumed that T_i is constant, and that $R_2\dot{n}_2V_d$ and $R_c\dot{n}_2V_c$ are both much smaller than $\Delta\Pi$. Eqs. (14-19) can then be written as

$$q_1 + 12\,\dot{n}_t\,T_i{}^2 = 96\,\dot{n}_1\,T_1{}^2 + 12\,\dot{n}_2\,T_1{}^2 \tag{20}$$

$$q_2 + 12\,\dot{n}_2\,T_1{}^2 = 96\,\dot{n}_2\,T_2{}^2 \tag{21}$$

$$\dot{n} + \dot{n}_2 = \dot{n}_2 \tag{22}$$

$$\Pi_o\,(T_1{}^2 - T_2{}^2) = Z_1\,\eta_{od}\,\dot{n}_1\,V_d/T_1{}^2 \tag{23}$$

$$\Pi_o\,(T_1{}^2 - T_2{}^2) = g\,(\rho_d - \rho_c)\,\Delta h \tag{24}$$

q_1 and q_2 are now the externally-applied heating power (partially the result of external heat leaks).

Eq. (23) is the same as Eq. (13) indicating that $\dot{n}_1$ in this approximation is equal to the internal circulation.

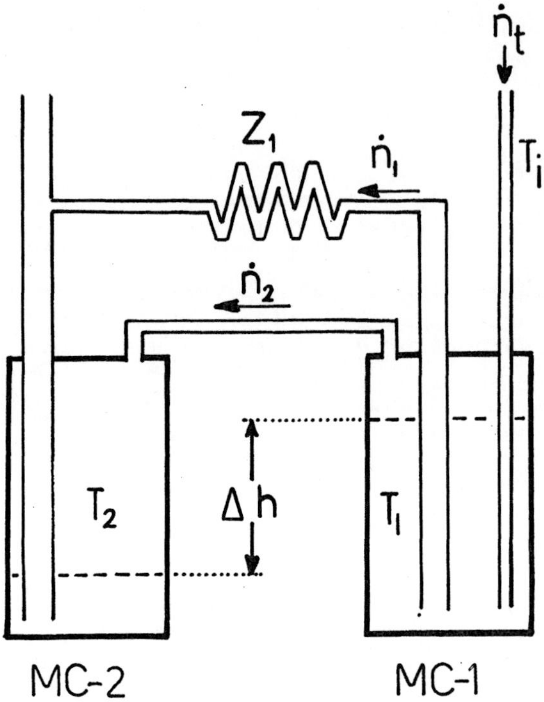

Figure 2. Flow of ^{3}He in a double mixing chamber. The symbols are explained in the text.

When $\dot{n}_t$ is lowered, $\dot{n}_2$ can theoretically become zero. In that case T_2 is very sensitive to external heat leaks and the double mixing chamber does not operate properly. By putting $q_1 = q_2 = 0$ and $\dot{n}_2 \approx 0$, Eqs. (20-23) show that this critical situation occurs when

$$7\ \Pi_o\ T_i^4 = 512\ Z_1\ \eta_{od}\ \dot{n}_t\ V_d. \tag{25}$$

The heights of the chambers have to be greater than the expected Δh in order to make sure that the phase boundaries are in the respective chambers. Numerical evaluation of Eqs. (21) and (24) gives

$$\frac{\Delta h}{\text{mm}} = 0.15\ \left(\frac{T_1}{\text{mK}}\right)^2 . \tag{26}$$

when Δh = 50 mm, MC-1 is at 18 mK.

From Eqs. (20-23) the cooling power of MC-2 for different $\dot{n}_t$ values can be calculated. To be able to compare the calculated cooling power with the measurements[4] we take $q_1 = 0$, $T_i = 36$ mK and $Z_1 = 14 \times 10^{12}$ m^{-3}.

In Fig. 3 the calculated dependences of T_1 and T_2 on $\dot{n}_t$ are given for $q_2 = 0$ and $q_2 = 200$ nW. At large heating powers and low flow rates $T_1 \approx T_2$ and the system behaves as a single mixing chamber. Also shown in Fig. 3 is the $q_2 - \dot{n}_t$ dependence for $q_2 = 1$ nW. From this curve it can be deduced that the cooling power is very small when n_t is only slightly larger than the critical value given by Eq. (25).

The dotted line in Fig. 3 gives the calculated T_M-$\dot{n}_t$ dependence representing the temperature of a single mixing chamber when it is heated with 200 nW with the same $T_i = 36$ mK.

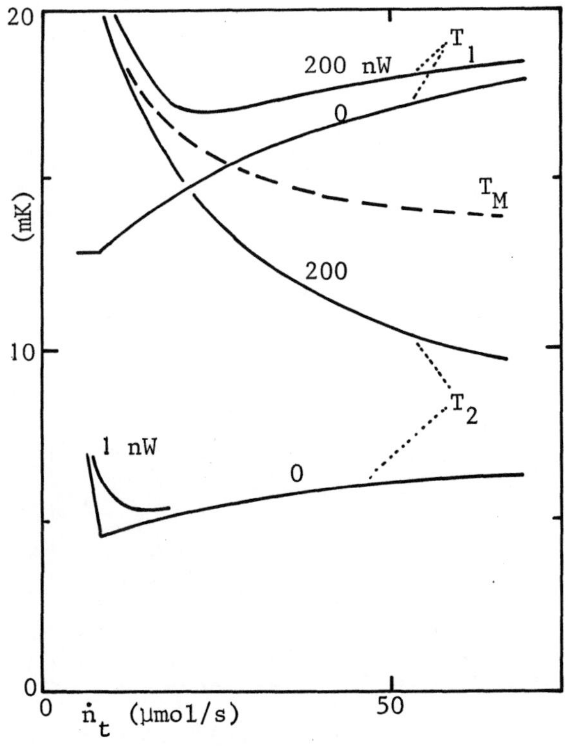

Figure 3. Calculated dependences of T_1 (for $q_2 = 0$ and 200 nW), T_2 (for $q_2 = 0$, 1 nW and 200 nW) and T_M (for $q = 200$ nW) on $\dot{n}_t$, for values of $T_i = 36$ mK and $Z_1 = 14 \times 10^{12}$ m^{-3}.

In our experiment $\dot{n}_t$ = 37 μmol/s. When the measured T_1, T_2 versus q_2 dependences are compared with the calculation, the agreement is fairly poor. E.g. for q_2 = 0 the calculated value of T_1 is 16.3 mK while the measured value if 14.0 mK. The agreement is better when R_1 in Eq. (17) is taken to be equal to $0.4 \times \eta_o Z_1/T_1^2$ instead of simply $\eta_o Z_1/T_1^2$ (Fig. 4). The flow impedance of the dilute outlet tube of MC-1 seems to be lower than expected from our simple model. A discrepancy of this kind was also reported by Schumacher.[9] An explanation may be that the temperature of the liquid in Z_1 is effectively 1.6 T_1. This is possibly one result of viscous heating in the tube which is equal to $\dot{n}_1 V_d \Delta p \approx \dot{n}_1 V_d \Pi_o T_1^2 \approx 43\, \dot{n}_1 T_1^2$ in SI-units. This amount of heat can warm the liquid to about 1.4 T_1, which is of the right order of magnitude.

The critical flow rate with $Z_1 = 5.7 \times 10^{12}\ m^{-3}$ is 19 μmol/s. In the experiment reported here, T_1 and T_2 were unstable at n_t = 16 μmol/s and stable at 25 μmol/s, indicating that the critical flow rate is indeed between 16 and 25 μmol/s.

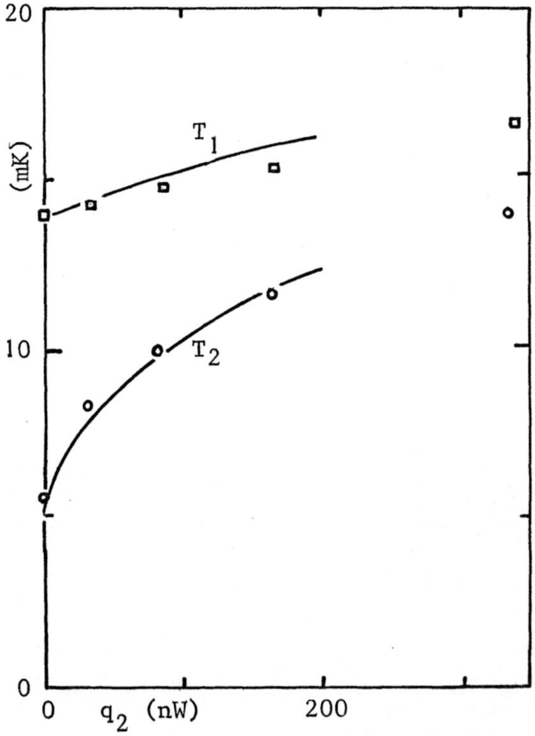

Figure 4. The lines give the calculated T_1 and T_2 dependences on q_2 for $\dot{n}_t$=37 μmol/s, T_i=36 mK and Z_1=5,6 x 10^{12} m^{-3}. The points represent the *measurements* for T_1(□) and T_2(o) with the same $\dot{n}_t$ and T_2 values, but with Z_1 = 14 x 10^{12} m^{-3}.

6. PURIFICATION

The ^3He circulated in the refrigerator contains a certain amount of ^4He. Part of the ^3He, arriving in MC-1, is dissolved in the ^4He. However, there is no ^3He in the dilute phase flowing into the second (and next) mixing chambers, because the dilute phase is heavier than the concentrated phase. Hence, MC-1 purifies the ^3He flowing to the other mixing chambers.

7. THREE MIXING CHAMBERS

The system of two mixing chambers can be extended by adding more mixing chambers.[4] At the moment we are studying the system of three mixing chambers. Our dilution refrigerator now has four step exchangers. With a single mixing chamber a temperature of 8.5 mK is reached at a flow rate of 26 μmol/s. With three mixing chambers the lowest magnetic temperature in MC-3 in the continuous mode was 3.0 mK while T_2 = 6.1 mK. The lowest temperature in the single cycle mode was limited to 2.8 mK by the quantity of concentrated ^3He, available at the beginning of the single cycle. This result shows that temperatures below 3.0 mK can be maintained in this system continuously if the incoming ^3He can be precooled properly.

8. THE LIMITING TEMPERATURE OF DILUTION REFRIGERATORS

The results of the group in Grenoble and the results with the cascade mixing chambers raise again the question whether dilution refrigerators have a fundamental low temperature limit. Wheatley *et al.*[9] showed that the limit of a refrigerator is intrinsically determined by viscous heating in the dilute-exit tube of the mixing chamber. From the heat balance in this tube a characteristic temperature T_o can be derived given by

$$\frac{T_o}{\text{mK}} = 4.4 \left(\frac{\text{mm}}{D_d}\right)^{1/3} \qquad (27)$$

and a characteristic length l_o given by

$$\frac{l_o}{\text{mm}} = 110 \, \frac{\mu\text{mol/s}}{\dot{n}} \left(\frac{D_d}{\text{mm}}\right)^{8/3} . \qquad (28)$$

The single cycle limit is 0.9 T_o. To maintain a certain temperature continuously, D_d has to be about a factor of 2 larger than the value deduced from the single cycle limit.[4] Hence

$$\frac{T_{cont}}{mK} \approx 5.0 \left(\frac{mm}{D_d} \right)^{1/3} \quad . \tag{29}$$

In order to reach 2 mK continuously with $\dot{n}$ = 1000 μmol/s, Eqs. (28) and (29) give $D_d \approx$ 16 mm and l_o = 180 mm as typical dimensions of the dilute outlet tube. For lower temperatures even larger machines will be necessary, both with respect to volume and to circulation rate. We have the impression that 2 mK will be practically the limit.

9. CONCLUSION

Dilution refrigerators now make it possible to reach temperatures in the 3 mK region continuously. It can be expected that the temperature will be further reduced to 2 mK in the near future.

ACKNOWLEDGEMENTS

We thank G. M. Coops for valuable discussion during the preparation of the manuscript.

References

(1) P. Das, R. de Bruyn Ouboter and K. W. Taconis, "Proc. 9th Int. Conf. on Low Temp. Phys.,"(Plenum Press, London, 1965), p. 1253.

(2) T. O. Niinikoski, Proc. 6th Int. Cryogenic Eng. Conf., (IPC Science and Techn. Press. Guildford, 1976) p. 102.

(3) G. Frossati and D. Thoulouze, ibid. p. 116.

(4) A.Th.A.M. de Waele, A. B. Reekers and H. M. Gijsman, ibid. p. 112.

(5) N. H. Pennings, R. de Bruyn Ouboter and K. W. Taconis, Physica, 81 B, (1976) 101; and N. H. Pennings, thesis, Leiden (1976).

(6) F. A. Staas, A. P. Severijns and H.C.M. van der Waerden, Phys. Lett. 53 A , 327, (1975).

(6) (cont.) G. Frossati, G. Schumacher and D. Thoulouze, "Proc. 14th Int. Conf. on Low Temp. Phys.," (North-Holland Publ. Comp., Amsterdam, 1975) Vol. 4, p. 13.

(7) T. O. Niinikoski, Nucl. Instrum. Methods 97, 95, (1971).

(8) J. D. Siegwarth and R. Radebaugh, Rev. Sci. Instrum. 43, 197, (1972), F. A. Staas, K. Weiss and A. P. Severijns, Cryogenics 14, 253, (1974).

(9) J. C. Wheatley, O. E. Vilches and W. R. Abel, Physics 4, 1, (1968).

(10) G. Frossati, communication at the Europhysics Study Conf. on Dilution Refrigeration, Lancaster 1976.

(11) G. Schumacher, ibid.

(12) A.Th.A.M. de Waele, A. B. Reekers and H. M. Gijsman, Physica 81 B, 323, (1976).

LIST OF CONTRIBUTORS

Adams, E. D., University of Florida

Ahonen, A. I., Helsinki University of Technology, Finland

Anderson, P. W., Princeton University

Band, W. T., Manchester University, England

Bernat, T. P., Louisiana State University

Biem, W., Institut für Theoretische Physik, Universität Giessen, West Germany

Britton, C. V., University of Florida

Buchholtz, L. J., Stanford University

Calder, I. D., University of Alberta, Canada

Chainer, T., Rutgers University

Chela-Flores, J., Centro de Física, I.V.I.C., Caracas

Cowan, B. P., Sussex University, Brighton, England

Cross, M. C., Bell Laboratories

deWaele, A.Th.A.M., Eindhoven University of Technology, Eindhoven, The Netherlands

Diehl, H.-W., Institut für Theoretische Physik, Universität Giessen, West Germany

Dynes, R. C., Bell Laboratories

Edwards, D. O., The Ohio State University

Ekholm, D. T., University of Massachusetts

Fairbank, W. M., Stanford University

Feder, J. D., The Ohio State University

Fetter, A. L., Stanford University

Flint, E. B., University of Florida

Franck, J. P., University of Alberta, Canada

Giannetta, R. W., Cornell University

Gijsman, H. M., Eindhoven University of Technology, Eindhoven, The Netherlands

Glaberson, W. I., Rutgers University

Glyde, H. R., University of Ottawa, Canada

Gould, C. M., Cornell University

Greywall, D. S., Bell Laboratories

Hall, H. E., Manchester University, England

Hallock, R. B., University of Massachusetts

Ho, T-L, Cornell University

Hook, J. R., Manchester University, England

Hu, C-R, Texas A&M University

Ihas, G. G., The Ohio State University

Ketterson, J. B., Northwestern University

Khanna, F. C., Chalk River Nuclear Laboratories, Canada

Kojima, H., Rutgers University

Kokko, J., Helsinki University of Technology, Finland

Krumhansl, J. A., Cornell University

Kumar, P., University of Southern California

Kummer, R. B., Bell Laboratories

Kurkijärvi, J., Research Institute for Theoretical Physics, Finland

Lee, D. M., Cornell University

Leiderer, P., Physik-Department E 10 der Technischen Universität München, West Germany

Lounasmaa, O. V., Helsinki University of Technology, Finland

Main, P. C., Manchester University, England

Maki, K., University of Southern California

Marston, P. L., Stanford University

McMillan, W. L., University of Illinois

Menn, K., Institut für Theoretische Physik, Universität Giessen, West Germany

Mermin, N. D., Cornell University

Meyer, H., Duke University

Mezhov-Deglin, L. P., Acad. of Sciences of USSR, Chernogolovka, Moscow District, USSR

Mochel, J. M., University of Illinois

Mueller, R. M., University of Florida

Narayanmurti, V., Bell Laboratories

Nayak, V. S., The Ohio State University

Nosanow, L. H., National Science Foundation

Osheroff, D. D., Bell Laboratories

Ostermeier, R. M., Rutgers University

Owers-Bradley, J. R., Sussex University, Brighton England

Paalanen, M. A., Helsinki University of Technology, Finland

Palmer, R. G., Princeton University

Pelizzari, C. A., Argonne National Laboratory

Rainer, D., Helsinki University of Technology, Finland

Reekers, A. B., Eindhoven University of Technology, Eindhoven, The Netherlands

Richards, M. G., Sussex University, Brighton, England

Richardson, R. C., Helsinki University of Technology, Finland

Roach, P. R., Argonne National Laboratory

Roach, P. D., Northwestern University

Roe, D., Duke University

Ruppeiner, G., Duke University

Rutledge, J. E., University of Illinois

Sandiford, D. J., Manchester University, England

Saslow, W. M., Texas A&M University

Schoepe, W., Helsinki University of Technology, Finland

Serene, J. W., State University of New York

Siggia, E. D., University of Pennsylvania

Sköld, K., Argonne National Laboratory

Smith, E. N., Cornell University

Soda, T., Tokyo University of Education, Japan

Stwalley, W. C., University of Iowa

Sungaila, Z., Argonne National Laboratory

Takano, Y., Helsinki University of Technology, Finland

Tam, C. P., The Ohio State University

Telschow, K. K., University of Massachusetts

Thomson, A. L., Sussex University, Brighton, England

Tsakadze, S. Y., Institute of Physics, Georgian Academy of Sciences, Tbilissi

Webb, R. A., Argonne National Laboratory

Wheatley, J. C., University of Southern California at San Diego

Woo, C-W, Northwestern University

Yefimov, V. B., Acad. of Sciences of USSR, Chernogolovka, Moscow District, USSR

Zinov'eva, K. N., Academy of Sciences of the USSR, Moscow

SUBJECT INDEX